Cuban Counterpoint

TOBACCO AND SUGAR

By FERNANDO ORTIZ

Translated from the Spanish by HARRIET DE ONÍS
Introduction by BRONISLAW MALINOWSKI
Prologue by HERMINIO PORTELL VILÁ
New Introduction by FERNANDO CORONIL

DUKE UNIVERSITY PRESS *Durham and London 1995*

© 1947 Alfred A. Knopf, Inc.
First printing in paperback by Duke University Press 1995
Second printing, 1999
All rights reserved
Printed in the United States of America on acid-free paper ∞
Library of Congress Cataloging-in-Publication Data appear
on the last printed page of this book.

CONTENTS

[v

ILLUSTRATIONS

Introduction to the
Duke University Press Edition
Transculturation and the Politics of Theory:
Countering the Center, Cuban Counterpoint

"*Un solo palo no hace monte*"[1]—CUBAN PROVERB

For without exception, the cultural treasures he surveys have an origin which he cannot contemplate without horror. They owe their existence not only to the efforts of the great minds and talents of those who created them, but also to the anonymous toil of their contemporaries. And just as such a document is not free of barbarisms, barbarism taints also the manner in which it was transmitted from owner to owner.—WALTER BENJAMIN

Las obras literarias no están fuera de las culturas sino que las coronan y en la medida en que estas culturas son invenciones seculares y multitudinarias hacen del escritor un productor que trabaja con las obras de innumerables hombres. Un compilador, hubiera dicho Roa Bastos. El genial tejedor, en el vasto taller histórico de la sociedad americana.[2]
—ANGEL RAMA

When *Cuban Counterpoint: Tobacco and Sugar* (1940) was published in Cuba, Fernando Ortiz (1881–1969) had established himself as one of the country's most influential public intellectuals. Like so many Latin American and Caribbean men of his position and generation, Fernando Ortiz had left his native land to receive a European education. When he returned to Cuba, where he remained until his death, he dedicated his life to the study of its popular traditions. Taking Cuba as his central concern, he addressed issues of national culture and colonialism, supported democratic institutions, promoted cultural organizations and journals, and authored works in areas ranging from criminology to ethnology. In *Cuban Counterpoint*'s prologue, Herminio Portell Vilá, a Cuban historian, insists that there is actually no need to introduce an outstanding work by this prominent author, and expresses gratitude that he has been allowed to link his name to that of Don Fernando Ortiz.

Perhaps in Portell Vilá's insistence one may detect a critical commentary on the introduction of the book, authored by the most renowned anthropologist of the time, Bronislaw Malinowski. A Polish émigré to England, Malinowski was a leading figure in defining anthropology as a scientific discipline and in conceptualizing fieldwork as the core of its method. In his introduction, Malinowski, then at Yale University, enthusiastically praises Ortiz's ethnographic skills and unabashedly presents the book as an outstanding example of functionalism, the theory of social integration with which Malinowski was closely aligned. He notes in particular his admiration for Ortiz's neologism "transculturation," and vows to employ it himself in his subsequent work: "I promised its author that I would appropriate the new expression for my own use, acknowledging its paternity, and use it constantly and loyally whenever I had occasion to do so" (1947, ix). Before his death in 1942, Malinowski wrote a number of papers and prepared two books, published posthumously. In these works he only used "transculturation" twice.

While I too recognize the privilege of introducing a book that speaks for itself, I acknowledge that, in the words of an earlier contrapuntal author, we necessarily read it "not just as we please, but under circumstances not chosen by ourselves" (Marx 1963, 15). It is difficult to assess the impact of ideas, to trace their origins and circulation, the paths through which they enter disciplinary canons and collective understandings, and the contexts that mark their reception by different publics. It is undeniable that Ortiz's greatest work has received exceptional international recognition; his book was translated into English in 1947, and in 1954 Columbia University, on the occasion of its two-hundredth anniversary, conferred on Ortiz an honorary doctorate. Yet my sense is that, given the conditions shaping its international reception, Ortiz's book has been read in ways that have overlooked aspects of its significance and have left its critical potential undeveloped. By offering a reading of selected sections of the book and of Malinowski's introduction, this essay seeks to enter into a dialogue with these texts, not so much to introduce the book, as to make its introduction ultimately unnecessary.

"A written preface," writes Gayatri Spivak in her introductory essay to *Grammatology* (Derrida 1974), "provisionally localizes the place where, between reading and reading, book and book, the interinscribing of 'reader(s),' 'writer(s),' and language is forever at work" (xii). A reader of Ortiz's *Cuban Counterpoint: Tobacco and Sugar* can readily understand the notion that "the return to the book is also the abandoning of the book" (Derrida quoted by Spivak, ibid.), that each reader and each reading of the same book opens up a different book. As Spivak suggests, "the preface, by daring to repeat the book and reconstitute it at another register, merely enacts what is already the case: the book's repetitions are always other than the book. There is, in fact, no 'book' other than these ever-different repetitions." Ortiz would have welcomed a perspective that, while respecting the integrity of a cultural text, recognizes its provisionality and inconclusiveness, the contrapuntal play of text against text and of reader against author. If, indeed, a counterpoint among reader, writer, and language is forever at work, my text pays tribute to Ortiz by engaging in this transcultural exchange, as Ortiz's book does, in counterpoint with the historical conditions of its own making.

Conditions of Reception

The publication of Ortiz's book in 1940 occurred at a time of international and domestic upheaval which frames the concerns of the text and helps explain its allegorical character. Fascism had begun to engulf Western Europe and to challenge the principles, already shaken by World War I, considered fundamental to "Western civilization." In Cuba, the strongman Fulgencio Batista, who controlled the state through intermediaries, gained widespread support and was elected president of the country. By the time Ortiz received his honorary doctorate in 1954, the United States had emerged as the dominant global power and the arbiter of Latin American affairs, and Batista had returned to the presidency of Cuba in 1952 by way of a coup. Batista's dictatorship, backed by the United States, came to an end in 1959 when a guerrilla-led revolution overthrew him and led to the first socialist regime in the Americas.

Cuban Counterpoint has circulated, until recent years, in a world divided into socialist and capitalist camps and modern and backward nations. For Third World nations—and this seemingly indispensable category was also born in the 1940s—socialism and capitalism have commonly been regarded as competing strategies to achieve modernity. While in the Third World, socialism and capitalism have offered competing images of the future, they have shared the assumption that the future, whatever its particular political or economic form, must of necessity be "modern." Ortiz's book did not quite fit the terms of this polarized debate. It was unconventional in form and content, did not express explicitly the wisdom of the times or reiterate prevailing currents of thought, and it proposed neither unambiguous solutions nor a blueprint for the future. Rather than straightforwardly offering an argument, it worked tangentially through poetic allusion, brief theoretical comments, and a detailed historical interpretation.

Half a century later, this edition of *Cuban Counterpoint* addresses a world where cultural differences and political inequalities cannot be mapped in terms of old polarities. The Second World has drastically contracted and transformed, while the First World is decentering and diversifying. Third World "development" programs of neoliberal design are accelerating the fractures within and among the nations of the "periphery." A number of intimately related processes, in which globalizing forms of capital accumulation and communication are met both with transnationalizing and reconfigured nationalist responses, have unsettled certainties associated with the belief in modernity.

Particularly in those intellectual fields shaped by feminist theory, Gramscian marxism, and poststructuralism, an attempt has been made to develop new perspectives and to bring excluded problems under critical observation. In certain respects these efforts have helped counter the silencing and stereotyping of subaltern collectivities and revealed their role in the making and the contestation of histories and cultures long represented from the homogenizing perspective of those who hold dominant power. This new edition of *Cuban Counterpoint* enters the space opened by these collective achievements; as an example of engaged cultural analysis, it may contribute to expanding it.

Modernity and Postmodernity: A Counterpoint from the Margins

Cuban Counterpoint examines the significance of tobacco and sugar for Cuban history in two complementary sections written in contrasting styles. The first, titled "Cuban Counterpoint," is a relatively brief allegorical tale of Cuban history narrated as a counterpoint between tobacco and sugar. The second, "The Ethnography and Transculturation of Havana Tobacco and the Beginnings of Sugar in America," is a historical essay that adheres loosely to the conventions of the genre. It is divided into two brief theoretical chapters and ten longer historical ones which discuss sociological and historical aspects of the evolution of sugar and tobacco production. Although each chapter can be read as a unit, together they present Ortiz's understanding of Cuban history.[3] Through their counterpoint, the two sections reinforce each other. While each is complex, they place different demands upon the reader. The historical essay is imposing and requires the reader to assimilate a vast amount of information. The brief allegorical essay poses a more unusual challenge: it reads almost too easily.

The allegorical essay seems to pull us in two directions at once. It is as if, through his playful evocation of the pleasures associated with sugar's and tobacco's consumption, Ortiz wishes to seduce us into enjoying the text with sensuous abandon. And yet it is also as if, through the unfolding dramatic plot which compellingly recounts a story of colonial domination, Ortiz wishes us to read this text in the same way that he reads tobacco and sugar: as complex hieroglyphs that elude definitive decoding. Through the interplay of these two readings the essay may seem at once to stand by itself and to call for continuing reinterpretation.

At a time when debates about postmodernity and modernity affect the climate of discussion about Latin America, there may be a desire to receive *Cuban Counterpoint* as a postmodern text on the basis of its unusual formal organization and its evident distance from positivism. There is a certain risk in this appropriation, however, for it is likely to deflect attention from the book's significance as a historical interpretation that seeks to integrate, through innovative methods of investigation and narration, the interplay of cultural forms and material conditions. In my view, Ortiz's analysis of the complex articulation of stabilizing and disruptive forces throughout Cuban history both questions prevailing assumption about the existence of separate "premodern" and "modern" domains, and demystifies certain pretensions of modernity itself. In this respect, we may wish to see *Cuban Counterpoint* as offering a historical analysis that can contribute to the understanding of Latin America's deepening social crisis and the emergence of a world at once increasingly interrelated and fractured.

From my position as a Venezuelan anthropologist working in the United States, I wish to approach *Cuban Counterpoint* as a valuable book for these difficult times. I take this text as an invitation to question the conceits of modernity and postmodernity alike. Ortiz shows that the constitution of the modern world has entailed the clash and disarticulation of peoples and civilizations together with the production of images of integrated cultures, bounded identities, and inexorable progress. His counterpoint of cultures makes evident that in a world forged by the violence of conquest and colonization, the boundaries defining the West and its Others, white and dark, man and woman, and high and low are always at risk. Formed and transformed through dynamic processes of transculturation, the landscape of the modern world must constantly be stabilized and represented, often violently, in ways that reflect the play of power in society. If a postmodern vision offers the image of fragmentary cultural formations unmoored from social foundations as an alternative to the modernist representation of integrated cultures rooted in bounded territories, Ortiz's perspective suggests that the formation of this vision be understood in relation to the changing geopolitics of empires. Ortiz invites us to understand the micro-stories of postmodernity and the master narrativities of modernity in relation to their respective conditions of possibility, rather than regarding one as epistemologically superior and thus trading the certainties of one age for those of another.

Cuban Counterpoint helps show the play of illusion and power in the making and unmaking of cultural formations. If the self-fashioning of sovereign centers entails the making of dependent peripheries, Ortiz celebrates the self-fashioning of these peripheries, the counterpoint through which people turn margins into centers and make fluidly coherent identities out of fragmented histories. Like other Caribbean thinkers who left their homelands and figure as foundational figures of postcolonial discourse, Fernando Ortiz struggled against Eurocentrism, although within the political and cultural confines of his nation and of reformist nationalist thought. While Ortiz clearly valorizes forms of sociality embedded in certain traditions and is hesitant with respect to the specific form of the future, he does not root identity in the past. His utopia involves less of a rupture with the present than Fanon's, but like Fanon, Ortiz uses binary oppositions (black and white, West and non-West) in a way that recognizes the experiential value of these terms for people subjected to imperial domination, but that also refuses to imprison an emancipatory politics in them. His allegorical essay recognizes the play of desire in the construction of colonial oppositions, vividly revealing how the colonial encounter forged cognitive categories as well as structures of sentiment. Ortiz treats binary oppositions not as fixities, but as hybrid and productive, reflecting their transcultural formation and their transitional value in the flow of Cuban history.[4]

Against the imperial alchemy that turns a Western particularity into a model of universality, *Cuban Counterpoint* calls attention to the play of globally interconnected particularities. Given Malinowski's role in establishing the centrality of anthropology as a Western discipline of otherness, there is a certain irony in the counterpoint which occurs between Malinowski and Ortiz. If Malinowski was the metropolitan ethnographer whose "magic" was most responsible for creating the concept of cultures as islands standing outside the currents of history,[5] Ortiz, constructing a perspective from the periphery, viewed cultural boundaries as artifices of power traced precariously on the sands of history.

Ortiz's playful treatment of cultural forms as fluid and unstable in *Cuban Counterpoint* explains the temptation to see him as a postmodern ethnographer *avant la lettre*.[6] Yet we should not forget the significance, in Ortiz's life work, of his critique of Eurocentric categories, his respect for the integrity, however precariously achieved, of subaltern cultures, and the attentiveness with which he studied the material constraints within which people make their cultures. I believe Ortiz would endorse Marshall Sahlins's warning concerning certain currents in postmodern ethnography:

> Everyone hates the destruction rained upon the peoples by the planetary conquests of capitalism. But to indulge in what Stephen Greenblat calls the "sentimental pessimism" of collapsing their lives within a global vision of domination, in subtle intellectual ideological ways makes the conquest complete. Nor should it be forgotten that the West owes its own sense of cultural superiority to an invention of the past so flagrant that it should make natives blush to call other peoples culturally counterfeit. (1993, 381)

In Ortiz's works the concept of "transculturation" is used to apprehend at once the destructive and constructive moments in histories affected by colonialism and imperialism. Through his critical valorization of popular creativity, Ortiz shows how the social spaces where people are coerced to labor and live are also made habitable by them, how in effect power resides not only in the sugar mill, but in the rumba. As a liberal democrat who had seen the failure of liberal democracy in Cuba and elsewhere, Ortiz could find little hope in a democratic option for Cuba in 1940. He probably found even less promise in Marxism, given its formulaic application in Cuba. If Ortiz's distrust of theory is related to Gramsci's "pessimism of the intellect," his analysis of popular culture reflects its counterpart, "the optimism of the will." This optimism takes as its central object the life-affirming creativity with which Cuban popular sectors countered their violent history. As if inspired by Cuban popular traditions, Ortiz offered *Cuban Counterpoint* as his own response to the critical circumstances of his time.

After the revolution in 1959, many professionals left Cuba, but Ortiz remained, conducting research on Cuban culture until his death in 1969. Now, when the promise of a democratic society appears as a receding mirage and the market parading as Freedom haunts much of the world, we may wish to remember how Ortiz found strength in Cuban popular traditions and recognized in them exemplary forms of sociality and creativity.[7]

The Caribbean, formed by a history of colonialism and neocolonialism, cannot be studied without addressing the geopolitics of empires. *Cuban Counterpoint* offers a glimpse into this history which demystifies its ruling fantasies—notions of the authentic native, of separate pure cultures, of a superior Western modernity.[8] Listening to the dialogue between Malinowski and Ortiz today may allow us to participate in their understanding of this history and to trace links between the politics of social theory and the geopolitics of empire. My discussion of these texts is organized into three parts. First, I discuss some of the personal, cultural, and political circumstances in which the book was produced. Second, I offer a reading of the book which centers on its first part.[9] Third, I discuss the politics of theory, through an examination of the counterpoint between Malinowski and Ortiz, and argue for the need to distinguish between theory production and canon formation.

Circumstances

Cuban Counterpoint was published when sixty-year-old Fernando Ortiz was at the height of his creative activity and Cuba was at the end of a tumultuous decade marked by numerous ruptures: domestic upheavals arbitrated by the United States; sharp swings in its U.S.-dependent economy; the collapse of a revolutionary civilian regime in 1933; and the consolidation of the army's power under Fulgencio Batista.

Born in Cuba in 1881 to a Cuban mother and Spanish father, Ortiz spent his youth in Minorca, Spain where he completed high school. While he went to Cuba in 1895 to study law, the turbulent conditions created by the War of Independence led him to return to Spain to complete his studies. In 1900 he obtained a bachelor's degree in law (*licenciatura*) in Barcelona and the next year a doctorate in Madrid. He returned to Cuba briefly in 1902, but left again to serve as Cuba's consul in Italy and Spain until 1906, when he was appointed public prosecutor for the city of Havana. He resigned from this position in 1916 when he was elected to a seven-year term in Cuba's House of Representatives. Throughout his life he combined academic pursuits with public service. He served several terms in Congress, where he sought to implement liberal political reforms, and in 1923 he headed the Committee of National Civic Restoration, created to combat the rampant corruption of the period.

During the short-lived revolutionary government led by Grau San Martín in 1933, which the United States refused to recognize, Ortiz was invited to join the cabinet and to offer a plan of reconciliation. While he did not join the regime, he made a proposal for national political unity built on the participation of major groups in Grau's government.[10] The plan was accepted by all; even Sumner Welles, the U.S. Ambassador hostile to Grau, found it a "reasonable compromise" (Aguilar 1972, 189). But Ortiz's formula ultimately failed "because of mutual suspicion and past resentment and the internal fragmentation of almost every group involved." With the loss of support for the civilian revolution, the military, headed by Batista, gained control of the state. Unable to rule without U.S. support, Grau went into exile in January 1934. Batista then ruled Cuba by way of puppet presidents until 1940, when he was elected president. This defeat of domestic progressive forces took place at the same time as the Spanish Republic's demise and the advance of fascism in Europe. Concerned with these developments, Ortiz organized "La Alianza Cubana por un Mundo Libre" in 1941 (The Cuban Alliance for a Free World).

Written in this political context, *Cuban Counterpoint* was the product of a career which sought, from multiple angles, to interpret Cuban society, analyze the sources of its "backwardness," and valorize the distinctive aspects of its culture. His scholarly work was marked at once by a continuity of concerns and a shift in perspectives. Ortiz's first book, *Los Negros Brujos* (1906), was a treatise of criminal anthropology focused on Afro-Cubans, their "superstitions," and their deviance. Framed by evolutionary positivist theories ascendant at the time and adhering to the biological reductionism of the Italian criminologist Cesare Lombroso, who prefaced the book, it analyzed the conditions which promoted criminality and "backward" beliefs among practitioners of "brujería" (sorcery) in Cuba. Illustrated with photos of the heads of Afro-Cuban criminals, it exemplified the biological notion of race widely accepted in Europe and the United States, and the assumption that those of African descent were a source of social disruption and stagnation.[11]

However, by 1910 Ortiz had begun to develop a sociological approach to race, one which emphasized cultural rather than biological factors as the basis of social progress. Concerned as before with Cuba's "backwardness," Ortiz felt Cubans must recognize their inferiority if they were to advance: "We are inferior, and our greatest inferiority consists, without doubt, in not acknowledging it, even though we frequently mention it" (1910, 27). But he explained that this inferiority was not due to "our race . . . but our sense of life, our civilization is much inferior to the civilization of England, of America, of the countries that today rule the world." He argued that Cubans could be civilized or uncivilized, like all men, "like those who are victorious, like those more backward ones who still splash around in the mud of barbarism" (28). Distancing himself from biological essentialism, he argued that people were physically alike; what Cubans needed, he argued, was "not a brain to fill the skull, but ideas to flood it and to wipe out its drowsiness. . . . We only lack one thing: civilization."[12] The "civilization" that Cubans needed, however, was European. Ortiz, in effect, transcoded biological signs into cultural signs, adopting at this time a racially marked evaluation of civilizing progress.[13]

The cause of Cuba's backwardness, of the corruption of its politicians, of the precariousness of its institutions, was attributed to the influence of the sugar industry by historian Ramiro Guerra y Sánchez in his highly influential *Azúcar y población en las Antillas* (1927), published at a time of heightened authoritarianism under President Gerardo Machado. This study profoundly affected a generation of intellectuals involved in the struggle for political reform and honesty in public life. In Guerra y Sánchez's analysis, Afro-Cubans figured as victims of the giant sugar factories that dominated the Cuban economy rather than as a source of Cuban culture.

For this reason, it is likely that Spengler's widely read work *The Decline of the West* (1918, translated into Spanish in 1923) exercised a greater influence on Ortiz. Its depiction of multiple paths leading toward historical development encouraged many Latin American intellectuals during the interwar period to view their societies as occupying not a lower stage in the unilinear development of Western civilization, but a unique position in a different historical pattern, one informed by its greater spiritual qualities and by the revitalizing mixture of races (Skurski 1994). Latin America no longer had to be seen as an incomplete version of Europe, but as an alternative to it.

The journal *Revista de Occidente*, founded by the Spanish philosopher José Ortega y Gasset in 1923, made German philosophy and historiography available to a generation of Cuban intellectuals for whom it provided intellectual resources with which to redefine Cuban identity (González Echevarría 1977, 52–60). According to Cuban novelist Alejo Carpentier, the journal became "our guiding light." It helped forge new links between Cuba and Spain, like the Spanish-Cuban Cultural Institute over which Ortiz presided (53). Spain, because of its own marginality within Europe, became a conduit for German thought, especially that of Spengler, which provided a compelling vision of historical diversity. As González Echevarría suggests,

> Spengler offers a view of Universal history in which there is no fixed center, and where Europe is simply one more culture. From this arises a relativism in morals and values: no more acculturation of blacks, no need to absorb European civilization. Spengler provided the philosophical ground on which to stake the autonomy of Latin American culture and deny its filial relation to Europe. (56)[14]

The shift in Ortiz's evaluation of the Afro-Cuban population and his concern for establishing the foundations of Cuban nationhood can be better understood in the light of this influence. Ortiz's alternative conception of Latin American development revalorizes popular and regional cultures but maintains a modified evolutionary framework (evidenced, for example, in his conception of cultural stages presented in the introduction to the second half of *Cuban Counterpoint*). Thus, while he keeps a notion of levels of cultural development, and in this respect reproduces certain biases concerning primitive and advanced civilizations, he significantly revalorizes contemporary Latin American cultures. Thus, in his *Africanía de la música folklórica en Cuba*, he seeks to establish the universal value of African and Afro-Cuban music, and to relativize European music as the standard of accomplishment. Ironically, to achieve this aim, Ortiz invokes the authoritative words of a European intellectual, Marcel Mauss: "Our European music is but *a* case of music, it is not *the* music" (1950, 331; my translation).

Cuban Counterpoint

Cuban Counterpoint, like Guerra y Sánchez's book, was published in
the context of a strongman's consolidation. But Ortiz's book, in contrast, is a highly metaphorical interpretation of Cuban history. Its
framework is not positivist, but literary, and it is modeled after the
work not of Lombroso, Guerra y Sánchez, or even Spengler, but of
the medieval Spanish poet Juan Ruiz, the archpriest of Hita.

The core of Ortiz's book is its first section. It creates a playful counterpoint between sugar and tobacco that is modeled, according to Ortiz, on Ruiz's allegorical poem, "Pelea que tuvo Don Carnal con Doña
Cuaresma," and inspired by Cuban popular traditions: the antiphonal
liturgy prayers of both whites and blacks, the erotic controversy in
dance measures of the rumba, and the versified counterpoint of
guajiros and the Afro-Cubans curros (1947, 4).[15] While Ruiz set
Carnival and Lent in a contest against each other, Ortiz engaged
sugar and tobacco in a theatrical interaction structured around their
contrasting attributes.

Ortiz's use of allegory not only draws on this long established literary form, but speaks as well to an allegorical literary tradition influential in Latin America's republican era. In the foundational novels
of the nascent republics, as Doris Sommer argues, national political
conflicts and the romantic ties of a couple from differing origins mirror each other, charting a resolution to divisions in the polity and
family through the formation of desire for the nation (1991).[16] As
icons of opposing regions and ways of life, the pair integrates the
fractured social collectivity.

In *Cuban Counterpoint* Ortiz offers an interpretation of Cuba's
social evolution narrated through the actions of sugar and tobacco,
products he introduces as "the two most important personages in the
history of Cuba" (4). Throughout the book he emphasizes their contrasting properties:

> Sugar cane lives for years, the tobacco plant only a few months . . . .
> The one is white, the other dark. Sugar is sweet and odorless; tobacco
> bitter and aromatic. Always in contrast! Food and poison, waking
> and drowsing, energy and dream, delight of the flesh and delight of
> the spirit, sensuality and thought, the satisfaction of an appetite and
> the contemplation of a moment's illusion, calories of nourishment and
> puffs of fantasy, undifferentiated and commonplace anonymity from
> the cradle and aristocratic individuality recognized wherever it goes,

medicine and magic, reality and deception, virtue and vice. Sugar is *she*; tobacco is *he*. Sugar cane was the gift of the gods, tobacco of the devils; she is the daughter of Apollo, he is the offspring of Persephone. (6)

He also establishes their profound impact on Cuban society and culture:

In the economy of Cuba there are also striking contrasts in the cultivation, the processing, and the human connotations of the two products. Tobacco requires delicate care, sugar can look after itself; the one requires continual attention, the other involves seasonal work; intensive versus extensive cultivation; steady work on the part of a few, intermittent jobs for many; the immigration of whites on the one hand, the slave trade on the other; liberty and slavery; skilled and unskilled labor; hands versus arms; men versus machines; delicacy versus brute force. The cultivation of tobacco gave rise to the small holding; that of sugar brought about the great land grants. In their industrial aspects tobacco belongs to the city, sugar to the country. Commercially the whole world is the market for our tobacco, while our sugar has only a single market. Centripetence and centrifugence. The native versus the foreigner. National sovereignty as against colonial status. The proud cigar band as against the lowly sack. (6–7)[17]

The contrasts of tobacco and sugar thread throughout the book and present themselves as a series of oppositions. However they have unexpected alignments that destabilize notions of fixed polarity: indigenous/foreign; dark/light; tradition/modernity; unique/generic; quality/quantity; masculine/feminine; artesan production/mass production; seasonal time/mechanical time; independent producers/monopoly production; generates middle classes/polarizes classes; "native" autonomy/Spanish absolutism; national independence/foreign intervention; world market/U.S. market.

These contrasts, while first described in Lombrosian fashion as deriving from the "biological distinction" between tobacco and sugar (4), unfold not as fixed qualities, but as themselves hybrid products. While tobacco is seen as male, its biological variety is seen as female. Tobacco is variously linked to the native (as an indigenous plant), to the European (as cultivated by white small holders), to the uniquely Cuban (as a transcultural product); it is related to the satanic, to the sacred, and to the magical. Although in its finished form it is an icon of Cuba's identity, it symbolizes foreign capital's control as well.

Linked to the violent history of indigenous, white colonist, and slave labor, tobacco has become a unique Cuban creation, leading Ortiz to call it "mulatto." Similarly, sugar's contrasts also change, as both a modernizing and an enslaving force identified with foreign domination as well as with Afro-Cuban labor. Their qualities are contradictory and multiple, carrying with them the marks of their shifting histories. In an encompassment of diverse attributes typical of the baroque, tobacco and sugar incorporate multiple meanings and transform their identities. As paradigmatic metaphors, they acquire new meanings by being placed within a syntagmatic structure through which they express a changing historical flow.

Yet this apparent mutability, which historicizes racial categories and productive relations, is stabilized by Ortiz's tendency to naturalize gender and to use common values associated with the masculine and the feminine as standards for valorization.[18] Tobacco tends to be masculine and to represent the more desirable features in Cuban culture; sugar, in contrast, stands for the feminine and represents the most destructive features of foreign capitalism. Quality and uniqueness become, in this alignment, strongly associated with masculinity and the national, whereas quantity and homogeneity are, in seeming paradox, linked to femininity and the international. Yet this paradox points to his representation of the character of capitalism on the periphery. While capitalism is powerful and therefore masculine, capitalism in the periphery is dependent and therefore feminine. As a dependent fragment in an expanding system of international capitalist relations, Cuba appears as feminized even as it is modernized, neocolonized rather than developed. The feminine becomes the sign of weakness and of seduction. As a result, Cuba, as a ground for national identity appears contradictorily as essential and as constructed, as a metaphysical entity and as a historical product.

As metaphorical constructs condensing a multiplicity of meanings, tobacco and sugar stand for themselves, as agricultural products, as well as for their changing conditions of production. Tobacco represents a native plant from which is made a product of great individuality and uniqueness, but also relations of production marked by domestic control over the labor process, individual craftsmanship, and the flexible rhythms of seasonal time. Sugar, on the other hand, represents not only a generic product derived from an imported plant, but also stands for industrial capitalist relations of production that reduce people to commodities, homogenize social relations and products, and subject labor to the impersonal discipline of machine production and to the fixed routines of mechanical time.

Symbols both of commodities and of productive relations, tobacco and sugar become defined reflexively by the conditions of production which they represent. This reciprocal interplay between products and their generative historical contexts constitutes a second counterpoint. As both products come under the impact of the capitalist forces, they become less differentiated and their attributes converge. They represent not only distinctive qualities or identities, but also their mutability under changing conditions.

Thus, the social identities of tobacco and sugar emerge from the interplay between their biological makeup and their productive relations. Except for certain aspects of their gendering, there is little that remains essential about them, for their biological attributes are mediated by human activity and modified by evolving patterns of production and consumption. Thus, in the last pages of the book's allegorical section, just when Ortiz identifies sugar with Spanish absolutism and tobacco with Cuban nationalism, he clarifies: "But today (1940), unfortunately, this capitalism which is not Cuban by birth or by inclination, is reducing everything to the same common denominator" (71). He returns to this idea in the conclusion of this section:

> We have seen the fundamental differences that existed between them (tobacco and sugar) from the beginning until machines and capitalism gradually ironed out these differences, dehumanized their economy, and made their problems more and more similar. (93)

The playful construction of contrasts between tobacco and sugar meets its counter in the sobering image of capital's growing domination of Cuban society. As Ortiz states in the conclusion of the book, in the face of this domination "many peoples and nations may find in tobacco their only temporary refuge for their oppressed personalities" (309). Yet Ortiz's work offers no predictions and seeks no closure.[19] Instead, the first section suggests a utopian solution in the form of a fairy tale. Asserting that "there was never any enmity between sugar and tobacco," Ortiz constructs a historical possibility which envisions the marriage, à la family romance, of Cuba's central actors, much as he had proposed a unifying alliance to resolve the political crisis of 1933:

> Therefore it would be impossible for the rhymesters of Cuba to write a "Controversy between Don Tobacco and Doña Sugar," as the

roguish archpriest would have liked. Just a bit of friendly bickering,
which should end, like the fairy tales, in marrying and living happily
ever after. The marriage of tobacco and sugar, and the birth of alcohol, conceived of the unholy ghost, the devil, who is the father of
tobacco, in the sweet womb of wanton sugar. The Cuban Trinity:
tobacco, sugar and alcohol. (93)

It is poetic justice that in the end, tobacco and sugar, these impersonators who had taken the license to borrow so many human attributes, reciprocate by becoming models of the generative powers of the
Cuban people.

This utopian allegory, however, bears the marks of its intellectual
origins within an elite discourse of reformist nationalism. It conjures
up the "unity of a collectivity" (Jameson 1981, 291) by means of a
trope of the liberal imagination with deep roots in Latin American
fiction: a fruitful marriage, compromise and fusion, rather than conflict or transformation. Ortiz envisioned national unity attained by
making the productive relations established under colonialism the
basis of Cuban culture. Yet alcohol, like tobacco and sugar, could not
escape the expanding grip of monopoly capital or stimulate more than
a transient illusion of community within a fractured nation. Ortiz's
utopia was imagined within the confining landscape of a neocolonial
commodity-producing society, revealing once again, how utopia and
ideology set the limits of each other in the battle over history.

Transculturation

This suggestion of utopia is followed by a sober second section where
Ortiz presents an amply documented study of the evolution of tobacco
and sugar production in Cuba. The historical discussion begins with
an unusual introduction of seven pages divided into two chapters. The
first, titled "On Cuban Counterpoint," is a two paragraph chapter in
which he explains that the second section is intended to give the preceding "schematic essay" supporting evidence. Yet he warns against a
simplistic reading of the first section:

It [the first section] makes no attempt to exhaust the subject nor does
it claim that the economic, social, and historical contrasts pointed out
between the two great products of Cuban industry are all as absolute
and clear-cut as they would sometimes appear. The historic evolution
of economic-social phenomena is extremely complex, and the variety
of factors that determine them cause them to vary greatly in the course
of their development. (97)

The second chapter is titled "On the Social Phenomenon of 'Transculturation' and its Importance in Cuba." As if to mark his distaste for "theory," Ortiz begs the reader's permission for introducing the neologism "transculturation." His introduction of this term comes in two parts. First, he explains that he is "employing for the first time" the term "transculturation," and invites others to follow him: "I venture to suggest that it might be adopted in sociological terminology, to a great extent at least, as a substitute for the term acculturation, whose use is now spreading." He explains that acculturation is being used to describe the process of transition from one culture to another and its manifold social repercussions, but asserts that transculturation is a more fitting term. Transculturation provides a larger conceptual framework within which to place the unpredictable features of Cuban society; it helps us understand

> the highly varied phenomena that have come about in Cuba as a result of the extremely complex transmutations of culture that have taken place here, and without a knowledge of which it is impossible to understand the evolution of the Cuban folk, either in the economic or in the institutional, legal, ethical, religious, artistic, linguistic, psychological, sexual, or other aspects of its life. (98)

Making his strongest claim, Ortiz asserts, "The real history of Cuba is the history of its intermeshed transculturations." Whose transculturation makes up "the real history of Cuba"?—that of tobacco and sugar? But sugar and tobacco now recede to the background of the narrative. Ortiz instead reviews the array of "human groups" that have populated the island over its history, from Indians to contemporary immigrants. He gives close attention to the violent conditions in which African slaves were forced to become part of these vast processes.

> [T]here was . . . the transculturation of a steady human stream of African Negroes coming from all the coastal regions of Africa along the Atlantic from Senegal, Guinea, the Congo, and Angola and as far away as Mozambique on the opposite shore of that continent. All of them snatched from their original social groups, their own cultures destroyed and crushed under the weight of the cultures in existence here, like sugar cane ground in the rollers of the mill.

Following the forced African migration "began the influx of Jews, French, Anglo-Saxons, Chinese, and the peoples from the four quarters of the globe. They were all coming to a new world, all on the way to a more or less rapid process of transculturation" (102). According to Ortiz,

> There was no more important human factor in the evolution of Cuba than these continuous, radical, contrasting geographic transmigrations, economic and social, of the first settlers, the perennially transitory nature of their objectives, and their unstable life in the land where they were living, in perpetual disharmony with the society from which they drew their living. Men, economies, cultures, ambitions were all foreigners here, provisional, changing, "birds of passage" over the country at its cost, against its wishes, and without its approval. (101)

After making this point Ortiz elaborates the concept of transculturation by contrasting it to the English term "acculturation." Evidently referring to the way the concept has been actually used in anthropological studies (rather than to the way it has been formally defined), he argues that acculturation implies the acquisition of a culture in a unidirectional process. Instead, transculturation suggests two phases, the loss or uprooting of a culture ("deculturation") and the creation of a new culture ("neoculturation"). He thus places emphasis on both the destruction of cultures and on the creativity of cultural unions. Giving credit to Malinowski's school for this idea, he says that cultural unions, like genetic unions between individuals, lead to offsprings that partake of elements of both sources, and yet are different from them.[20]

Ortiz insists that the concept of transculturation is indispensable for an understanding of Cuba, "whose history, more than that of any other country of America, is an intense, complex, unbroken process of transculturation of human groups, all in a state of transition" (103). Yet, far from restricting this term to Cuba, he argues that for analogous reasons, transculturation is fundamental for understanding the history of "America in general." He left it for others to apply this concept to societies in which native peoples remained an important sector of the population. It was through his analysis, more than through his brief formal definition, that Ortiz showed his understanding of transculturation.

Counterfetishism

While in the first section of *Cuban Counterpoint* we are told that "the most important personages of Cuban history" are sugar and tobacco, in the second section we learn that "the real history of Cuba" is made up of "the intermeshed transmigrations of people." What is the significance of this apparent contradiction, of this shift from commodities to people as the central characters of Cuban history?

Perhaps it is related to the strange effect the book produces upon its readers. The more Ortiz tells us about tobacco and sugar, the more we feel we learn about Cubans, their culture, musicality, humor, uprootedness, their baroque manner of refashioning their identities by integrating the fractured meanings of multiple cultures. Imperceptibly, we likewise begin to understand the social forces that have conditioned the ongoing construction of Cuban identities within the context of colonial and neocolonial relations. How is it that a book about two commodities produces this effect?

The mystery of this effect, and the apparent contradiction between these two views, is perhaps resolved by realizing that Ortiz treats tobacco and sugar as highly complex metaphorical constructs that represent at once material things and human actors. Moreover, by showing how these things/actors are defined by their social intercourse under specific conditions, he illuminates the forces shaping the lives of the real actors of Cuban history—of Africans "like sugar cane ground in the rollers of the mill," or of Cuban nationalists turned into interventionists, like the tobacco of the foreign-controlled cigarette industry.

Ortiz, in my view, uses the fetish power of commodities as a poetic means to understand the society that produces them.[21] Without making reference to Marx, he shows how the appearance of commodities as independent entities—as potent agents in their own right—conceals their origins in conflictual relations of production and confirms a commonsense perception of these relations as natural and necessary. The meaningful misrepresentation that occurs when social relations appear encoded as the attribute not of people, but of things, transforms commodities into opaque hieroglyphs, whose mysterious power derives from their ability to misrepresent and conceal reality, and whose multiple meanings can only be deciphered through social analysis.

By constructing a playful masquerade of tobacco and sugar, Ortiz links the fetish to the poetic and transgressive possibilities of the carnivalesque. Using the idiom of fetishized renderings of Cuban culture, he presents a counterfetishistic interpretation that challenges essentialist understandings of Cuban history. In this respect, his work resonates with Walter Benjamin's treatment of fetishism. Unlike others members of the Frankfurt school, who were primarily concerned with demystifying the fetish in the service of reality, Benjamin sought to apprehend how the fetish commands the imagination, at once revealing and appreciating its power of mystification. By treating tobacco and sugar not as things but as social actors, Ortiz in effect brings them back to the social world which creates them, resocializes them as it were, and in so doing illuminates the society that has given rise to them. The relationships concealed through the real appearance of commodities as independent forces become visible once commodities are treated as what they are, social things impersonating autonomous actors.

As the narrative unfolds, tobacco and sugar indeed become historical personages; they appear as social actors with political preferences, personal passions, philosophical orientations, and even sexual proclivities. It becomes clear that tobacco and sugar, far from mere things, are changing figures defined by their intercourse with surrounding social forces. By turning them into full-fledged social actors, Ortiz has shown that they can appear as autonomous agents only because they are in fact social creatures, that is, the products of human interaction within the context of capitalist relations of production. Like Marx, who by personifying Madame la Terre and Monsieur le Capital in *Capital* highlighted their fetishization in capitalist society (1981, 969), Ortiz simultaneously presents sugar and tobacco as consummate impersonators and unmasks them.

The book's conclusion makes this clear. It follows an account of how Havana tobacco was accepted in Europe and its cigars "became the symbol of the triumphant capitalist bourgeoisie," and how the democratic cigarette eventually replaced the cigar, affecting in turn how these commodities were produced:

> But cigars and cigarettes are now being made by machines just as economy, politics, government, and ideas are being revised by machines. It may be that many peoples and nations now dominated by the owners of machines can find in tobacco their only temporary refuge for their oppressed personalities. (1947, 309)

Tobacco, the symbol of Cuban independence, of exceptional skill and unique natural factors, appears now as an increasingly homogenized mass product controlled by foreign interests, like sugar. But at this stage the issue is not to dissolve once again the sharp contrasts established earlier between tobacco and sugar, but to unmask these pretentious actors as mere creatures of human activity. The point is not so much that they are the products of machines, but of machines producing under specific social relations. At the end of the book, the owners of the machines emerge as the leading actors, for they dominate the structure and aims of production.

Similarly, at the conclusion of the first section, as the counterpoint of tobacco and sugar comes to the end, the pursuit of money and power emerges as a major force structuring the pattern of Cuban transculturations ever since the conquest. Ortiz approvingly quotes Ruiz's verses on the powers of money, the commodity that stands for all commodities, the universal fetish:

> Throughout the world Sir Money is a most seditious man
> Who makes a courtesan a slave, a slave a courtesan
> And for his love all crimes are done since this old earth began. (81)

The chase after money and power in Cuba had helped fashion a social world which trapped them in subordinate relation to external conditions beyond their control. As Portell Vilá explains,

> A difference of half a cent in the tariff on the sugar we export to the United States represents the difference between a national tragedy in which everything is cut, from the nation's budget to the most modest salary, even the alms handed to a beggar, and a so-called state of prosperity, whose benefits never reach the people as a whole or profit Cuba as a nation. (Ortiz 1947, xix)

Just as money could "make a courtesan a slave," it made tobacco, the emblem of distinction, into a mass product like sugar. At the end, Cubans, with no control over the winds of history, appear as "birds of passage," transient creatures with fluid identities.

Without referring to parties, groups, or personalities, Ortiz depicts the dynamics of neocolonial Cuba, the malleable loyalties and identities of its major actors, the provisional character of its arrangements and institutions, the absence of control over its productive relations. He had seen how no political principle was secure; noninterventionists asked the U.S. ambassador to intervene, pro-civilians allied themselves with the military, advocates of honesty became masters of corruption. In Ortiz's narrative no names need to be mentioned, for tobacco and sugar act as a mirror in which one could see reflected familiar social identities.

By casting commodities as the main actors of his historical narrative, Ortiz at once displaces the conventional focus on human historical protagonists and revalorizes historical agency. Acting as both objects and subjects of history, commodities are shown to be not merely products of human activity, but active forces which constrain and empower it. Thus historical agency comes to include the generative conditions of agency itself. As a critique of reification, Ortiz's counterfetishism questions both conservative interpretations that reduce history to the actions of external forces, and humanist and liberal conceptions that ascribe historical agency exclusively to people. His counterfetishism encompasses a critique of Western humanism's essentialization of the individual and its hierarchization of cultures. As Paul Eiss argues, Ortiz enacts not only a counterfetishism, but a counterhumanism. "In addition to unmasking human social relations hidden in the apparent activity of commodities, Ortiz unmasks the agency of commodities which is hidden in apparently human agencies or characteristics. Ortiz's counterhumanism not only constitutes a brilliant spatial critique, but also challenges stable narratives and identities of colonial and neocolonial history" (1994, 35). Transculturation thus breathes life into reified categories, bringing into the open concealed exchanges among peoples and releasing histories buried within fixed identities.

Counterpoints of Theory: Malinowski and Ortiz

Given Malinowski's international academic reputation, it is understandable that Ortiz, whose prestige was local, would welcome the opportunity to have Malinowski introduce his book. He regarded transculturation as a critical concept that countered prevailing anthropological theory by directing attention to the conflictual and creative history of colonial and neocolonial cultural formations; it offered the possibility of recasting not solely Cuban history, but that of "America in general." In all likelihood he felt that the endorsement of a metropolitan authority of Malinowski's stature would help gain him recognition.

Ortiz acknowledges, with a tone of formal correctness, Malinowski's "approbation" of his new term. At the close of his introduction to *Cuban Counterpoint*'s second half, in which he presents his concept of transculturation, Ortiz adopts an impersonal tone and uses the passive voice when recounting the granting of approval by this intellectual authority: "When the proposed neologism, transculturation, was submitted to the unimpeachable authority of Bronislaw Malinowski, the great figure in contemporary ethnography and sociology, it met with his instant approbation" (1947, 103). Resuming his characteristically direct narration, Ortiz explains that "under his eminent sponsorship, I have no qualms about putting the term in circulation." Yet, he makes no further comment on Malinowski's introduction.

Malinowski is more explicit. He recounts in his introduction that he visited Cuba in 1939 and was pleased to meet Ortiz, whose work he had admired. He was enthusiastic about his plan to introduce the term transculturation as a replacement for the prevailing terms relating to cultural contact ("acculturation, diffusion, cultural exchange, migrations or osmosis of culture"). After stating that he promised Ortiz he would use it in the future, Malinowski recounts that "Dr. Ortiz then pleasantly invited me to write a few words with regard to my 'conversion' in terminology, which is the occasion for the following paragraphs" (lvii).

Just as it is reasonable to assume that Ortiz hoped to receive international validation through Malinowski's authority, we may surmise that Malinowski sought to consolidate his own reputation and that of functionalism, by supporting while aligning with his own theoretical position the work of a noted anthropologist from the margins. The introduction reflects the tension between these two aims. At one level, Malinowski highlights the importance and originality of the book. He praises Ortiz's style and mastery of ethnographic materials, offers an appreciative exegesis of the book's argument, and recognizes the validity of the term transculturation. Moreover, he supports plans for the creation in Cuba of an international research center. In brief, the introduction expresses strong support for the work of a peripheral ethnographer by a metropolitan anthropologist, an unusual and significant gesture.[22]

At another level, however, the introduction assimilates Ortiz's project into Malinowski's own, blunting its critical edge and diminishing its originality. This assimilation takes place through three related moves. First, Malinowski aligns Ortiz's transculturation with his own ideas concerning cultural contact; second, he defines Ortiz as a functionalist without evidence that this is the case; third, he reads *Cuban Counterpoint*, literally as a book on tobacco and sugar, as material objects, without attending to their complex cultural structure as commodities and to the critical use Ortiz makes of this complexity. Let me explain these points further.

First, Malinowski presents as Ortiz's the notion that "the contact, clash, and transformation of cultures cannot be conceived as the complete acceptance of a given culture by any one 'acculturated' group" (lx) and supports it by quoting two statements from a 1938 article he wrote on cultural contact in Africa. These quotes, focusing on cultural "ingredients" and "typical phenomena of cultural exchanges" —"schools and mines, Negro places of worship and native courts of justice, grocery stores and country plantations"—present the idea that cultural contact affects both cultures and results in new cultural realities. While a sound idea, this formulation is well within the framework presented in the memorandum on "acculturation" by Redfield, Linton, and Herskovits, considered the definitive statement on the subject at the time.[23]

During the 1930s, as the subjects of anthropological study could no longer be kept within the "slots" of the "primitive" and the "traditional" where they had been contained,[24] British and American anthropology renewed their theoretical concern with issues of cultural diffusion and contact. Through studies of "culture contact" in England and "acculturation" in the United States, anthropologists not only redefined the objects of anthropological study, but also addressed issues of contemporary relevance. In a review of acculturation studies, Ralph Beals remarks that "The obvious utility of acculturation studies for the solution of practical problems was also a factor in their early popularity" (1955, 622). According to him, this sense of "utility" was not unrelated to the exercise of state power in the colonies and at home:

> The beginnings of interest in contact situations in Great Britain, France, and Holland coincided with the rise of a new sense of responsibility toward colonial peoples, while in the United States the great development of acculturation studies coincided with the Depression era and its accompanying widespread concern with social problems. (622)

At this time British anthropology shifted its focus, geographically, from the Pacific to Africa, and thematically, from the study of pristine cultures to the study of "cultural contact." Malinowski was best known for his pathbreaking *Argonauts of the Western Pacific* (1922), a study of the Trobriand Islanders in which he defined anthropology as the study of culture as an integrated whole suspended in time. He adapted to this shift in focus, however, and defined as "the real subject-matter of field-work . . . not the reconstruction of a pre-European native of some fifty or a hundred years ago" but "the changing Melanesian or African" (1935, 480). His reason was the native's contact with the Western world: "He has already become a citizen of the world, is affected by contacts with the world-wide civilization, and his reality consists in the fact that he lives under the sway of more than one culture."

Malinowski had come to recognize not only that cultural contact entailed the transformation of both cultures leading to the formation of a new one, but also that colonial politics needed to be included in the ethnography of African societies. He had developed a three-column approach to the study of contact situations whereby data was placed in the appropriate column, thus mapping out the three phases of culture change: impinging culture; receiving culture; compromise and change. This model is refined in his book, *The Dynamics of Cultural Change*, edited by Phyllis M. Kaberry and published posthumously in 1945.[25] Yet, in this work, composed of notes and papers written between 1936 and his death in 1942, he continues to treat social change taxonomically, without attending to its historical or cultural depth. Malinowski's "Method of the Study of Culture Contact" in this book is evaluated by Beals as "a rather mechanical organization for analytical purposes." The three columns now have become five:

(A) white influences, interests and intentions; (B) process of culture contact and change; (C) surviving forms of tradition; (D) reconstructed past; and (E) new forces of African spontaneous reintegration or reaction. The discussion and illustrations utilized adhere rigidly to Malinowski's classical functionalist approach, save for a slight and grudging bow to time elements in the category of the "reconstructed past." The "method" is also focused directly upon Africa and upon administrative problems. (1955, 630)

This five column approach was only a partial improvement over his previous scheme. The two new columns—"reconstructed past" and "new forces of spontaneous African reaction"—did little to add a historical dimension. According to Beals, "Malinowski, on the whole remained intransigent to the end concerning history, despite his inclusion of a column for the 'Reconstructed Past' in his outline tables. This indeed he really viewed not as a 'reconstructed past' but as a past remembered" (631). His disciple Lucy Mair stated that "of the specimen charts which are published in *The Dynamics of Culture Change*, only that on warfare contains entries under both of these heads, and his comment on the reconstruction of the past is that, though of interest for the comparative study of warfare, it is 'of no relevance whatever for the application of anthropology' " (1957, 241).

Functionalism's attempt to address cultural change in colonial Africa did not lead to a reevaluation of its assumptions, but to the domestication of colonial history: conflict was contained within integration and transformation within reproduction. Thus, "savage" cultures remained safely subsumed by European "civilization." We will never know what kind of book Malinowski would have written had he lived longer, but given his taxonomic treatment of change in the works that comprise *The Dynamics of Culture Change*, it is understandable that he did not engage Ortiz's work or even mention the term "transculturation."

Second, Malinowski, rather than distancing himself from Ortiz, sought to include him in his camp. In the introduction to *Cuban Counterpoint* he refers to Ortiz as a functionalist three times. Given Malinowski's aversion to history, his insistence on casting Ortiz as a functionalist is telling, particularly since he presents Ortiz's extensive use of history as an expression of functionalist principles: "Like the good functionalist that he is, the author of this book resorts to history when it is really necessary" (lxii).

Apparently Malinowski believes history became "really necessary" for Ortiz as a tool to study changing patterns of tobacco and sugar production, independent of colonialism or imperialism. He construes these products as a source of pride for Ortiz, who,

> Cuban by birth and by citizenship, is justly proud of the role his country has played in the history of sugar, through the vast production of its centrals, and in that of smoking, through having developed in its vegas the best tobacco in the world. . . . He describes the triumphal march of tobacco all over the face of the globe and determines the profound influence exerted by sugar on the civilization of Cuba, its principal effect having been, perhaps, to occasion the importation from Africa of the many and uninterrupted shiploads of black slave workers. (lxiii)

Ortiz refers to tobacco and sugar as two sources of pride for Cuba, as "the country that produced sugar in the greatest quantity and tobacco

in the finest quality" (92–93). Yet his analysis suggests how prob-
lematic the sense of national identification through these two commodities was. As we have seen, Ortiz had sought to find in the
reflections of a sugar crystal a history of colonial domination. Ortiz's
pride was not in the volume of sugar produced in Cuba but in the
creation of a culture in Cuba that countered the degradation of this
history; the quality of its tobacco served as a metaphor of Cuba's
unique culture.

"Was Ortiz really a functionalist?" asks Cuban historian Julio Le
Riverend, in an introduction to a Venezuelan edition of this book
published in 1978.[26] He answers his own question by indicating that
Ortiz repeatedly asserted that he was not. Le Riverend presents Ortiz
as a thinker familiar with classical and contemporary social theory
who had read Comte, Marx, and Durkheim, as well as many contemporary thinkers, among them Malinowski. According to Le Riverend, Ortiz systematically avoided theoretical discourse and showed
an increasing preference for a historical approach (1978, xx–xxiii); he
was an eclectic intellectual who resisted procrustean labels. Ortiz
must have recognized the irony of accepting his public presentation
as a functionalist in return for the intellectual acknowledgement of
a book that sought to counter metropolitan anthropology and the
imperial imposition of labels on Cuba.

Third, Malinowski's treatment of sugar and tobacco as mere things,
divorced from their cultural and political significance, serves to obscure the metaphorical character of the book and blunt its critical
edge. It must be remembered that Malinowski saw himself as no
ordinary anthropologist, but as one who combined literary sensitivity
with theoretical ambitions—he aspired to be the "Conrad of anthropology" (Stocking 1983, 104). Yet there is little indication that Malinowski appreciated the literary qualities of *Cuban Counterpoint*, its
unconventional structure, its allegorical character, or its originality
as an engaged ethnography produced by a native anthropologist involved in the political struggles of his nation. He reads transculturation as a technical term that expresses a certain dynamism in cultural
exchanges, not as a critical category intended to reorient both the
ethnography of the Americas and anthropological theory. In Malinowski's introduction there is little receptivity to a reading of *Cuban
Counterpoint* as a critical intervention in Cuban historiography, or,
least of all, as a text that could develop metropolitan anthropology.

Theoretical Transculturation: Travelling
Theory/Transcultural Theory

Many Cuban intellectuals have recognized Ortiz's wide range of accomplishments; Juan Marinello, echoing a term coined by Ortiz's secretary, Rubén Martínez Villena, called Ortiz "the third discoverer of Cuba" (after Columbus and Humboldt).[27] While many intellectuals have paid tribute to Ortiz,[28] few have directly engaged or developed his work outside Cuban circles. Perhaps the most remarkable exception is Uruguayan literary critic Angel Rama. His *Transculturación narrativa en la America Latina*,[29] whose title pays tribute to Ortiz, begins with an appreciative discussion of his work and shows its relevance for Rama's own attempt to examine Latin American narratives from a Latin American perspective.[30] Using Ortiz's concept, Rama offers a critical examination of the anthropological and literary work of José Maria Arguedas, a Peruvian ethnologist and writer who committed suicide after dedicating his life to revalorizing and integrating the Quecha and Hispanic cultural traditions that make up his nation. For Rama, transculturation facilitates the historical examination of Latin American cultural production in the context of colonialism and imperialism. Perhaps through the influence of Rama's work, Ortiz's ideas have received some recognition in literary criticism and cultural studies.[31]

Among anthropologists, however, Ortiz's presence is marginal. Ralph Beals, in an overview of "acculturation studies," offers the following evaluation: "In his preface to the work (*Cuban Counterpoint*), Malinowski is enthusiastic about the new term, but one finds no serious consideration of the reciprocal aspects of culture contact in any of his own publications. 'Transculturation' has had some use by Latin American writers, and, were the term 'acculturation' not so widely in use, it might profitably be adopted" (1955, 628). Yet Mexican anthropologist Gonzalo Aguirre Beltrán criticizes the concept transculturation on etymological grounds and argues that the term "created more confusion." (1957, 11). Brazilian anthropologist Darcy Ribeiro's monumental synthesis of Latin American cultural formations includes a critical discussion of the concept of acculturation in terms that resemble Ortiz's position, but despite Ribeiro's erudite references to a large number of authors, he does not mention Ortiz, and retains the term acculturation (1971, 24; 37–39).[32] Neither are Ortiz's books mentioned in Mexican anthropologist Nestor García Canclini's important work *Culturas híbridas* (1989). In a heated debate among prominent Latin Americanists in the United States about approaches to the history of non-European peoples, Ortiz's ideas were not considered (Taussig 1989; Mintz and Wolf 1989). In a review of Caribbean anthropology, Michel-Rolph Trouillot, in contrast, mentions Ortiz's work in the context of his discussion about the development of a historically oriented anthropology and situates him in relation to the work of other Caribbean anthropologists, such as Price-Mars in Haiti and Pedreira in Puerto Rico (1992, 29).

The 1944 and 1957 editions of the *International Encyclopedia of the Social Sciences*, edited by Seligman and Johnson, do not mention Ortiz. The 1968 edition edited by David L. Sills, despite its proclaimed goal of being more international in practice, reproduces this silence. As a commentator states,

> Despite the stated aim of the editors of the new encyclopedia to recognize the main international contributions to the development of anthropology and to overcome the Anglocentric character of the work produced by Seligman, the names of the founders of Afro-american studies, Nina Rodríguez and Fernando Ortiz, do not appear in the six hundred biographic entries (my translation; Ibarra, 1990).

Even as a marginal note, Ortiz's presence may be ephemeral. In the introduction to the first edition of Malinowski's *The Dynamics of Culture Change*, Phyllis Kaberry comments that Malinowski rarely used the term acculturation for he preferred the phrase culture contact, but she also states that once he had "advocated the adoption of a term coined by Don Fernando Ortiz, namely, 'transculturation'" (1945, vii). Kaberry, in a footnote, gives the source of this reference as the introduction to *Cuban Counterpoint* and adds that "Malinowski also employed this term in his article 'The Pan-African Problem of Culture Contact.'" However, she does not elaborate. It is remarkable that in her introduction to the second edition of Malinowski's book, published a decade and a half later, she modified this section. Her reference to Fernando Ortiz is dropped without explanation, and together with it the reference to Malinowski's article in which he had used Ortiz's transculturation (1961). This may explain why Wendy James, in an informative assessment of Malinowski's increasingly critical opinion of colonial powers, strikingly titled "The Anthropologist as Reluctant Imperialist," makes no mention of this article (1973, 41–69).[33] Through these silences that appear in works analyzing the development of Malinowski's ideas, Ortiz's influence on his thought is erased. Emerging anthropological canons appear exclusively as Malinowski's own.

How are authors from the periphery recognized at the center? Edward Said's discussions of the complicity between imperialism and knowledge have played a pathbreaking role in unsettling Euro-centric representations and in valorizing the work of authors from the periphery. Paradoxically, his comments on nonmetropolitan anthropology reveal the intricate mechanisms through which its marginalization is often unwittingly reinscribed. In a talk delivered to a professional gathering of anthropologists in 1987, he considers the limits of peripheral anthropology and suggests that imperial power is so dominant in the periphery that anthropologists working at the center must recognize the special responsibility they have:

> To speak about the "other" in today's United States is, for the contemporary anthropologist here, quite a different thing than say for an Indian or Venezuelan anthropologist: the conclusion drawn by Jurgen Golte in a reflective essay on "the anthropology of conquest" is that even non-American and hence "indigenous" anthropology is "intimately tied to imperialism," so dominant is the global power radiating out from the great metropolitan center. To practice anthropology in the United States is therefore not just to be doing scholarly work investigating "otherness" and "difference" in a large country; it is to be discussing them in an enormously influential and powerful state whose global power is that of a superpower (213).

Indeed, Jurgen Golte, a German anthropologist specializing in Andean studies, does not condemn anthropology in its entirety as a discipline bound to imperialism (Golte 1980). Rather, his argument implies that Latin American anthropology cannot escape from this complicity because it originates not in the European Enlightenment—presumably the foundation of metropolitan anthropology—but in European imperialism:

> Anthropology in Latin America is the instrument of the dominant classes in their relationship with the exploited classes. It forms part of a cultural context derived from bourgeois European thought which reifies potentially exploitable human groups. There are few indications that permit us to see a potential significance for anthropology in the context of the liberation of those who have been its objects, since Latin American anthropology as a discipline had its origin not in the tradition of the European Enlightenment but in the tradition of European imperialism (391).

Paradoxically, the epigraph of Golte's article is a quote from José Maria Arguedas, written in Quechua, in which Arguedas criticizes the colonial attitudes of the "doctors" working in the *Instituto de Estudios Peruanos*. Instead of seeing in Arguedas an example of a different kind of anthropology, as Angel Rama does, Golte treats his statement as a confirmation that no critical anthropology can exist in Latin America:

(Arguedas) learned anthropology in order to put his knowledge at the service of the Quechua peoples but failed to reach his goal. For him anthropology was not suitable for expressing and appreciating the Quechua worldview. Poetry, and the novel proved more valuable to him, but even with these he lost hope as the Quechua tradition and experience were being rapidly annihilated. He committed suicide in January 1970. (386–87)

Undoubtedly, Arguedas struggled with the tools he received and tried to adapt them to his own purposes, not always successfully. Like Ortiz, Arguedas rejected prevailing assumptions about progress and "acculturation," and sought instead to explore the dynamics of cultural transformation underpinning the formation of cultures in Latin America (1977). In a statement written before his death, on the occasion of receiving the literary prize "Inca Garcilaso de la Vega," Arguedas states, "I am not an acculturated person; I am a Peruvian who proudly, as a happy devil, speaks in Christian and in Indian, in Spanish and in Quechua" (my translation; 1971, 297). He explains that a principle guiding his life work was the effort to view Peru as an infinite source of creativity, a country endowed with such extraordinarily diverse and rich traditions, with such imaginative myths and poetry, that, "From here to imitate someone is quite scandalous" (my translation; 298).[34]

While Golte evidently intends to demonstrate Latin American anthropologists' complicity with imperialism in order to critique it, his blanket dismissal unwittingly completes the scandal of imperialism. An examination of the relationship between anthropology and imperialism should make visible not only the complicity, but also the contrapuntal tension between the two. While Said does not distance himself from Golte's opinion, his plea for reading cultures contrapuntally in *Culture and Imperialism* (1993) opens a space for a more nuanced evaluation of cultural formations involving centers and peripheries. He argues for the interactional, or contrapuntal, constitution of cultural identities. "For it is the case that no identity can ever exist by itself and without an array of opposites, negatives, oppositions: Greeks always require barbarians and Europeans, Africans, Orientals, etc." (52). He returns to this idea in his conclusion: "Imperialism consolidated the mixture of cultures and identities on a global scale. But its worst and most paradoxical gift was to allow people to believe that they were only, mainly, exclusively, white, or Black, or Western, or Oriental" (336). Against this paradoxical gift, Said ends by offering, with a sense of urgency, a contrapuntal perspectivism:

> Survival in fact is about the connections between things, in Eliot's phrase, reality cannot be deprived of the "other echoes (that) inhabit the garden." It is more rewarding—and more difficult—to think concretely and sympathetically contrapuntally about others than only about "us" (336).

A contrapuntal perspective, by illuminating the complex interaction between the subaltern and the dominant, should make it difficult to absorb one into the other, completing, however unwittingly, the work of domination.

It is significant that two critics of imperialism developed, independently of each other and fifty years apart, a contrapuntal perspective for analyzing the formation of cultures and identities. While Said derived his notion of counterpoint from Western classical music, Ortiz was inspired by Cuban musical and liturgical popular traditions. Perhaps a contrapuntal reading of Said and Ortiz points to a counterpoint between classic and popular music, and beyond that, to one between the cultures of Europe, Africa, and America.

Ortiz's presence as an echo in Said's garden, however, makes also more visible the need to understand the systemic and yet little known operations through which centers and margins are reproduced.[35] One may be tempted to see in the silences surrounding Ortiz, even as his ideas have had an impact on writing and analysis, a confirmation of Dipesh Chakrabarty's argument that Third World histories are written with reference to First World theoretical canons and thus to regard this as yet another proof that social theory is an attribute of the center (1992). Yet Ortiz's work complicates this view. His understanding of the relational nature of cultural formations undermines the distinction between First and Third worlds that seems so central to authors related to the Subaltern Studies project.

For instance, Gyan Prakash's suggestive proposal for writing post-Orientalist histories depends on a fundamental distinction between First and Third worlds (1990). While Prakash rejects foundationalist historiography, he ultimately brings "third world positions" as a slippery strategic foundation that guarantees "engagement rather insularity" (403). By anchoring the writing of history in the "Third World," he thus hopes to counter the possibility that a postfoundational historiography may lead to the aestheticization of the politics of diversity (407). Yet if the justification for this form of strategic foundationalism is its political efficacy, it must also account for the political consequences of categories that may polarize and obscure contests often fought on more varied terrains.

By examining how cultures shape each other contrapuntally Ortiz shows the extent to which their fixed and separate boundaries are the artifice of unequal power relations. A contrapuntal perspective may permit us to see how the Three Worlds schema is underwritten by fetishized geohistorical categories which conceal their genesis in inequality and domination; more important, it may help develop nonimperial categories which challenge rather than confirm the work of domination.[36] The issue is not that the categories we have at our disposal, such as the "First" or "Third" worlds, should not be used, for it is evident that in certain contexts they are not only indispensable but also efficacious, but that their use should attend to their limits and effects. Arguments about theory production and subaltern historiography polarized in terms of the Three Worlds schema run the risk of reinscribing the hierarchical assumptions which underpin it.

In "Travelling Theory" Said discusses how theory travels through a study of the migration of Lukacsian Marxism after WWI to France and England and its transformation in the works of Lucien Goldman and Raymond Williams (1983). While James Clifford considers Said's essay to be "an indispensable starting place for an analysis of theory in terms of its location and displacements, its travels," he also states that "the essay needs modification when extended to a postcolonial context" (1989, 184). He objects to Said's delineation of four stages of travel (origin; distance traversed; conditions for reception or acceptance; and transformation and incorporation in a new place and time), because, in his words, "these stages read like an all-too-familiar story of immigration and acculturation. Such a linear path cannot do justice to the feedback loops, the ambivalent appropriations and resistances that characterize the travels of theories, and theorists, between places in the 'First' and 'Third' worlds." Clifford complements Said's view of "linear" theoretical travel within Europe, with a conception of the "non-linear complexities" of theoretical itineraries between First and Third worlds (184–85).

While there are significant differences in the way theory travels between different regions of the world, perhaps one may push Clifford's argument further and propose that all theoretical travel is defined by "non-linear complexities," by processes of "transculturation" rather than "acculturation." The dichotomy between linear theoretical travel within Europe and non-linear theoretical travel elsewhere anchors theory production at the center. Thus, Clifford argues that Marx, who came from backward Germany, was "modernized" by moving to Paris and London. "Could Marx have produced Marxism in the Rhineland? Or even in Rome? Or in St. Petersburg? It is hard to imagine, and not merely because he needed the British Museum and its blue books. Marxism had to articulate the 'center' of the world—the historically and politically progressive source" (181). From a different perspective one could ask: could Marx have produced Marxism if he had not grown up in the Rhineland and kept it close to his concerns? Could Smith or Proudhon have produced Marxism? Marxism entailed a non-linear transculturation of intellectual formations involving not only "backward" and "modern" locations of European culture, but also dominant and subaltern perspectives within it. This suggests that all theoretical travel is inherently transcultural, but that the canonization of theory entails the retrospective erasure and "linearization" of traces and itineraries.

In my view, canons, not theories, are imperial attributes. While theoretical production—broadly understood as self-critical forms of knowledge—takes place in multiple forms and sites, disciplinary canons and the canonization of their creators largely remains the privilege of the powerful. Yet even canons, despite their hardness, are inhabited by subaltern echoes.[37] There is no reason to assume that theory travels whole from center to periphery, for in many cases it is formed as it travels through the interaction between different regions. The recognition of the existence of a dynamic exchange between subaltern and dominant cultures, including subaltern and metropolitan anthropologies, may lead to the realization that much of what today is called "cultural anthropology" may be more aptly addressed as "transcultural anthropology."

Transcultural Anthropology: Thinking
Contrapuntally about Malinowski

If transculturation is a two-way process, could we find Ortiz's echo in Malinowski's growing acceptance of a more dynamic conception of culture change? Could it be that through his effort to contain *Cuban Counterpoint*, Malinowski was affected by Ortiz's ideas? Through a contrapuntal reading of Malinowski and Ortiz I wish to trace links between the changing geopolitics of empire and the politics of social theory and see the formation of theoretical canons as the product of transcultural exchanges. It is well established that Malinowski's ideas had changed on the basis of his experiences in Africa and that he shared, with many European intellectuals during the interwar period, a growing disillusion with notions of progress. I am not interested here in the impossible task of estimating the extent of Ortiz's influence on Malinowski, only in noting its presence.

In the care that Malinowski took to contain *Cuban Counterpoint* within functionalist anthropology I note a certain ambivalence, a veiled desire to domesticate its power. I read this ambivalence as a tension between denial and disavowal, between totally repressing and fleetingly recognizing Ortiz's originality as an ethnographer—an originality that countered metropolitan assumptions about "home" and "abroad," "science" and "fiction," "civilized" and "savage cultures."

Three years earlier, Malinowski introduced *Facing Mt. Kenya* (1965), written by his former student, the independence leader Jomo Kenyatta, stating, "Anthropology begins at home . . . we must start by knowing ourselves first, and only then proceed to the more exotic savageries" (vii). Clearly, in the case of Ortiz and Kenyatta, home and the exotic savageries coincided; as anthropologists, beginning at home entailed adopting a self-reflexive view from within and from without. In the case of Malinowski, for whom home and science coincided at the center, there was no privileged place outside the center on which to stand in order to begin at home; a self-reflexive view risked undermining the conceits of home and science alike.

In Malinowski's case, the absence of a totalizing perspective may be seen in relation to Perry Anderson's argument that British bourgeois culture is organized around an "absent center," the lack of a total theory of itself. The void at the center generated, according to Anderson, a "pseudo-center—the timeless ego . . . the prevalence of psychologism."

A culture which lacks the instruments to conceive the social totality inevitably falls back on the nuclear psyche as a First Cause of society and history. This invariant substitute is explicit in Malinowski, Namier, Eysenck and Gombrich. It has a logical consequence. Time exists only as intermittence (Keynes), decline (Leavis), or oblivion (Wittgenstein). Ultimately, (Namier, Leavis, or Gombrich), the twentieth century itself becomes the impossible object (1968, 56).

Ortiz's totalizing historical narrative may have been particularly challenging to Malinowski, for *Cuban Counterpoint* constantly displaces and re-places home and exile, the national and the international, centers and peripheries, and shows how they are formed historically through constant interplay. "Historicity," as Michel-Rolph Trouillot argues in his review of Caribbean anthropology, "once introduced, is the nightmare of the ethnographer, the constant reminder that the groupings one needs to take for natural are human creations, changing results of past and ongoing processes" (1992). Ortiz's historical perspective sought not closure, but ruptures and openings. "Ortiz has never been able to encircle a subject of his study. In breaking the circle he seeks a problem's integral meaning, its significance in the world as a whole," according to a journalistic account titled "Mister Cuba" (Novás Calvo 1950). Ortiz's contrapuntal viewpoint also informed his practical concerns: "He has never thought of national affairs as separate from world affairs" (ibid.). Home and abroad, science and politics, self and other, were intimately related in Ortiz's historical work.

In *Argonauts of the Western Pacific*, Malinowski argues that an anthropological vision requires the perspective of a detached observer, capable of seeing the functioning of the whole society. "Exactly as a humble member of any modern institution, whether it be the state, or the church, or the army, is *of* it and *in* it, but has no vision of the resulting integral action of the whole, still less could furnish any account of its organization, so it would be futile to attempt questioning a native in abstract sociological terms" (1922, 12). Ortiz looks at Cuba, his "home," not from a detached archimedean point, but from within; his integral vision of the whole was developed by being *in* it and *of* it. As an intellectual from the periphery, developing a critical perspective from within does not preclude, but rather is conditioned by, a view from without. Yet his critical distance entails a critique of distance and of the view from afar. Thus detachment is not the opposite of commitment, but its necessary condition. Implicitly challenging the notion of the detached observer, Ortiz's work summons anthropologists, at the center or at the periphery, to recognize their positionality, the historicity of what Walter Mignolo has theorized as "the locus of enunciation" (1993). With particular urgency in postcolonial societies, this task involves taking a critical stance with respect to the available standpoints. Ortiz's work reflects a creative struggle to construct, rather than merely to occupy, a critical locus of enunciation.

It is difficult for me to imagine that Malinowski did not even glimpse the significance of Ortiz's achievement. If I am right in perceiving a tension in Malinowski's introduction between repressing and fleetingly seeing Ortiz's originality, perhaps we may see this tension on the two occasions when Malinowski used the term transculturation.

The first appears in Malinowski's attempt to lay the foundations of functionalist anthropology in his *A Scientific Theory of Culture and Other Essays* (1944). In the second chapter, "A Minimum Definition of Science for the Humanist," he states that theory finds a source of inspiration and correction in practical concerns: "Finally, in all this the inspiration derived from practical problems—such as colonial policy, missionary work, the difficulties of culture contact, and transculturation—problems that legitimately belong to anthropology, is an invariable corrective of general theories" (14). Malinowski does not acknowledge the paternity of this concept, or explain its significance.[38] While in later chapters he surveys developments in anthropological theory and identifies the authors related to them, he does not mention Fernando Ortiz or his work and ideas.

Ironically, on the basis of Malinowski's minimal use of the term transculturation in this book, the Oxford English Dictionary credits Malinowski as introducing the word; needless to say, it does not distinguish the term from acculturation, which was Ortiz's intent in coining it:

Transculturation(f. Trans- 3 + CULTURE *sb.* + -ATION) = Acculturation.
  1941 B. MALINOWSKI *Sci. Theory of Culture* (1944) ii. 14 Practical problems—such as . . . the difficulties of culture contact and transculturation—problems that legitimately belong to anthropology. 1949 Psychiatry XII. 184. This paper . . . has shown that the process of transculturation is not really a process of adaptation to a culture

but to a political situation. 1970 R. STAVENHAGEN in I. L. Horo-
witz *Masses in Lat. Amer.* vii. 287. We use the terms "transculturation" and "acculturation" interchangeably (1991, 2096).[39]

In striking contrast to Malinowski's minimal use of transculturation
in this canon-setting book, in the last article he wrote before his death
in 1942, "The Pan-African Problem of Culture Contact" (1943),[40] he
mentions the term several times, fully crediting Fernando Ortiz:

> We shall, in a moment, have a closer look at the general principles of
> this cultural transformation—or transculturation, as we might call
> it—following the great Cuban scholar, Dr. Fernando Ortiz, whose
> name may well be mentioned here, for he is one of the most passionate
> friends of the Africans in the New World and a very effective spokes-
> man of their cultural value and sponsor of their advancement (650).

In this article Malinowski takes an unusually strong critical stance
with respect to "the onslaught of white civilization on native cultures." In response to this onslaught, he states that "the anthropologist
should immediately register that a great deal of African culture was
destroyed or undermined in the process."

> The African lost a great deal of his cultural heritage, with all the
> natural privileges which it carried of political independence, of per-
> sonal feedom, of congenial pursuits in the wide, open spaces of his
> native land. He lost that partly through the predatory encroachment
> of white civilizations, but largely through the well-intentioned at-
> tempts of his real friends. At the same time, he did not gain any
> foothold in white citizenship in the social and cultural world of
> European settlers, officials, and even missionaries and educators—a
> foothold the promise of which was implicit in the very fundamental
> principles of Christianity and education alike (651).

In response to the ravages of colonialism, Malinowski makes an extraordinary proposal: the establishment in Africa of "an equitable
system of segregation, of independent autonomous development"
(665).

How to explain Malinowski's exceptional use of Ortiz's transculturation, his emotional denunciation of colonial destruction, his strong
critique of white civilization, his proposal for empowering Africans
at this time? We may find a clue to this puzzle in the way Malinowski justified his proposal for a system of "autonomous" African
development.

Speaking as a European, and a Pole at that, I should like to place here as a parallel and paradigm the aspirations of European nationality, though not of nationalism. In Europe we members of oppressed or subject nationalities—and Poland was in that category for one hundred and fifty years, since its first partition, and has again been put there through Hitler's invasion—do not desire anything like fusion with our conquerors and masters. Our strongest claim is for segregation in terms of full cultural autonomy which does not even need to imply political independence. We claim only to have the same scale of possibilities, the same right of decision as regards our destiny, our civilization, our careers, and our mode of enjoying life.

In this unusual statement, Malinowski places himself in the text, but this time not as an impartial observer standing on an archimedeal point outside history, as in his early texts, or as a concerned anthropologist, as in some of his later writings, but as a positioned historical actor, a kindred victim of history's atrocities. A decentered and fragmented Europe seems to have enabled Malinowski to locate himself *in it*, to *be of it*, to speak *from it*. It is as if at the zenith of his life, the advance of fascism in Europe, the occupation of Poland, the destruction of his own "home," had made him receptive to the claims and experience of other oppressed groups. At that moment he was able to acknowledge Ortiz, to fulfill the promise he once made to him.

At a time when no place can be safe from history's horrors or innocent of its effects, we may wish to establish our affiliation with the Malinowski of 1942, rather than with the canonical figure of *Argonauts*. Malinowski's acknowledgement suggests how Ortiz's ideas helped him view cultural transformations from a nonimperial perspective and support the claims of subject peoples. In the spirit of Ortiz's work, we may honor his memory by suspending belief in his individual authorship, and remembering *Cuban Counterpoint* as a text in which "cultural treasures," as Walter Benjamin and Angel Rama recognized, cease to owe their existence exclusively to the work of elites and become, as products of a common history, the achievement of popular collectivities as well. Reflecting Ortiz's own counterpoint with these collectivities, *Cuban Counterpoint* celebrates the popular imagination and vitality that inspired this work: the "antiphonal prayer of the liturgies of both whites and blacks, the erotic controversy in dance measures of the rumba and . . . the versified counterpoint of the unlettered guajiros and the Afro-Cubans curros" (1947, 4).

Fernando Coronil

Notes

I dedicate this essay to the memory of Oriol Bustamante and Sara Gómez, *amigos cubanos*. This essay incorporates comments of readers of a previous article on Ortiz (1993): John Comaroff, Roberto Da Matta, Paul Friedrich, Roger Rouse, Rafael Sanchez, David Scobey, and Rebecca Scott. This version benefitted from discussions after a lecture at Duke University and at the Affiliations faculty workshop, University of Michigan, and survived the scrutiny of participants in my Occidentalism graduate seminar: Marty Baker, Laurent Dubois, Paul Eiss, Javier Morillo, Colleen O'Neal, David Pederson, and Norbert Ross. George Stocking provided perceptive comments concerning my argument about the links between Ortiz and Malinowski. Julie Skurski helped at every stage. My gratitude to all.

1. "One tree does not make a forest." *Monte* has multiple meanings and associations in Cuba, e.g., "forest," "mountain," "woods," "bush." See Lydia Cabrera's *El Monte* (1975), which is dedicated to Fernando Ortiz.

2. "Literary works do not exist outside of cultures, but crown them. To the degree that these cultures are the multitudinous creations of centuries, the writer becomes a producer, dealing with the work of innumerable others: a sort of compiler (Roa Bastos might have said), a brilliant weaver in the vast historical workshop of American society." (From Angel Rama [1982]. My thanks to John Charles Chasteen for providing these translations.)

3. For the 1963 edition of this book, published by the Consejo Nacional de Cultura in La Habana, Ortiz added twelve chapters (over two hundred additional pages). He reorganized the second part by dividing it into two sections, "Historia, etnografía y transculturación del tabaco habano," which discusses tobacco production in thirteen chapters (III–X; XIX–XXII, and XXV), and "Inicios del azúcar y de la esclavitud de negros en Americas," which examines aspects of the evolution of sugar production in ten chapters XI–XVIII; XXIII–XXIV). The section on tobacco includes (as Chapter II) the theoretical introduction to the second part, where Ortiz presents the term "transculturation." There is no indication that he regarded this addition as definitive; in some respects, it only highlights the open-ended character of the book. Except for few modifications, he left the first section unchanged. The original Spanish and English editions included the prologue by Herminio Portell Vilá; in fact, Portell Vilá's name precedes Malinowski's in the Cuban edition (but not in the English version). In the 1963 edition, Portell Vilá's prologue is not included, but it appears again in an edition published by the La Universidad de las Villas in 1983.

4. These comments only intend to suggest certain links between the work of Ortiz and that of other postcolonial intellectuals. Benita Parry discusses the role of binary oppositions in Fanon and other postcolonial thinkers in an illuminating discussion of colonial discourse (1987). Homi Bhabha has brought up the important dimension of desire through a Lacanian reading of ambivalence in colonial situations (1985).

5. For an illuminating discussion of Malinowski's impact on anthropology, see Stocking (1983).

6. It should be noted, however, that this temptation has not led to the inclusion of Ortiz in contemporary discussions about alternative forms of ethnographic writing. For references that defined this discussion, see Marcus and Fisher (1986) and Clifford and Marcus (1986).

7. Ortiz's detailed examination of Afro-Cuban musical traditions in *Africanía de la música folklórica de Cuba* (1950) and *Los bailes y el teatro de los negros en el folklore de Cuba* (1951) valorizes some aspects of the social dimensions of these musical forms (participative, improvisational, playful, democratic, etc.), without, however, paying sufficient attention to the gender inequalities reinscribed in them. These studies, more than his theoretical discussion, make evident the significance of his notion of "transculturation." They also develop this concept in interesting directions. For example, Ortiz makes creative use of Marett's concept of "horizontal and vertical transvalorization" in his discussion of the transculturation of high and low musical traditions (1950, ix–xix).

8. Many Caribbean writers have shown the connection between colonialism and the Caribbean (one can think here of writers as diverse as José Martí, C. L. R. James, Eric Williams, Nicolás Guillén, Frantz Fanon, Aime Cesaire, Alejo Carpentier, George Lamming, Michelle Cliff, etc.). For a perceptive discussion of this connection in relation to anthropology, see Trouillot (1992, 22). While the Caribbean offers a powerful vantage point from which to look at colonialism, it needs to be complemented with perspectives from societies in which native populations were not destroyed, and remain, through processes of transculturation necessarily different from those that have taken place in the Caribbean, active forces in the present.

9. These two sections draw freely from a previous article of mine (1993).

10. His solution was to "keep Grau as provisional president while changing the structure of the government so as to include representatives of all important political groups, thus working toward a genuine 'national' government" (Aguilar 1972, 189). See also Louis A. Pérez (1986).

11. For a discussion of Ortiz's frame of mind at this time, which shows the burden of evolutionary, positivist, and racist scientific ideologies on his thinking, see Aline Helg (1990). Helg also argues that Ortiz's work at the beginning of the century helped consolidate dominant forms of racial prejudice (250, 1990). In an interesting discussion of "the conceptual horizon" within which Ortiz began his work in Cuba, Maria Poumier suggests that the race war of 1912, in which several thousand Afro-Cubans were massacred, helped consolidate certain silences about the discourse on race in Cuba at a time when Cuba's "whitening" was regarded as a necessary condition for its regeneration and unification as a nation. She situates Ortiz's silence with respect to this war as a sign of his ambiguous relationship with dominant views. Poumier also echoes the opinion of Cuban experts who believe, on the basis of the extraordinary ethnographic information presented in Ortiz's works, that at one point he must have become an initiate of the Abakuá, an Afro-Cuban men's society, and that his obligations to its strict codes of secrecy may have constrained him from speaking on current issues of race. Helg's and Poumier's informative articles contribute to a fuller understanding of Ortiz's thought by specifying the context of its initial formation. As their works make evident, an evaluation of Ortiz's entire corpus must avoid the danger both of deifying Ortiz as a "discoverer" of Cuba and champion of liberal and socialist nationalist projects, and of vilifying him as a ventriloquist of racial prejudice. These polarizing positions only simplify Ortiz's role in forming Cuban nationalist discourse and reduce the complex dynamic between his ideas and dominant ideologies to a few tenets. The challenge is to appreciate at once Ortiz's striking transformation *and* its limits, given the origins of his work

in racist ideologies and his position as an elite intellectual under changing political regimes. My thanks to Rebecca Scott for sharing these sources.

12. "No cerebro que llene el cráneo, sino ideas que lo inunden y limpien su modorra . . . Sólo nos falta una cosa: civilización." (my translation).

13. I am indebted to Julie Skurski for this formulation.

14. Ironically, this denial of a "filial relation to Europe" took place through filial links to Europe, making these enduring ties shape the intellectual landscape on which Afro-Cubans were imagined as a source of culture, not an obstacle to it. This irony reminds me of a cartoon by Quino, the Argentinian humorist, depicting a discussion between the Mafalda and Libertad, two politically concerned young girls, in which one of them asserts: "The problem with Latin Americans is that we always imitate others. We should be like North Americans, who don't imitate anyone."

15. The term "curros" referred to freed Afro-Cubans who roamed the streets of La Habana during the first half of the nineteenth century and were considered part of the underworld. Ortiz examined this topic in *Los negros curros*, a posthumous book (1986).

16. On the ambiguities which marked this process, see Skurski (1994). On allegory in the literature of the periphery, see the exchange between Fredric Jameson (1986) and Aijaz Ahmad (1987).

17. Ortiz built on a tradition of thought, both popular and academic, that established connections between tobacco and sugar and the formation of Caribbean nations. Hoetink argues that Pedro F. Bonó, in an essay written in the 1880s about the Dominican Republic, was the first scholar in the Caribbean to evaluate the social impact of tobacco and to develop an argument about its "democratic character." He states that this argument was "repeated and elaborated with great literary and scientific erudition by the Cuban scientist, Fernando Ortiz (1980, 5) (my translation). Bonó argued that tobacco was the basis of democracy in the Dominican Republic because it promotes economic stability among agricultural workers and landholders (Rodriguez 1964, 199). My thanks to Robin Derby for these references.

18. I am indebted to Roxanna Duntley Matos and Norbert Otto Ross for this observation.

19. Ortiz's approach contrasts with that of other critics of the period. Guerra y Sánchez, in the preface to the second edition of his *Azúcar y población en las Antillas* (1927), notes that his book had originally predicted the sugar crisis of 1930, which was determined by the "historical laws" that govern this industry, and had demonstrated how these events followed "with mathematical precision" from his account of Cuban history.

20. It is not clear who Ortiz has in mind, if Malinowski or some of his followers. Harriet de Onís translated "la escuela de Malinowski" as "Malinowski's followers." I prefer the term "school." My preceding reference to the genetic union between individuals is also a more literal translation of Ortiz's text than Onís's "reproductive process." George Stocking has suggested that Ortiz may have read Malinowski's "Methods of Study of Cultural Contact" (1938) (personal correspondence).

21. There is a vast literature on this subject. Most contemporary thinkers build upon the insights of Marx and Durkheim. For a discussion of anthropological perspectives on totemism and fetishism, see Terence Turner (1985). See also a discussion of commodities by Appadurai (1988) and Ferguson (1988).

22. According to George Stocking, Malinowski was in the habit of co-opting both people and concepts for his functionalist movement. He perceptively notes that while "too 'strong' a reading of his preface runs the risk of making too much of small things . . . positively valued, this is precisely the point of de- or re-constructionist readings (to make much of silences, or contradictions, or implications)" (personal communication).

23. They viewed acculturation as a two-way process mutually affecting the groups in contact; "Acculturation comprehends those phenomena which result when groups of individuals having different cultures come into continuous first-hand contact, with subsequent changes in the original cultural patterns of either or both groups . . . " (1936, 10). Malinowski's critique of Herskovits's "acculturation" shows not only their differences— which center on the role of applied anthropology, which Malinowski defended—but also on their shared understandings. Malinowski's agreement concerning the need to take into account the temporal or historical dimensions in the study of change shows how far he in fact is from Ortiz's position (1939).

24. See Trouillot's suggestive discussion of the "savage slot" in anthropological theory (1991); I am borrowing his term here.

25. Kaberry explains that this book is based on materials on this subject from Malinowski's 1936–1938 seminars at the London School of Economics, his 1941 seminar at Yale University, and other papers and articles. Kaberry acknowledges that although she was familiar with Malinowski's ideas, the book she edited was "not the book that he would have written" (1961, vi).

26. This edition, published in 1978, was part of the "Biblioteca Ayacucho" series. Among the editors of this series was Angel Rama, an admirer of Ortiz's work. The introduction by Julio Le Riverend is included in the 1983 edition of Ortiz's book published by the Editorial de Ciencias Sociales in La Habana.

27. Marinello made this statement, which in certain respects highlights Ortiz's links to "outsiders" and to an outsider's perspective, in a section of *Casa de las Américas* dedicated to Ortiz after his death in 1969 (which included articles by Nicolas Guillén, José Luciano Franco, José Antonio Portuondo, and Miguel Barnet. I have mentioned the introduction to the 1940 edition by H. Portell Vilá and the introduction to the Venezuelan edition of *Cuban Counterpoint* by Julio Le Riverend. Working within the context of U.S. universities see Cuban authors Antonio Benítez-Rojo's book, *Repeating Island*, dedicated to Ortiz on the occasion of *Cuban Counterpoint's* fiftieth anniversary (Benítez-Rojo closely aligns Ortiz with postmodernism), and Gustavo Pérez Firmat's *The Cuban Condition: Translation and identity in Modern Cuban Literature*. In an interesting critique of Todorov's and Naipaul's ideas, José Piedra (1989) uses the term "transculturation," without explicitly acknowledging Ortiz's work, to propose a transformative approach to colonial encounters. Coincidentally, Piedra supports his argument by comparing Malinowski's *Argonauts of the Western Pacific* to his *A Diary in the Strict Sense of the Term*, noting that in the diary one can detect traces of transcultural exchanges between Malinowski and the Trobrianders that are suppressed in his monological scholarly text (xx; for a discussion of this work see Coronil 1989).

28. Ibarra refers to works by Alfred Metraux, Alfonzo Reyes, Roger Bastide, Jean Price Mars, Melville Herskovits (1949). C. L. R. James refers to Ortiz's work in glowing terms: "it is the first and only comprehensive study of the

West Indian people. Ortiz ushered the Caribbean into the thought of the twentieth century and kept it there" (1963:395).

29. "Narrative Transculturation in Latin America," an important book as yet untranslated into English.

30. Rama appreciatively states that transculturation expresses "a Latin American perspectivism, including in that which it may incorrectly interpret" (1982:33). The reference to "incorrect interpretation" concerns Aguirre Beltrán's etymological critique of Ortiz's term (1957).

31. For example, Ortiz's ideas appear in the theoretical essays on cultural transformation by Bernardo Subercaseaux (1987) and George Yudice (1992). Mary Louise Pratt's *Imperial Eyes Travel Writing and Transculturation* (1992), credits Ortiz for the term transculturation and develops her conception of a linguistics of contact into a suggestive analysis of cultural transformation in "contact zones."

32. Ribeiro refers to Ortiz only in relation to specific aspects of Cuban ethnography.

33. I am grateful to Riyad Koya for bringing this article to my attention and for his helpful comments concerning Malinowski's affiliations and ideas.

34. "Imitar desde aquí a alguien resulta algo escandoloso."

35. The reality of Ortiz's absence in contrast to the expectation of his presence became clearer when I delivered a version of this essay at Duke University. While Walter Mignolo had assumed before the talk that Said's contrapuntal perspectivism in *Culture and Imperialism* built upon Ortiz's ideas, Fredric Jameson believed that Claude Levi-Strauss, given his *Mythologiques*, must have known and been influenced by Ortiz's work. After the lecture, Mignolo and I examined Said's book, and Jameson checked Levi-Strauss's works; in neither case did we find any reference to Ortiz. Yet Jameson conjectured that Levi-Strauss must have known about *Cuban Counterpoint*, given Malinowski's introduction to the book, his own work in Brazil, and his travels and contact in South America. He exclaimed that if Malinowski's partial recognition of Ortiz's ideas is disturbing, Levi-Strauss's silence is "thunderous."

36. I develop this idea in "Beyond Occidentalism: Towards Non-Imperial Geohistorical Categories," *Critical Inquiry* (forthcoming).

37. There are, of course, counter canons and alternative modalities of establishing the significance of ideas and authors. This schematic formulation seeks to suggest a way of thinking about the role of power in the relationship between theory production and canon formation. In the Angloamerican world there is an important literature on canon formation. For broad ranging discussions of the role of canons in literary theory which relativize Eurocentric standards from a Latin American perspective, see Navarro (1985), Mignolo (1991), and Pastor (1988).

38. According to Huntington Cairns, the editor of this posthumous book, Malinowski had revised the first two hundred typed pages of the manuscript (1944, vii). Thus it is clear that Malinowski not only wrote, but revised the paragraph where he mentions transculturation.

39. I am grateful to Colleen O'Neal for bringing this reference to my attention. The OED does not consider translated texts, and thus *Cuban Counterpoint* could not figure as a source of the term. Thus, what may have been an isolated error reflects a systematic exclusion of the contribution of non-English writers to the English language. My thanks to Bruce Mannheim and Charles Bright for these observations.

40. This is the article that disappeared as a citation in Kaberry's introduction to the second edition of Malinowski's *The Dynamics of Culture Change*.

Works Cited

Aguilar, Luis. 1972. *Cuba 1933: Prologue to Revolution*. Ithaca: Cornell University Press.

Aguirre Beltrán. Gonzalo, 1957. *El proceso de aculturación*. Mexico: Universidad National Autónoma de México.

Ahmad, Aijaz. 1987. "Jameson's Rhetoric of Otherness and the 'National Allegory.'" *Social Text* 17 (Fall): 3–25.

Anderson, Perry. 1968. "Components of the National Culture." *New Left Review* 50 (July-August): 3–57.

Appadurai, Arjun, ed. 1988. *The Social Life of Things: Commodities in Cultural Perspective*. Cambridge: Cambridge University Press.

Arguedas, José María. 1971. *El zorro de arriba y el zorro de abajo*. Buenos Aires: Editorial Losada.

———. 1977. *Formación de una cultura nacional indoamericana*. Mexico: Siglo Veintiuno.

Beals, Ralph. 1955. "Acculturation." *Anthropology Today: An Encyclopedic Inventory*. Edited by Alfred L. Kroeber. Chicago: The University of Chicago Press.

Benítez-Rojo, Antonio. 1992. *The Repeating Island: The Caribbean and the Postmodern Perspective*. Durham, N.C.: Duke University Press.

Bhabha, Homi K. 1985. "Signs Taken for Wonders: Questions of Ambivalence and Authority under a Tree Outside Delhi, May 1817." *Critical Inquiry* 12(1): 144–65.

Cabrera, Lydia. 1975. *El monte*. Miami: Ediciones Universal.

Canclini, Néstor García. 1989. *Culturas híbridas: Estrategias para entrar y salir de la modernidad*. México: Grijalbo.

Chakrabarty, Dipesh. 1992. "Postcoloniality and the Artifice of History: Who Speaks for 'Indian' Pasts?" *Representations* 37 (Winter): 1–26.

Clifford, James, 1989. "Notes on Travel and Theory." *Inscriptions* 5: 177–188.

Clifford, James, and Marcus, George E. (eds). 1986. *The Poetics and Politics of Ethnography*. Berkeley: University of California Press.

The Compact Oxford English Dictionary, 1989. Second Edition, Oxford: Clarendon Press.

Coronil, Fernando, 1989. "Discovering America—Again: The Politics of Selfhood in the Age of Postcolonial Empires." In "Colonial Discourses" edited by Rolena Adorno and Walter Mignolo, *Dispositio* 14 (36–38): 315–31.

———. 1993. Challenging Colonial Histories: Cuban Counterpoint/Ortiz's Counterfetishism." *Critical Theory, Cultural Politics, and Latin American Narrative*. Edited by Steven M. Bell, Albert H. LeMay, and Leonard Orr.

——— (forthcoming). "Beyond Occidentalism: Towards Non-Imperial Geohistorical Categories."

Derrida, Jacques. 1974. *Of Grammatology*. Baltimore: The Johns Hopkins University Press.

Eiss, Paul. 1994. "Politics of Space, or Political Emptiness?" Paper for the seminar, "Occidentalism and Capitalism," University of Michigan.

Ferguson, James. 1988. "Cultural Exchange: New Developments in the Anthropology of Commodities." *Cultural Anthropology*, 3(4): 488–513.

Firth, Raymond. 1957. *Man and Culture. An Evaluation of the Work of Bronislaw Malinowski*. London: Routledge and Kegan Paul.

Golte, Jurgen. 1980. "Latin America: The Anthropology of Conquest." *Anthropology: Ancestors and Heirs*. Edited by Stanley Diamond. New York: Moulton Publishers.

González Echevarría, Roberto, 1977. *Alejo Carpentier: The Pilgrim at Home*. Ithaca: Cornell University Press.

———. 1985. *The Voice of the Masters: Writing and Authority in Modern Latin American Literature*. Austin: University of Texas Press.

Guerra y Sánchez Ramiro. 1927. *Azúcar y población en las Antillas*. Havana Cultural S.A.

Helg, Aline. 1990. "Fernando Ortiz our la pseudo-science contre la sorcellerie Africaine á Cuba." *La pensée métisse. Croyances africaines et rationalité occidentale en questions*. Paris: Presses Universtaires de France.

Herskovits, Melvile. 1938. *Acculturation: The Study of Culture Contact*. Locust Valley, N.Y.: Augustin.

Hoetink, H. 1980. "El Cibao 1844–1900: su aportación a la formación social de la República. *Eme Eme. Estudios Dominicanos* 8(48): 3-19.

Ibarra, Jorge. 1990. "La herencia científica de Fernando Ortiz." *Revista Iberoamericana* 56.

James, C. L. R. 1963. *The Black Jacobins. Toussaint L'Ouverture and the San Domingo Revolution*. New York: Vintage Books.

James, Wendy. 1973. "The Anthropologist as Reluctant Imperialist." *Anthropology and the Colonial Encounter*. Edited by Talal Asad. Atlantic Highlands: Humanities Press.

Jameson, Fredric. 1981. *The Political Unconscious*. Ithaca: Cornell University Press.

———. 1986. "Third-World Literature in the Era of Multinational Capital," *Social Text* 15 (Fall): 65–88.

Kaberry, Phyllis. 1945. "Introduction." *The Dynamics of Culture Change*, by Bronislaw Malinowski. New Haven: Yale University Press.

———. 1961. "Introduction." *The Dynamics of Culture Change*, by Bronislaw Malinowski. New Haven: Yale University Press.

Kenyatta, Jomo. 1965. *Facing Mt. Kenya*. New York: Vintage Books.

Mair, Lucy. 1957. "Malinowski and the Study of Social Change." *Man and Culture: An Evaluation of the Work of Bronislaw Malinowski*. Edited by Raymond Firth. London: Routledge & Kegan Paul.

Malinowski, Bronislaw. 1922. *The Argonauts of the Western Pacific*. London: Routledge.

———. 1935. *Coral Gardens and Their Magic*. 2 vols. Bloomington, Ind.

———. 1938. Introductory Essay, "Methods of Study of Cultural Contact in Africa." International African Institute Memorandum, vi–xxxviii.

———. 1939. "The Present State of Studies in Culture Contact. Some Comments on an American Approach." *Africa* 12(1).

———. 1943. "The Pan-African Problem of Culture Contact." *American Journal of Sociology* 48(6): 649–65.

———. 1944. *A Scientific Theory of Culture and Other Essays*. Chapel Hill: The University of North Carolina Press.

———. 1945. *The Dynamics of Culture Change*. New Haven: Yale University Press.

Marcus, George E., and Fisher, Michael M. J. 1986. *Anthropology as Cultural Critique*. Chicago: The University of Chicago Press.

Marx, Karl. 1963. *The 18th. Brumaire of Louis Bonaparte.* New York: International Publishers.

———. 1981. *Capital.* Vol. 3. New York: Vintage Books.

Marinello, Juan. 1969. Untitled. *Casa De Las Americas* 10(55–57): 4.

Mignolo, Walter. 1991. "Canons A(nd) Cross-Cultural Boundaries (Or, Whose Canon Are We Talking About?)" *Poetics Today* 12(1): 1–28.

———. 1993. "Colonial and Postcolonial Discourse: Cultural Critique or Academic Colonialism." *Latin American Research Review* 28(3): 120–34.

Mintz, Sidney. 1985. *Sweetness and Power: The Place of Sugar in Modern History.* New York: Penguin Books.

Mintz, Sidney, and Eric Wolf. 1989. "Reply to Michael Taussig." *Critique of Anthropology 9:1.*

Mounier, María. 1993. "Le champ de vision du Troisiem Decouvreur de Cuba: Fernando Ortiz (1881–1969)." *Seminario de Historia de WBA Comunications.* Bellaterra: Universitat Autónoma de Bareclona.

Navarro, Desiderio. 1985. "Otras reflexiones sobre eurocentrism y antieurocentrismo en la teoría literaria de la América Latin y Europa. *Casa de las Américas* 150 (May–June): 68–78.

Novás Calvo, 1950. "Mister Cuba." *The Americas, a Review of Latin American News* 12(6): 6–8, 48.

Ortiz, Fernando. 1906. *Los negros brujos (apuntes para un estudio de etnología criminal).* Madrid: Libreria Fernando Fé.

———. 1910. La reconquista de América: reflexiones sobre el *panhispanismo.* Paris: Sociedad de Ediciones Literarias y Artísticas.

———. 1940. *Contrapunteo cubano del tabaco y el azúcar.* Havana: Jesús Montero. (new editions: New York, 1947 and 1970; Las Villas, 1963; Barcelona, 1973; Caracas, 1978; La Habana, 1983.)

———. 1947. *Cuban Counterpoint. Tobacco and Sugar.* New York: A. A. Knopf.

———. 1965 (1950). *Africanía de la música folklórica de Cuba.* Havana: Editora Universitaria.

———. 1951. *Los bailes y el teatro de los negros en el folklore de Cuba.* Havana: Ministerio de Educación.

———. 1963. *Contrapunteo cubano del tabaco y el azúcar.* Havana: Empresa Consolidada de Artes Gráficas.

———. 1986. *Los negros curros.* Havana: Editorial de Ciencias Sociales.

Parry, Benita. 1987. "Problems in Current Theories of Colonial Discourse." In *Oxford Literary Review.* 9.

Pastor, Beatriz. 1988. Polémicas en torno al canon: implicaciones filosóficas, pedagógicas y políticas. *Casa de las Américas* 171 (November–December): 78–87.

Pérez Firmat, Gustavo, 1989. *The Cuban Condition: Translation and Identity in Modern Cuban Literature.* Cambridge: Cambridge University Press.

Pérez, Louis A. 1986. *Cuba under the Platt Amendment, 1902–1934.* Pittsburgh: University of Pittsburgh Press.

Prakash, Gyan. 1990. "Writing Post-Orientalist Histories of the Third World: Perspectives from Indian Historiography." *Comparative Studies in Society and History* 32(2): 383–408.

Pratt, Mary Louise. 1986. "Fieldwork in Common Places." *Writing Culture: The Poetics and Politics of Ethnography.* Edited by James Clifford and George E. Marcus. Berkeley: University of California Press.

———. 1987. "Toward a Linguistics of Contact." *The Linguistics of Writing:*

Arguments Between Language and Literature. Nigel Fabb, Derek Attridge, Alan Durant, and Colin MacCube. New York: Methuen.

———. 1992. "Imperial Eyes." *Travel Writing and Transculturation.* London and New York: Routledge.

Rama, Angel. 1982. *Transculturación Narrativa en América Latina.* Mexico: Siglo XXI.

Redfield, Robert, Ralph Linton and Melville H. Herskovits. 1936. "Memorandum for the Study of Acculturation." *American Anthropologist.* 38: 230–233.

Ribeiro, Darcy. 1971. *The Americas and Civilization.* New York: E. P. Dutton and Co.

Riverend, Julio Le. 1978. "Ortiz y sus contrapunteos." Introduction. *Contrapunteo Cubano del Tabaco y el Azucar* by Fernando Ortiz. Caracas: Biblioteca Ayacucho.

Rodriguez, Demorizi. 1964. *Papeles de Pedro F. Bonó.* Santa Domingo: Editora Gritora del Caribe.

Sahlins, Marshall. 1993. "Good Bye Tristes Tropes: Ethnography in the Context of Modern World History. *Assessing Cultural Anthropology.* Edited by Robert Borofsky. New York: McGraw Hill.

Said, Edward. 1983. *The World, the Text and the Critic.* Cambridge: Harvard University Press.

———. 1989. "Representing the Colonized: Anthropology's Interlocutors." *Critical Inquiry* 15: 205–25.

———. 1993. *Culture and Imperialism.* New York: Alfred A. Knopf.

Sills, David L., ed. 1968. *International Encyclopedia of the Social Sciences.* New York: Macmillan.

Skurski, Julie. 1994. "The Ambiguities of Authenticity in Latin America: *Doña Bárbara* and the Construction of National Identity" *Poetics Today* 15(4): 59–81.

Sommer, Doris. 1991. *Foundational Fictions: The National Romances of Latin America.* Berkeley: University of California Press.

Spivak, Gayatri Chakravorty. 1974. "Translator's Preface." *Of Grammatology,* by Jacques Derrida. Baltimore: The Johns Hopkins University Press.

Stocking, George. 1983. "The Ethnographer's Magic. Fieldwork in British Anthropology from Tylor to Malinowski." *Observers Observed: Essays on Ethnographic Fieldwork.* Washington, D.C.: American Anthropological Association.

Subercaseaux, Bernando. 1987. "La apropriación cultural en el pensamiento latinoamericano." *Mundo* 1:29–37.

Taussig, Michael. 1989. "History as Commodity in Some Recent American (Anthropological) Literature." *Critique of Anthropology* 9(1): 7–23.

Trouillot, Michel-Rolph. 1991. "Anthropology and the Savage Slot." *Recapturing Anthropology.* Edited by Richard G. Fox. Santa Fe: School of American Research Press.

———. 1992. "The Caribbean Region: An Open Frontier in Anthropological Theory." *Annual Review of Anthropology* 21:19–42.

Turner, Terence. 1985. Animal Symbolism, Totemism, and the Structure of Myth." *Animal Myths and Metaphors in South America.* Edited by Gary Urton. Salt Lake City: University of Utah Press.

Yúdice, George. 1922. "Postmodernity and Transnational Capitalism in Latin America." *On Edge. The Crisis of Contemporary latin American Culture.* Edited by George Yúdice, Jean Franco, and Juan Flores. Minneapolis: University of Minnesota Press, 1992.

Introduction

I HAVE known and loved Cuba ever since the days of a prolonged stay in the Canary Islands during my early years. To the Canary Islanders Cuba was the land of promise, where they went to make money and then return to their homes on the slopes of Mount Teide or around Gran Caldera, or else to settle in Cuba and return to their native islands only for a holiday, humming Cuban songs, parading the Creole mannerisms and customs they had picked up, and relating the wonders of that beautiful land where the royal palm queens it, and the sugar-cane fields and the bottom lands where the tobacco grows spread their verdure as far as the eye can see. After establishing these contacts with Cuba in my youth, my ties with the island were strengthened later when I became acquainted with the name of Fernando Ortiz and his work in the field of sociology. His research into the African influences in Cuba, his investigation of the economic, social, and cultural aspects resulting from the interplay of influences between Africans and Latin Americans, always impressed me as being model works in their field.

So when at last I met Fernando Ortiz during my first visit to Havana in November 1939, it was a source of pleasure and profit to me to take greater advantage of his time and patience than is generally considered permissible on such short acquaintance. As might have been expected, we often discussed those most interesting of phenomena: the exchanges of cultures and the impact of civilizations on one another. Dr. Ortiz told me at the time that in his next book he was planning to introduce a new technical word, the term *transculturation,* to replace various expressions in use such as "cultural exchange," "acculturation," "diffusion," "migration or osmosis of culture," and similar ones that he considered inadequate. My instant response was the enthusiastic acceptance of this neologism. I promised its author that I would appropriate the new expression for my own use, acknowledging its paternity, and use it constantly and loyally whenever I had occasion to do so. Dr. Ortiz then pleasantly invited me to write a few words with regard to my "conversion" in terminology, which is the occasion for the following paragraphs.

[lvii

There is probably nothing more misleading in scientific work than the problem of terminology, of the *mot juste* for each idea, of finding the expression that fits the facts and thus becomes a useful instrument of thought instead of a barrier to understanding. It is evident that quarreling over mere words is but a waste of time; what is not quite so apparent is that the imp of etymological obsessions often plays mischievous tricks on our style—that is to say, on our thoughts—when we adopt a term whose component elements or basic meaning contains certain false or misleading semantic implications from which we cannot free ourselves, and thus the exact sense of a given concept, which in the interests of science should always be exact and unequivocal, becomes confused.

Take, for example, the word *acculturation,* which not long ago came into use and threatened to monopolize the field, especially in the sociological and anthropological writings of North American authors. Aside from the unpleasant way it falls upon the ear (it sounds like a cross between a hiccup and a belch), the word *acculturation* contains a number of definite and undesirable etymological implications. It is an ethnocentric word with a moral connotation. The immigrant has to *acculturate* himself; so do the natives, pagan or heathen, barbarian or savage, who enjoy the benefits of being under the sway of our great Western culture. The word *acculturation* implies, because of the preposition *ad* with which it starts, the idea of a *terminus ad quem.* The "uncultured" is to receive the benefits of "our culture"; it is he who must change and become converted into "one of us."

It requires no effort to understand that by the use of the term *acculturation* we implicitly introduce a series of moral, normative, and evaluative concepts which radically vitiate the real understanding of the phenomenon. The essential nature of the process being described is not the passive adaptation to a clear and determined standard of culture. Unquestionably any group of immigrants coming from Europe to America suffers changes in its original culture; but it also provokes a change in the mold of the culture that receives them. Germans, Italians, Poles, Irish, Spaniards always bring with them when they transmigrate to the nations of America something of their own culture, their own eating habits, their folk melodies, their musical taste, their language, customs, superstitions, ideas, and temperament. Every change of culture, or, as I shall say from now on, every transculturation, is a process in which something is always given in return for what one receives, a system of give and take. It is a process in which both parts of the equation are modified, a process from which a new reality emerges, transformed and complex, a reality that is not a mechanical agglomeration of traits, nor even a mosaic, but a new phenomenon, original and independent. To describe this process the word *trans-culturation,* stemming from Latin roots, provides us with a term that does not contain the implication of one certain culture toward which the other must tend, but an exchange between two cultures, both of them active, both contributing their share, and both co-operating to bring about a new reality of civilization.

In Dr. Ortiz's excellent analysis (Part II, Chapter ii) he points out clearly and convincingly that even the first Iberian settlers of Cuba, those who arrived shortly after its discovery by Christopher Columbus, did not bring with them to that West Indian island their Spanish culture in its totality, complete and intact. Dr. Ortiz indicates how this new choice of those settlers, motivated by different reasons and objectives, operated a change in them by the very fact of their migration to the New World. The make-up of the new society was determined from its beginning by the fact that the settlers were strained through the sieve of their own ambitions, of the various motives that caused them to leave their homeland and took them to another world where they were to live. These were people who, like the Pilgrim Fathers of Anglo-Saxon America, were not only seeking another land in which to re-establish the peace of their homes, but also had deep-seated reasons for forsaking their native land.

It would be as preposterous to suggest that the Spaniards who settled in Cuba became "acculturated"—that is, assimilated—to the Indian cultures, as it would be to affirm that they did not receive from the natives very tangible and definite influences. It will suffice to read this account of tobacco and sugar to realize how the Spaniards acquired from the Indians one of the two basic elements of the new civilization they were to develop in Cuba during the four centuries of their domination, and how the other was brought in by them to this island of America from across the ocean. There was an exchange of important factors, a *transculturation,* in which the chief determining forces were the new habitat as well as the old traits of both cultures, the interplay of economic factors peculiar to the New World as well as a new social organization of labor, capital, and enterprise.

In the course of his study Dr. Ortiz shows how the different waves of Mediterranean culture (Genoese, Florentine, Jewish, and Levantine) each brought with it its own contribution to the give and take of the process of transculturation. Negroes also reached Cuba, first from Spain itself, which before the discovery of the West Indies already included among its population great masses of African Negroes, and then directly from the different countries of Africa. And so, century after century, came successive arrivals of immigrants, French, Portuguese, Anglo-Saxon, Chinese . . . down to the recent influx of Spaniards after the late civil war and of Germans who took refuge there in their flight from Hitlerism. The author of this book emphasizes the importance of studying in all these cases both aspects of this contact and looking upon this phenomenon of integration as a transculturation—that is to say, as a process in which all the new elements are fused, adopting forms that are already established while at the same time introducing exotic touches of their own and generating new ferments.

With the kind indulgence of the reader, I can substantiate my complete accord with Fernando Ortiz by quoting from an earlier work of my own. On different occasions I have emphasized that the contact, clash, and transformation of cultures cannot be conceived as the complete acceptance of a given culture by any one "acculturated" group. Writing about the contacts between Europeans and Africans on the Dark Continent, I attempted to show how the two races "exist upon elements taken from Europe as well as from Africa . . . from both stores of culture. In so doing both races transform the borrowed elements and incorporate them into a completely new and independent cultural reality." [1]

I also suggested at the time that the result of the exchange of cultures cannot be regarded as a mechanical mixture of borrowed elements. "The developments resulting from exchanges of cultures are completely new cultural realities which must be studied in the light of their own significance. Moreover the typical phenomena of cultural exchanges (schools and mines, Negro places of worship and native courts of justice, grocery stores and country plantations) are all subject to the contingencies of the two cultures which flank them on both sides throughout their formation and development. It is a fact that these typical social phenomena are conditioned by the interests, the objectives and the impact of Western civilization; but

---

[1] *Methods of Study of Culture Contact in Africa.* Memorandum XV, International Institute of African Languages and Cultures, 1938, p. xvii.

they are also determined by the cultural reality of the Africans' reserves. Therefore it becomes apparent once more that we must take into account at least three phases in this constant interplay between the European and African cultures. The changes brought about in this way cannot be foretold or postulated, no matter how careful the scrutiny of the ingredients making up the two parent cultures. Even if we knew all the 'ingredients' that are to go into the formation of a school or a mine, a Negro church or a native court of justice, we could not foresee or foretell what the development of the new institution would be, for the forces that create such institutions and determine their course and development are not 'borrowed' but are sired by the institution itself." [2]

These quotations prove how completely my approach to the problem coincides with the analysis made by Fernando Ortiz in this volume. And I do not need to add that the fact gives me pride.

I think that with the foregoing I have complied with Dr. Ortiz's desire. Now, it would be as impertinent as it is unnecessary for me to comment upon the value and merits of this book. The intelligent reader will take account of the wealth of sober scientific labor and searching social analysis underlying the brilliant outward form of the essay, the fascinating play of words and the ingenious setting forth of contrasts and differences in this *Counterpoint*. In clear and vivid language, employing documentation as sound as it is unpedantic, Dr. Ortiz gives us first the initial definition of what he means by "counterpoint" between sugar and tobacco. Then he sets about translating his brilliant phrases into concrete and descriptive information. We see how the ecological conditions of Cuba made the island an ideal place for sugar and tobacco. This last point really requires no documentation or special pleading: the words Cuba and Havana are synonymous with the delights, the virtues, and the vices of the smoker. We all know that the luxury, the enjoyment, the æsthetics, and the snobbishness of smoking tobacco are associated with these three syllables: Havana.

The author then proceeds to give a brief résumé of the chemistry, the physics, the technique, and the art of the production of those two commercial products. As befits a true "functionalist" who knows full well that the æsthetics and psychology of the sensory impressions must be taken into account together with the habitat and the technology, Dr. Ortiz proceeds to study the beliefs, superstitions, and

[2] Ibid., p. xxiv.

cultural values that touch upon the substance as well as the acts of smoking and sweetening. In the vein of Voltaire, the author slyly considers the supposed malignancy and Satanic associations of the diabolical weed. The religious and mystical attributes of tobacco are a theme of this book that will make it of special value to anthropologists.

Returning to the contrasts between these two vegetable products, the author points out the difference between the skill and care demanded in the cultivation, harvesting, selection, and manufacture of tobacco, and the rough nature of the agricultural, industrial, mechanical, and mercantile steps in the preparation of sugar. We come close to the soil of Cuba, and make the acquaintance of the tobacco-growers and cigar-makers as well as of the slaves and peons who work in the canefields and the sugar mills. In all lovers of good tobacco and those who have known the sweetness of the sugar of Cuba, these intimate pictures of the Cuban landscape where these products are raised will arouse a keen personal interest. In the passages describing the raising of tobacco, the technique of its cultivation, its harvesting, its curing, and its final preparation, there is a wealth of narrative charm and analysis as well as important information that will be of keen interest to those who devote themselves to anthropology and economy and will fascinate the ordinary reader.

Fernando Ortiz belongs to that school or tendency of modern social science known today by the name of "functionalism." No one sees more clearly than he that the economic and ecological problems of works and skills are fundamental in the industries dealt with in this book; but the author is also clearly aware of the fact that the psychology of smoking, its æsthetics, the beliefs and the emotions associated with each of the finished products described by him, are important factors in their consumption, their distribution, and their manufacture. Reading the paragraphs in which he describes the fine art of cigar manufacture, the personal devotion of the growers and cigar-makers to their constant task of improvement and selection to endow with sensual beauty the material object that satisfies the habit, one might almost say the vice, of the habitual smoker, there came to my mind over and over again the best definition that has been given of beauty: *"La beauté n'est que la promesse du bonheur"* (Stendhal).

Like the good functionalist he is, the author of this book resorts to history when it is really necessary. The chapters dealing with the different methods of working the land, depending on whether it is for sugar or tobacco, with the differences in the systems of labor, whether by free workmen, slaves, or hired laborers, and, finally, those having to do with the varying political implications of the two industries are written as much from the historical point of view as from the functional. A number of the most important historical facts have been more fully documented in the supplementary chapters that comprise the second part of the book.

With reference to the political implications inherent in the basic problem of this book, Dr. Ortiz has refrained from any unwarranted judgments. Nevertheless, as regards this aspect, I hope to see the book translated into English and read by the students, the politicians, and of course the general public of the United States. Fernando Ortiz, Cuban by birth and by citizenship, is justly proud of the role his country has played in the history of sugar, through the vast production of its centrals, and in that of smoking, through having developed in its vegas the best tobacco in the world. The author reminds us that it was none other than Christopher Columbus himself who carried tobacco from Cuba to give it to the Old World and who brought sugar to these West Indian islands. He describes the triumphal march of tobacco all over the face of the globe and determines the profound influence exerted by sugar on the civilization of Cuba, its principal effect having been, perhaps, to occasion the importation from Africa of the many and uninterrupted shiploads of black slave workers. And the author points out, too, how in the fortunes of Cuba tobacco and sugar have been closely intertwined with the fabric of its relations with foreign nations.

In both these principal aspects of its economic life Cuba is at present becoming more and more closely linked to the United States. The disastrous events now taking place in Europe tend to make this connection even closer and more exclusive. But the same observations I made above in analyzing the phenomenon of transculturation could be repeated here if we transfer them to a slightly different field. Interdependence is mutual. Cuba, together with Mexico, is the closest of the Latin-American nations in which the "good-neighbor policy" should be set up with all the intelligence, foresight, and generosity of which statesmen and even the captains of finance of the United States are sometimes capable.

In my long conversations with Don Fernando we meditated on the problem of why great North American institutions of teaching and investigation have been established in China, in Syria, on the Bosporous or the shores of the Pacific, and not in the countries of Latin America. If some of the great and richly endowed cultural foundations of the United States were to contribute to the creation of institutes of economic and social investigation in these different countries it would do much, very much, to foster a better mutual understanding and greater economic co-operation between the nations of this hemisphere. If my understanding of these problems is correct and unbiased, Cuba, from this point of view, is the outstanding spot of Latin America, the most suited for establishing there a clearing house of information, ideas, influences, and cultural movements which would contribute effectively to goodwill and mutual understanding.

This book is a masterpiece of historical and sociological investigation, as admirably condensed and documented as it is free of sterile, pedantic erudition. To be sure, several of its sections, and even many of its paragraphs, could be used as guides to works of investigation in the field of ethnography. Those who work in those institutes of economic and social research whose creation Fernando Ortiz proposed at the Eighth American Scientific Congress, recently held in Washington (May 1940), and which the assembly unanimously agreed to recommend, especially with regard to the national institute to be set up in Cuba, could well initiate their activities with topics as profoundly complex and significant as the role of sugar and tobacco in the economy, ethnography, and sociology in the present and future of the Cuban people. As an outline for the development of the work of such investigations the present book is ideal. With these scientific efforts of study and analysis of the objective realities through which the complex social phenomena of peoples reveal themselves, the understanding between the Americas would become greater, more perfect, and more fruitful the goodwill of the North Americans toward Cuba, the most important and closest of her island neighbors of Latin America. It is obvious that here, as in every phase or phenomenon of transculturation, the influences and understanding would be mutual, as would the benefits.

BRONISLAW MALINOWSKI

Yale University, July 1940

By Way of Prologue

THE presence of these words of introduction of mine to a further contribution from the pen of Fernando Ortiz to his basic studies of Cuban national themes can be explained only as one more opportunity he has given me to allow my humble name to appear alongside his in his work of investigation, evaluation, interpretation, and dissemination of our culture, a work that is without doubt the most original and fruitful, the most universal in scope and nationally useful that Cuba has produced in its whole history.

There is therefore nothing startling in my sincere statement of belief that prologues to Fernando Ortiz's books are superfluous, in view of the author's international repute, the quality of his work, and the fact that they come from the hand of the person who in our days completely fulfills Martí's dictum on Domingo del Monte, to whom he referred as "the most real and useful Cuban of his day." This work, like the others that flow from his pen in furtherance of the ends of culture, so clearly bears this out that no further recommendation or judgment, beyond the author's outstanding merits, is needed to commend it to the attention and interest of the thoughtful reader.

Writing a prologue to a book of Fernando Ortiz's is a task that honors the prologizer. It gives him an opportunity to link a name of more or less significance to the intellectual achievements of a privileged mind equipped with an active, creative erudition that is astounding in its profoundness and variety. Aside from the bonds of the old and warm friendship that exists between us, my only right to set down these words of introduction to Dr. Ortiz's stimulating and highly original study of the parallel roles played by sugar and tobacco in the economy of Cuba lies, perhaps, in the fact that I, too, have on various occasions considered this fundamental problem which Dr. Ortiz has examined wittily and precisely, entertainingly and with scientific exactitude in these pages, which are essential to an understanding of our national evolution.

Century-old economic and political errors that sprang up independently of the will of the Cuban people as the result of Spain's restrictive colonial policy have been responsible for the monstrous increment of the sugar industry, which today weighs so heavily and with such disastrous effects on the life of our country. In the suggestive pages of this searching economic study, whose implications Dr. Ortiz has tried in vain to hide behind its whimsical title, he proves in detail and incontrovertibly that sugar cane, the industry that exploits it, the system that has developed around it, and so on, represent something foreign to our country, completely accidental, like a parasitic body, which although attached to us for centuries, still serves foreign rather than national interests, as though its loyalty to its other-world origin made it impossible for it ever to shed its characterizing traits of exploitation, unfair privilege, and protectionism.

In the past century as in the present the economists serving the interests of the sugar-producers, and others who have not wished to go to the heart of the problem or lacked the capacity to do so, have developed the theory that there is an identity between these interests and those of the nation, that sugar and Cuba are synonymous. This is the prevalent idea with regard to our country in Europe and in certain American countries, to the point where, by association of ideas, at the mere mention of the name of Cuba it almost seems as though they were tasting a lump of sugar, after the manner of Pavlov's dogs. In the eighteenth century Benjamin Franklin referred to Cuba and the other islands of the West Indies as "the sugar islands," and the nickname of "the sugar-bowl of Santo Domingo" was commonly applied to the French colony of Haiti, the principal source of supply of sugar for the European markets before Cuba made up its mind to replace Haiti, after the latter's industry had been ruined by its war of independence, in the dubious honor and very relative advantage of being the "sugar-bowl of the world."

The attempted identification of the sugar interests with those of the nation is completely artificial and is the result of misconceptions and selfishness. The Cuban sugar industry has never been self-sufficient; it has always lived on the favor and sacrifices of others, like some huge parasite sucking out juices more vital than those that come from the cane under the pressure of grinders and crushers through which, in excruciating torture, pass the sweet stalks and the happiness of those whose destiny it is to work with them.

Before emancipation it was the slave whose labor produced sugar. On his physical and moral suffering the great parasite and the exploiters, who never share their profits when things are going well and who always demand new sacrifices the moment their earnings begin to decline, battened. When some fifty years ago the system of slavery, which had been very largely at the service of the sugar-producers, was abolished, it was replaced by the growing proletarization of the Cuban people in the same capacity. Under the domination of this industry today we Cubans drag out our weary existence. A difference of half a cent in the tariff on the sugar we export to the United States represents the difference between a national tragedy in which everything is cut, from the nation's budget to the most modest salary, even the alms handed to a beggar, and a so-called state of prosperity, whose benefits never reach the people as a whole or profit Cuba as a nation.

The arbitrary resolve of the secretary of agriculture of the United States to suspend the system of sugar quotas, the no less unjust decision of the president of that country to increase the duty on sugar several mills, send a chill of terror into the heart of every Cuban, for this means a shrinkage of income, increased unemployment, fewer working days, a decrease in purchases abroad, and the consequent falling off of customs duties, which supply almost fifty per cent of the government's funds. There are those who selfishly confuse patriotism with the price of sugar, and who, blinded by their ignorance and ambition, like to talk with others of their own sort of what a good thing it would be if Cuba formed part of the United States so sugar could enter that country duty-free. . . .

The last Cuban revolution, which continued from 1895 to 1898 and brought to an end Spanish sovereignty over the island, destroyed for a time the domination of the sugar industry over national life. Gómez and Maceo's march from east to west, which swept like an avalanche of flame over the canefields of the islands and the sugar mills, which had been warned to stop grinding, turned the machinery, the system of communications, and the plantations all to ashes. The revolution had a golden opportunity to reorganize the economic life of the nation, giving up the system of a cash crop —that is to say, a crop for export, which gives us a lower standard of living than we should have and for which it is responsible because it obliges us to spend this money, which hardly warms our pockets, on a multitude of imported articles that increase the cost of living of our people unnecessarily. The United States, through the government imposed on Cuba by military intervention, shored up and re-established for its own benefit the system that had been overthrown. This was the most serious harm, among many others, to which the occupation of the island by the United States army gave rise, inasmuch as the irresponsible and arbitrary petty officials who for four years did as they pleased in Cuba purposely avoided the cardinal problem, which was the reorganization of Cuba's economy to free it from the domination of sugar, and fostered the revival of the great parasite on Cuba's national life, so that through it Cuba would continue to be effectively dominated by the United States. It was in this way that the economic aims of the revolution of 1895 were frustrated with victory almost in sight, and everything still remains to be done in exterminating this dangerous monster if Cuba is to survive. . . .

Tobacco, with its whole complicated process of planting, harvesting, preparation, and distribution, is the other extreme of this parallelism which Dr. Ortiz examines in this thoughtful study. The plant and the manner of preparing its aromatic leaves for smoking are indigenous to us, and it was here that Europeans first saw and learned smoking. The tobacco industry is Cuban in origin and it has almost always been essentially Cuban. The preparation of tobacco demands special care, skill, a technique all its own, and the intervention of experts in the different branches, which makes it the joint effort of growers, sorters, strippers, cigar-makers. These workers are progressive, informed, alert, and well organized to protect their rights. For many years during the eighteenth century, during the past century, and down to our own times the tobacco-grower and the tobacco worker have been the representatives of Cuban nationalism. They initiated or co-operated in the movements for independence and were the most capable spearhead of the Cuban proletariat. The invasion of overseas capitalism and the growing mechanization of the tobacco industry have wrought a profound change in its composition and workings and have occasioned a crisis, familiar to us but not irreparable like that of sugar, which affects plantings, prices, and sales.

In the case of tobacco, however, Cuba can defend itself better than where sugar is concerned, for, as José Manuel Cortina intelligently points out, we have a natural monopoly. A considerable number of countries produce tobacco in larger quantities than Cuba, but none of them equals it in the quality of the leaf, which is unique and which makes a Havana cigar a privileged product. This is due to the special soil of Vuelta Abajo or Manicaragua or Guantánamo as well as to the special methods of cultivation worked out by our growers, and also to the introduction of certain measures of planned economy, which date back to the eighteenth century. The captain general, Marquis de la Torre, seeing that the excess of production of tobacco of inferior grade was undermining the demand for Cuban tobacco on the world market and causing great stocks to pile up in the warehouses of Seville, Cádiz, and elsewhere, ordered the compulsory plowing under of those fields that were producing tobacco of poor quality and encouraged its cultivation in regions that produced a rich, fragrant leaf of superior grade. This was planned economy; thanks to it the bases of a natural selection to which Cuban tobacco owes the fame it has enjoyed in the world for a century and a half were established.

This whole historical and economic process, with its profound political connotations for Cuba, is set forth by Dr. Ortiz in this original study. The various references to the archpriest of Hita, with his witty and pertinent observations, point up the fact that the outstanding Cuban of his day, in this comparison between Doña Azúcar and Don Tabaco, is most ably pursuing an educational and patriotic purpose that deserves the praise of all Cubans willing to think independently on the subject of vested interests.

HERMINIO PORTELL VILÁ

Havana, December 1939

CUBAN COUNTERPOINT

Tobacco and Sugar

I

CUBAN COUNTERPOINT

Cuban Counterpoint

C ENTURIES ago a jovial Spanish archpriest, a famous poet of the Middle Ages, personified Carnival and Lent and made them speak in unforgettable verses, cleverly putting into the affirmations and rebuttals of the satirical contest between them their contrasting ethics and the ills and benefits that each has conferred upon mankind. This allegorical dialogue by the priest Juan Ruíz, *"Pelea que ovo Don Carnal con Doña Quaresma"* in his *Libro de Buen Amor,* redounded to the glory not only of his name but of the parish of Hita, whose fame rests exclusively upon that of the genial composer of rustic love songs and every manner of unabashed, mocking verse.

Perhaps that famous controversy imagined by this great poet of the Middle Ages might serve as a literary precedent to permit me now to personify dark tobacco and "high yellow" sugar, and let them, in the guise of a fable, uphold their vying merits. But lacking, as I do, authority either as poet or as priest to conjure up creatures of fantasy and lend them human passions and superhuman significance, all I can do is to set down, in drab prose, the amazing contrasts I have observed in the two agricultural products on which the economic history of Cuba rests.

These contrasts are neither religious nor moral, as were those rhymed by the poet priest between the sinful dissipations of Carnival and the purifying Lenten-tide abstinences. Tobacco and sugar are opposed to each other in the economic as in the social field, and even strait-laced moralists have taken them under consideration in the course of their history, viewing the one with mistrust and the other with favor. Moreover, the contrasting parallelism between tobacco and sugar is so curious, like that between the two characters in the archpriest's dialogue, that it goes beyond the limits of a merely social problem and touches upon the fringes of poetry. A poet might be able to give us in robust verses a *Pelea de Don Tabaco y Doña Azúcar*—a "Controversy between Don Tobacco and Doña Sugar." This type of

[3

dialogued composition which carries the dramatic dialectic of life into the realm of art has always been a favorite of the ingenuous folk muses in poetry, music, dance, song, and drama. The outstanding examples of this in Cuba are the antiphonal prayers of the liturgies of both whites and blacks, the erotic controversy in dance measures of the rumba, and in the versified counterpoint of the unlettered *guajiros* and the Afro-Cuban *curros.*

A typical folk ballad, or one of those ten-line stanza poems in the vernacular of the *guajiros* or *curros,* whose disputants were the masculine tobacco and the feminine sugar might be of educational value in schools and song festivals, for in the study of economic phenomena and their social effects it would be hard to find more eloquent lessons than those afforded by Cuba in the startling counterpoise between sugar and tobacco.

The contrast between tobacco and sugar dates from the moment the two came together in the minds of the discoverers of Cuba. At the time of its conquest, at the beginning of the sixteenth century, by the Spaniards who brought the civilization of Europe to the New World, the minds of these invaders were strongly impressed by two gigantic plants. The traders arriving from the other side of the ocean had already fixed the greedy eyes of their ambition on one; the other they came to regard as the most amazing prize of the discovery, a powerful snare of the devil, who by means of this unknown weed stimulated the senses as with a new kind of alcohol, the mind with a new mystery, the soul with a new sin.

Out of the agricultural and industrial development of these amazing plants were to come those economic interests which foreign traders would twist and weave for centuries to form the web of our country's history, the motives of its leaders, and, at one and the same time, the shackles and the support of its people. Tobacco and sugar are the two most important figures in the history of Cuba.

Sugar and tobacco are vegetable products of the same country and the same climate, but the biological distinction between them is such that it brings about radical economic differences as regards soil, methods of cultivation, processing, and marketing. And the amazing differences between the two products are reflected in the history of the Cuban nation from its very ethnological formation to its social structure, its political fortunes, and its international relations. (See Part II, Chapter i.)

The outstanding feature of our economic history is in reality this multiform and persistent contrast between the two products that have been and are the most typical of Cuba, aside from that period of brief duration at the beginning of the sixteenth century when the conquistadors' gold-mining activities and the cultivation of yucca fields and stock-raising to supply cassava bread and dried meat for the conquerors' expeditions took pre-eminence. Thus a study of the history of Cuba, both internal and external, is fundamentally a study of the history of sugar and tobacco as the essential bases of its economy.

And even in the universal history of economic phenomena and their social repercussions, there are few lessons more instructive than that of sugar and tobacco in Cuba. By reason of the clarity with which through them the social effects of economic causes can be seen, and because few other nations besides ours have presented this amazing concatenation of historical vicissitudes and this radical contrast, this unbroken parallelism between two coexisting orders of economic phenomena, which throughout their entire development display highly antithetical characteristics and effects, it is as though some supernatural teacher had purposely selected Cuba as a geographic laboratory in which to give the clearest demonstrations of the supreme importance of the basic economy of a nation in its continuous process of development.

The posing and examination of this deep-seated contrast which exists between sugar and tobacco, from their very nature to their social derivations, may throw some new light upon the study of Cuban economy and its historical peculiarities. In addition it offers certain curious and original instances of *transculturation* of the sort that are of great and current interest in contemporary sociological science. (See Part II, Chapter ii.)

Tobacco and sugar are both products of the vegetable kingdom that are cultivated, processed, and sold for the delectation of the mouth that consumes them.

Moreover, in the tobacco and sugar industry the same four factors are present: land, machinery, labor, and money, whose varying combinations comprise the history of these products.

But from the moment of their germination in the earth to their final human consumption tobacco and sugar behave in ways almost always radically opposed.

Sugar cane and tobacco are all contrast. It would seem that they were moved by a rivalry that separates them from their very origins. One is a gramineous plant, the other a solanaceous; one grows from cuttings of stalk rooted down, the other from tiny seeds that germinate. The value of one is in its stalk, not in its leaves, which are thrown away; that of the other in its foliage, not its stalk, which is discarded. Sugar cane lives for years, the tobacco plant only a few months. The former seeks the light, the latter shade; day and night, sun and moon. The former loves the rain that falls from the heavens; the latter the heat that comes from the earth. The sugar cane is ground for its juice; the tobacco leaves are dried to get rid of the sap. Sugar achieves its destiny through liquid, which melts it, turns it into syrup; tobacco through fire, which volatilizes it, converted into smoke. The one is white, the other dark. Sugar is sweet and odorless; tobacco bitter and aromatic. Always in contrast! Food and poison, waking and drowsing, energy and dream, delight of the flesh and delight of the spirit, sensuality and thought, the satisfaction of an appetite and the contemplation of a moment's illusion, calories of nourishment and puffs of fantasy, undifferentiated and commonplace anonymity from the cradle and aristocratic individuality recognized wherever it goes, medicine and magic, reality and deception, virtue and vice. Sugar is *she;* tobacco is *he*. Sugar cane was the gift of the gods, tobacco of the devils; she is the daughter of Apollo, he is the offspring of Persephone.

In the economy of Cuba there are also striking contrasts in the cultivation, the processing, and the human connotations of the two products. Tobacco requires delicate care, sugar can look after itself; the one requires continual attention, the other involves seasonal work; intensive versus extensive cultivation; steady work on the part of a few, intermittent jobs for many; the immigration of whites on the one hand, the slave trade on the other; liberty and slavery; skilled and unskilled labor; hands versus arms; men versus machines; delicacy versus brute force. The cultivation of tobacco gave rise to the small holding; that of sugar brought about the great land grants. In their industrial aspects tobacco belongs to the city, sugar to the country. Commercially the whole world is the market for our tobacco, while our sugar has only a single market. Centripetence and centrifugence. The native versus the foreigner. National sovereignty as against colonial status. The proud cigar band as against the lowly sack.

Tobacco and sugar cane are two gigantic plants, two members of the vegetable kingdom which both flourish in Cuba and are both perfectly adapted, climatically and ecologically, to the country. The territory of Cuba has in its different zones the best land for the cultivation of both plants. And the same happens in the combinations of the climate with the chemistry of the soil.

Even though all sugar is alike, Cuba possesses certain special conditions for its cultivation. The climate for cane is that determined by the isothermal lines of 68° rather than by mere tropical location. Broadly speaking, the sugar-producing zone of the world lies between latitude 22° North, the position of Havana, and 22° South, that of Rio de Janeiro. The whole of the West Indies lies within this geographical region; but Cuba, because of its proximity to the northern limit, and because of the effects of the adjacent winter cold, has advantages over the other islands. In no other part of the world do sun, rainfall, land, and winds collaborate as they do there to produce sugar in those little natural sugar mills of the cane stalks. The hot rainy season is very favorable for the rapid growth of the cane, and it rains heavily in Cuba. If "the cane prepares its sugar in the sweat of its leaves," to use the phrase of Alvaro Reynoso, we might say that the torrential rains bring to the cane the treasure of calories which is the gift of its father, the sun. If the latter grows angry and withholds the rain, the cane is stunted and impoverished. At the same time the pleasant winter season, without frost but with cold snaps, hastens the crystallization of the saccharose and in Cuba guarantees the vegetation cycle of the cane, its growth and its maturity. Nature in Cuba has given sugar cane a perfect annual cycle of growth and production, which affords it a situation of privilege there.

As for Cuban tobacco, being, as it is, the best in the world, it is unnecessary to analyze the advantages of soil and climate. From the excellence of the plant one can infer that of its natural conditions of production. A noteworthy poet, Narciso Foxá, said of the tobacco of his country: "A special gift conferred upon Cuba."

Sugar cane and tobacco are typical plants of the tropics, lasciviously hot-blooded, hating the cold, given to prodigious growth in stalk and leaves, with a tendency to "grow rank," as the Cuban countryman puts it.

Cane and tobacco do not concentrate all their strength in spears and ears, like wheat and corn, plumed like conquerors vainglorious of their lineage. Nor like the yucca and the potato, humble inhabitants of the earth, do they hide their wealth in the ground as in a miser's pot. But for the plant of wheat, corn, yucca, or potato, their consumption by man is their complete destruction. Each of these plants as it gives its fruit to mankind gives him its life and its posterity as well. If man wishes the plant he despoils to reproduce itself, to give further fruits, he must forgo part of its yield, must save grains from the ear of wheat or corn, some of the tubercules of the roots, and only in this way can the miracle of creation be repeated for the future. Not so with sugar cane or tobacco, more generous by far; each plant gives to man not only its complete usefulness but at the same time its unbroken continuity.

Cane and tobacco yield their sought-for wealth in such a way that they can make a present of it without depriving themselves of the roots or seeds that are the means by which they perpetuate the possibility of their favors. Sugar cane, after offering up its juicy stem to the very last of its precious joints, will shoot up again from its own fertile roots and reproduce its rich stalks year after year, as long as it has the help of earth and sun. Tobacco, after each plant has given every one of its aromatic leaves to the harvester, offers him its myriad seeds to ensure the repetition of its gifts the following year. The difference between the two plants lies in the fact that the cane comes up again from its own roots whereas tobacco is born anew from its seeds which it flourishes aloft. (See Part II, Chapter iii.)

Tobacco is born, sugar is made. Tobacco is born pure, is processed pure and smoked pure. To secure saccharose, which is pure sugar, a long series of complicated physiochemical operations are required merely to eliminate impurities—bagasse, scum, sediment, and obstacles in the way of crystallization.

Tobacco is dark, ranging from black to mulatto; sugar is light, ranging from mulatto to white. Tobacco does not change its color; it is born dark and dies the color of its race. Sugar changes its coloring; it is born brown and whitens itself; at first it is a syrupy mulatto and in this state pleases the common taste; then it is bleached and refined until it can pass for white, travel all over the world, reach all mouths, and bring a better price, climbing to the top of the social ladder.

"In the same box there are no two cigars alike; each one has a different taste," is a phrase frequent among discerning smokers, whereas all refined sugar tastes the same.

Sugar has no odor; the merit of tobacco lies in its smell and it offers a gamut of perfumes, from the exquisite aroma of the pure Havana cigar, which is intoxicating to the smell, to the reeking stogies of European manufacture, which prove to what levels human taste can sink.

One might even say that tobacco affords satisfaction to the touch and the sight. What smoker has not passed his hand caressingly over the rich *brevas* or *regalías* of a freshly opened box of Havanas? Do not cigar and cigarette act as a catharsis for nervous tension to the smoker who handles them and holds them delicately between lips and fingers? And what about chewing tobacco or snuff? Do they not titillate their users' tactile sense? And, for the sight, is not a cigar in the hands of a youth a symbol, a foretaste of manhood? And is not tobacco at times a mark of class in the ostentation of brand and shape? At times nothing less than a *corona corona,* a crowned crown. Poets who have been smokers have sung of the rapt ecstasy that comes over them as they follow with eyes and imagination the bluish smoke rising upward, as though from the ashes of the cigar, dying in the fire like a victim of the Inquisition, its spirit, purified and free, were ascending to heaven, leaving in the air hieroglyphic signs like ineffable promises of redemption.

Whereas sugar appeals to only one of the senses, that of taste, tobacco appeals not only to the palate, but to the smell, touch, and sight. Except for hearing, there is not one of the five senses that tobacco does not stimulate and please.

Sugar is assimilated in its entirety; much of tobacco is lost in smoke. Sugar goes gluttonously down the gullet into the intestines, where is is converted into muscle-strengthening vigor. Tobacco, like the rascal it is, goes from the mouth up the turnings and twistings of the cranium, following the trail of thought. *Ex fumo dare lucem.* Not for nothing was tobacco condemned as a snare of the devil, sinful and dangerous.

Tobacco is unnecessary for man and sugar is a requisite of his organism. And yet this superfluous tobacco gives rise to a vice that becomes a torment if it is denied; it is far easier to become resigned to doing without the necessary sugar.

Tobacco contains a poison: nicotine (see Part II, Chapter iv); sugar affords nourishment: carbohydrates. Tobacco poisons, sugar nourishes. Nicotine stimulates the mind, giving it diabolical inspiration; the excess of glucose in the blood benumbs the brain and even causes stupidity. For this reason alone tobacco would be of the liberal reform group and sugar of the reactionary conservatives; fittingly enough, a century ago in England the Whigs were regarded as little less than devils and the Tories as little less than fools.

Tobacco is a medicinal plant; it was so considered by both Indians and Europeans. Tobacco is a narcotic, an emetic, and an antiparasitic. Its active ingredient, nicotine, is used as an antitetanic, in cases of paralysis of the bladder, and as an insecticide. In olden times it was used for the most far-fetched cures; according to Father Cobo, "to cure innumerable ailments, in green or dried leaf form, in powder, in smoke, in infusion, and in other ways." Cuban folklore has preserved some of these practices in home remedies. Snuff was used as a dentifrice. At the beginning of the nineteenth century a very bitter-tasting variety, known as Peñalvar, was manufactured in Havana and exported to England for this purpose; it contained a mixture of powdered tobacco and a kind of red clay. Tobacco has always been highly prized for its sedative qualities, and was regarded as a medicine for the spirit. For this reason, if long ago the savages censed their idols in caves with tobacco to placate their fury with adulation, today one burns the incense of tobacco in the hollows of one's own skull to calm one's worries and breathe new life into one's illusions.

Sugar, too, has its medicinal side and is even a basic element of our physiological make-up, producing psychological disturbances by its deficiency as by its excess. For this reason, and because of their scarcity, sugar and tobacco were sold centuries ago at the apothecary's shop. But in spite of their old association on the druggist's shelves, tobacco and sugar have always been far removed. In the opinion of moralists tobacco was vicious in origin, and was abominated by them and condemned by kings as much as it was exalted by the doctors.

Tobacco is, beyond doubt, malignant; it belongs to that dangerous and widespread family of the Solanaceæ. In the old Eurasian world the Solanaceæ were known to inspire terror, torment, visions, and delirium. Mandragora produced madness and dreams and acted as an aphrodisiac. Atropa gave its name to one of the Fates. Belladonna gave the sinful blackness of hell to the pupils of beautiful women's eyes. Henbane was the narcotic poison of classic literature. The various daturas were the source of alkaloids that the Indians of Asia as well as those of America employed in their rites, spells, and crimes. In our New World this family of cursed plants was regenerated. Even though the Datura, of which the lowly Jimson weed is a species, still works its diabolical will here, inspiring the mystic frenzy of Aztecs, Quechuas, Zuñis, Algonquins, and other native tribes, America has paid its debt of sin with interest, bestowing on mankind other plants of the solanaceous family, but upright, edible members, such as the potato, which today is cultivated more extensively throughout the world than wheat; the tomato, the "love apple" of the French, whose juice is considered a stimulating wine today; and the pepper, that king of spices, which carries to all the globe the burning and vitamin-rich stimulus of the tropical sun of America.

But in addition to these exemplary plants with their nutritious, homely, respectable fruits, the Solanaceæ of America set afoot in the world that scamp of the family, tobacco, neither fruit nor food, sly and conceited, lazy and having no other object than to tempt the spirit. The moralists of Europe were fully aware of the mischief-making properties of that irresistible Indian tempter. Quevedo said in Spain that "more harm had been done by bringing in that powder and smoke than the Catholic King had committed through Columbus and Cortés." But those were rogues' days and nothing could be devised to halt this Indian tobacco which, like the Limping Devil, went roving all over the world because everywhere it found a longing for dreams and indulgence for rascalities.

In Europe tobacco became utterly degraded, the instrument of crime, the accomplice of criminals. In the eighteenth century there was a general fear of being poisoned by deadly poison mixed with snuff. "Perfumed snuff was at times the vehicle of poison," says the historian of tobacco, Fairholt. "In 1712 the Duke of Noailles presented the Dauphine of France with a box of Spanish snuff, a gift which pleased her mightily. The snuff was saturated with poison, and after inhaling it for five days the Dauphine died, complaining of a severe pain in her temples. This caused great excitement, and there was great fear of accepting a pinch of snuff, and likewise of offering it. It was generally believed that this poisoned snuff was used in Spain and by Spanish emissaries to get rid of political opponents, and also that it was employed by the Jesuits to poison their enemies. For this reason it was given the name of 'Jesuit snuff.' This fear persisted for a long time." In 1851 tobacco was guilty of murder. The Count of Bocarmé was put to death in Mons for poisoning his brother-in-law with nicotine that was extracted from tobacco for this purpose.

As though to heighten the malignity of tobacco, there is that special virus, or ultra-virus, which attacks it, and produces the dread disease known as mosaic. Sugar cane, too, suffers from a mosaic; but that which preys upon tobacco is produced by the first of the filtrable viruses, which was not only the first to be discovered, in 1857, but is the most infectious of all. It is stubbornly immune to ether, chloroform, acetone, and other similar countermeasures. There is something diabolical about this virus of tobacco mosaic. Its behavior is almost supernatural. It has not yet been ascertained whether it is a living molecule at the bottom of the life scale, or merely a macromolecule of crystallized protein. As though it had a double personality, the virus is as inert as distilled water, as inoffensive as a cherub, until it comes into contact with tobacco. But as soon as it penetrates the plant it becomes as active and malignant as the worst poison, like a mischievous devil in a vestry room. It almost seems as though it were in the essence of the tobacco that the virus finds the evil power by which it mottles the plant, dressing it up like a devil or a harlequin. The instant the tiniest particle of the infernal virus establishes contact with the protoplasm of tobacco, all its evil powers come to life, it infects every healthy plant, reproduces by the million, and in a few days a whole crop is stricken and destroyed by the virosis. As though the virulence of tobacco were the most deadly, when the Indians had to sleep in places infested by poisonous animals they were in the habit of spreading tobacco around themselves as a defense, for, as Father Cobo says, "it has a great malevolence against poisonous animals and insects" and drives them away as by magic.

Now to the traditional malignity of tobacco another and more cruel is being attributed: the power to cause cancer by means of the tars extracted from it. An Argentine doctor (Dr. Angel H. Roffo) smeared these tars on the skin of rabbits, and cancer resulted "in every case." This did not occur with the tars distilled from Havana tobacco, but even with these half of the cases experimented with developed cancer.

At the same time scientists are still studying the possibility that cancer may be produced by an ultra-virus, that is to say, one of those protein viruses which, although chemical compounds, behave with lifelike activity, multiplying when in contact with certain living organisms, growing and dying like living cells. A scientist (Dr. W. W. Stanley) who achieved fame by isolating certain viruses in the form of crystals, holds the belief that whether those viruses that are invisible even with the microscope are the cause of cancer or not, they hold the secret of those irritations of the tissues, and in them are to be found the governing factors of the vital process in all cells, whether normal or cancerous. The puzzling feature of this horrible disease, which seems to consist in a wild reproduction of living cells out of harmony with hereditary structural rhythms, and the no less puzzling phenomenon of this ultra-virus of tobacco mosaic, which also manifests itself as the unforeseen coming to life of certain molecules that suddenly lose their inertia on coming into contact with tobacco, and reproduce and proliferate madly, carrying the germs of life, add a new mystery to the nature of tobacco. Can it be that there is something in tobacco that is a powerful stimulant of life, that can make cells proliferate in this wild manner and give to inert molecules the vital power of reproductivity, just as its smoke stimulates the weary, guttering spirit so it may flame up anew and live with renewed vigor?

There is always a mysterious, sacral quality about tobacco. Tobacco is for mature people who are responsible to society and to the gods. The first smoke, even when it is behind one's parents' backs, is in the nature of a *rite de passage,* the tribal rite of initiation into the civic responsibilities of manhood, the test of fortitude and control against the bitterness of life, its burning temptations, and the vapors of its dreams. The Jivaro Indians of South America, as a matter of fact, use tobacco in the celebration of *kusupaní,* the ceremony that marks the coming of age of the youths of the tribe. Among certain Indians of America, like the Jivaros, and some of the Negro peoples of Africa, such as the Bantus, the spirit of tobacco is masculine, and only men may cultivate the plant and prepare it for the rites. Sugar, on the other hand, is not a thing for men, but for children in their tender infancy, something mothers give their little ones as soon as they can taste, like a symbolic omen of the sweetness of life. "With sugar or honey, everything tastes good," goes the old saying.

Tobacco was always a thing of consequence. It was the glory of the conquerors of the Indies, then the mariner's companion on his ocean voyages, and the comrade of old soldiers in distant lands, or settlers returning from America, of self-satisfied magnates, rich business men; and it became the seal and emblem of every man who was able to buy himself a pleasure and display it in defiant opposition to the conventionalisms that would put a check-rein upon pleasure.

In the fabrication, the fire and spiraling smoke of a cigar, there was always something revolutionary, a kind of protest against oppression, the consuming flame and a liberating flight into the blue of dreams. For this reason the reciprocal offering of tobacco is a fraternal rite of peace, like the swearing of blood brotherhood among savages or the firing of salvos between battleships. When Europe met America for the first time, the latter offered tobacco in sign of friendship. When Christopher Columbus stepped on American soil for the first time in Guanahaní on October 12, 1492, the Indians of the island greeted him with an offertory rite, a gift of tobacco: "Some dried leaves, which must be a thing highly esteemed among them, for in San Salvador they made me a present of them." To give leaves of tobacco or a cigarette was a gesture of peace and friendship among the Indians of Guanahaní, among the Tainos, and among others of the continent. Just as it is today among the whites of civilized nations. Smoking the same pipe, taking snuff from the same snuffbox, or exchanging cigarettes is a rite of friendship and communion like sharing a bottle of wine or a loaf of bread. It is the same among the Indians of America, the whites of Europe, and the Negroes of Africa.

Tobacco is a masculine thing. Its leaves are hairy, and as though weathered and tanned by the sun; its color is that of the earth. Twisted and enveloped in its wrapper as a cigar, or shredded and smoked in a pipe, it is always a boastful and swaggering thing, like an oath of defiance springing erect from the lips. In days gone by, the country women of Cuba, who shared with their men the joys and tasks of their rustic existence, smoked their home-made cigars, and not a few in the cities preserved these rural customs in their own homes. All through Europe certain highborn and emancipated ladies of the aristocracy smoked in the seductive intimacy of their boudoirs. Even the daughters of the Grand Monarch smoked, although Louis XIV himself abhorred tobacco. The custom spread, but then gradually disappeared until finally only the peasant women of certain countries continued to smoke a pipe. Among the upper classes some ladies went on smoking Havana cigars, but this was an eccentricity that occasioned much comment. In the present age, which has attenuated the social dimorphism between the sexes, women smoke perhaps more than their hardy mates. But even today they limit themselves to cigarettes, the babies of cigars, of embryonic masculinity, all wrapped in rice paper, and with gold tips, and even perfumed, sweetened, and perverted like effeminate youths. The women who smoke cigarettes today remind one of those exquisite abbés of the eighteenth century who mixed their snuff with musk, ambergris, rose vinegar, and other exotic perfumes. They do not smoke real cigars, *puros,* pure in content and in name, as they were invented by the Indians of Cuba, in their pristine simplicity, naked, unadorned, without the adulterations, mixtures, wrappings, perfume, and refinements of a decadent civilization. A cigar is smoked with "the five senses" and with meditation, which comes as sensation is transformed into thought and ideals; but one smokes a cigarette without thinking or reflecting, as a habit one has fallen into, which among women is a mark of smartness and frivolous coquetry.

If tobacco is male, sugar is female. The leaves of its stalk are always smooth, and even when burned by the sun are still fair. The whole process of sugar-refining is one continual preparation and embellishment to clean the sugar and give it whiteness. Sugar has always been more of a woman's sweetmeat than a man's need. The latter usually looks down upon sweets as a thing below his masculine dignity. But if where tobacco is concerned women invade man's field smoking cigarettes, which are the children of cigars, men return the compliment in their consumption of sugar, not in the form of sweets, syrups, or candy, but as alcohol, which is the offspring of the sugar residues.

There is no rebellion or challenge in sugar, nor resentment, nor brooding suspicion, but humble pleasure, quiet, calm, and soothing. Tobacco is boldly imaginative and individualistic to the point of anarchy. Sugar is on the side of sensible pragmatism and social integration. Tobacco is as daring as blasphemy; sugar as humble as a prayer. Don Juan, the scoffer and seducer, probably smoked tobacco, while the little novitiate Doña Inés must have munched caramels. Faust, that discontented philosopher, probably puffed at a pipe, while the gentle, devout Marguerite nibbled at sugar wafers.

Character analysts would classify sugar as a *pycnik,* tobacco as a leptosomae type. If sugar was the treat that Sancho, the gluttonous peasant, relished, tobacco might well have answered the purpose for Don Quixote, the visionary hidalgo. Sancho was too poor to get his fill of sugar; tobacco was too dear to reach La Mancha in time to delight the impoverished squire. But it is reasonable to believe that the one would have stuffed himself on cakes and the other would have seen visions and fabulous monsters in the puffs of smoke. And if Don Quixote had ever come upon a smoker puffing out smoke, he would have considered this one of his most fantastic adventures, which is what they tell happened in 1493 to one of the early users of tobacco, who as he was smoking in his home at Ayamonte in Cuba was believed to be possessed of a devil and was denounced by the officials of the Holy Inquisition, who refused to tolerate any smoke that was not that of incense or the faggots heaped about the stake.

Psychologists would say that sugar is an extrovert, with an objective, matter-of-fact soul, and that tobacco is an introvert, subjective and imaginative. Nietzsche might have called sugar Dionysian and tobacco Apollonian. The former is the mother of the alcohol that produces the sacred joy and well-being. In tobacco's spirals of smoke there are fallacious beauties and poetic inspirations. Perhaps old Freud wondered whether sugar was narcissiatic and tobacco erotic. If life is an ellipsis with its two foci in stomach and loins, sugar is food and nourishment while tobacco is love and reproduction.

In their origins sugar and tobacco were equally pagan, and still are by reason of their sensual appeal. In both their pagan beginnings go far back, even though they were unknown to the old gods and peoples of the Mediterranean world, who used bread and wine in their orgies, mysteries, and communions. Jehovah promised his people a land flowing with milk and honey, not with tobacco and sugar. The Hebrew knew neither sugar nor tobacco, nor did Jesus and his apostles, or the Christian faithful. These latter learned the taste of sugar from the Arabs during the crusades to Jerusalem, on the Moslem-held islands of Cyprus and Sicily, from the Moors in the gardens of Valencia or on the plains of Granada. The white peoples of the Middle Ages did not know tobacco, but sugar was familiar to them. The archpriest of Hita could gorge himself on sugared delicacies. In the fourteenth century he was writing (*Libro de Buen Amor,* stanza 1,337):

> *All kinds of sugar with these nuns are plentiful as dirt,*
> *The powdered, lump, and crystallized, and syrups for dessert;*
> *They've perfumed sweetmeats, heaps of candy—some with spice*
> *of wort,*
> *With other kinds which I forget and cannot here insert.*[1]

But the roguish clergyman knew nothing of tobacco and its delights.

The Christians discovered tobacco when they discovered the Indians of the New World, first in Cuba, then in the other islands of the West Indies, and then in the countries of the Spanish Main. (See Part II, Chapter v.) This was toward the end of the fifteenth century, at the very beginning of the modern era. And was not this moment marked by the discovery of a New World by the white men of Europe?

[1] Translation by E. K. Kane.

It would seem as though tobacco had lived in hiding, exercising its powers in the jungles of an unknown world, until civilization was ready to receive its stimuli with the arrival of the Renaissance and rationalism. (See Part II, Chapter vi.)

Tobacco is "the gracious plant that gives smoke, man's companion," wrote the Cuban José Martí. And with this steady company for every hour, even those of solitude and vigil, those hours of man's mysterious fecundations, he found in it consolation for his spirit, a spur for thought, and a ladder of inspiration. To Martí tobacco's role in history has been "to comfort the thoughtful and delight the musing architects of the air." America surprised Europe with tobacco, that genie who built castles in the air, and the sixteenth century was the century of Utopias, of the cities of dreams.

Tobacco smoke wafted the breath of a new spirit through the Old World, analytic, critical, and rebellious. In the end the smoke of the Indian tobacco proved itself more powerful by arousing the minds of men than that of the Inquisition's pyres hounding them mercilessly.

Tobacco and sugar were children of the Indies; but the latter was born in the East, the former in the West. One's name is of Sanskrit origin; the other still keeps its native savage nomenclature. In the far-off Indies they believe that sugar came to them as a gift of the gods with the dew from heaven, to nourish and sustain the joys of the flesh, and then goes into the earth and is absorbed by it after the body that consumed it has rotted away. In these Indies of ours it was believed that tobacco sprang from the earth through the action of the spirit of the caverns, and that after being burned in the human mouth and dissipated in trances of delight, its volatile essence rises carrying a message to the heavenly powers.

Despite the fact that both plants originated among heathen peoples, sugar was never frowned upon by the Church and was considered veritable ambrosia. But tobacco was regarded as an invention of the devil and was savagely persecuted to the lengths of excommunication and the gallows for its users. The devils are very astute, and to deceive the unsuspecting they often added to tobacco something to give it a sweet taste and an exotic perfume. Vanilla, mustard, anisette, caraway seed, and even molasses were all used to cover up the rascality of tobacco with the blessed cape of sweetness. Especially tobaccos of inferior quality, whose power of temptation was slight, and which were generally rejected as being unquestionably "infernal" if they were not disguised by some sweet, sugary flavor and given an odor that might be confused with that of sanctity. This was what the devil used to do with inexperienced and reluctant smokers, in the case of twist for pipes, and snuff; today he does it with ladies' cigarettes. Virtuous perfumes, sanctimonious flavors—perversions fomented by greed or inspired by Satan.

In the smoking of a cigar there is a survival of religion and magic as they were practiced by the *behiques,* the medicinemen, of Cuba. The slow fire with which it burns is like an expiatory rite. The smoke that rises heavenward has a spiritual evocation. The smoke, which is more pleasing than incense, is like a fumigatory purification. The fine, dirty ash to which it turns is a funereal suggestion of belated repentance. Smoking tobacco is raising puffs of smoke to the unknown, in search of a passing consolation or a hope which though fleeting beguiles for a moment. For this reason tobacco has been called "the anodyne of poverty" and the enemy of heartache.

> *Take a little tobacco*
> *And your anger will pass.*

These are the words Lope de Vega puts into the mouth of a Spaniard in the third act of *La mayor desgracia de Carlos V.* "When things go bad, try tobacco," says the old proverb, to express the hope-filled calm that settles over a man as the smoke of his cigar curls upward. Tobacco, according to the Cuban poet Federico Milanés, is:

> *The fragrant leaf that, turned to gentle smoke,*
> *Drives from man's brow the leaden cares away.*

For, as George Sand said, "It assuages grief and peoples solitude with a thousand pleasant images."

Even in the manner of lighting a cigar there is a sort of liturgical foretaste of mystery; whether it is by means of a spark struck from flint by steel, or by a phosphorous match whose scratched head bursts into flame. The devil has caused more wax to be employed in the tiny vestas burned in the rites of tobacco than the gods in the votive candles before their shrines. In this machine age the century-old liturgical traditions are dying out, and automatic lighters are being introduced for the smoker and electric bulbs for the churches. But in both cases the flickering flame of fire that ignites, illuminates, and burns like the spirit still persists. There is no trace of ritual observed in the consumption of sugar.

Sugar is the product of human toil, but it may be consumed by a beast; tobacco is rough and natural, but it has been set apart by Satan for the exclusive use of the being that calls himself the lord of creation, perhaps because he considers himself the last animal fashioned by the Maker and the only one having the right to sin.

There are those who might think that because of these concomitances between the devil and tobacco, the clergy might have been averse to its delights and cultivation, although, naturally, they did not refuse the tithes on the tobacco fields, which were carefully collected by the diligent tithers. There were probably churchmen who owned tobacco plantations; but, to the best of our knowledge, priests in Cuba did not count tobacconists' shops or cigar factories among their worldly goods, though it would be risky to deny the possibility of their having a share in such businesses, especially in these days of anonymous stockholders in great commercial enterprises which make possible easy and inconspicuous investments in the large tobacco companies. If the churchmen did not have plantations it was not because of fear of the devil, nor yet because of a distaste for the worldly attractions of trade or repugnance to owning slaves and treating them after the custom of the country. It is a well-known fact that from the beginning of the sixteenth century the clergy of these islands had numerous slaves for their service and their business undertakings, at times "more Negroes and plantations than the laity," according to a complaint made to the King in 1530 by the lawyers Espinosa and Zuazo. And there is no question that there were clergymen who raised cane, and were even plantation-owners, openly and aboveboard, for the Jesuits had several sugar refineries here, each with its complement of Negro slaves, obedient to the work bell and the crack of the overseer's whip. In any case, the churchmen quickly came to terms with tobacco, and the factories of Havana even made the finest-quality cigars especially for the clergy, as well as for the royal family. (See Part II, Chapter vii.)

While it was possible for sugar, no matter where it was grown, always to be equally sweet, it was never possible to produce anywhere else in the world tobacco comparable to that of Cuba or a cigar to rival a pure Havana. This explains that folk song of Andalusia referring to tobacco, in the form of a riddle:

> *In Havana I was born*
> *And in all the world known.*

The manufacture of sugar quickly achieved a uniformity of product as a result of the complete similarity of industrial processes. Almost all plants contain sugar, some in abundance, like sugar cane, beets, and many others; they are cultivated in many lands, and different methods are used to extract their juice and, from this, the more or less refined crystals; but in the end there is only one kind of sugar. All saccharoses are the same. Even in the canefield each variety of cane reproduces itself without variation every year, not only because the same root sends up new shoots each year, but through the canes themselves, which come up from their own cuttings if these are rooted down for new plantings. In this reproduction of sugar cane there can be no breeding, genetic crossings, or variations. But no matter what the initial juice-yielding capacity of the cane, the unity of the final product is always the same.

Uniformity has never been possible, and never will be, where tobacco is concerned. The botanical varieties that contain nicotine are few; but even within each variety, and even in tobacco itself, each field, each crop, each plant, and perhaps each leaf has its own unique quality. And since the reproduction of tobacco is by means of seed, which each plant produces in great abundance, it is not to be wondered at that each crop contains many variations, the result of infinite crossbreedings and mixtures, of the carefully studied selections of the planter and the strange mutations and chromosomic caprices of nature. One of the greatest and most difficult problems of tobacco-planters and manufacturers is to maintain an unchanging standard in the quality of their product, which has an established reputation.

The infinite and constant variety, natural or induced, is the secret of the success or failure of the tobacco industry, depending on the taste of the smoker. The planter and miller of sugar cane has no such problem, for he knows that, in the last analysis, all saccharose is the same, an amorphus mass, of similar granulation, without class or distinction.

The taste, the color, and the aroma of a cigar depend not only on its being of real tobacco, but on its being a Havana (which is the best in the world); on the region where it was grown (Vueltabajo, Semivuelta, Vueltarriba, Partidos, etc., if it is Cuban; or from Virginia, Java, Sumatra, Turkey, Egypt . . . or the devil's little acre); on its year, on the fertilizer that was used on the fields, on the weather conditions, on its fermentation, stripping, selection, stacking, leaf, wrapper, filler, blend, rolling, moisture, shape, packing, ocean shipping, the way it is lighted, the way it is smoked; in a word, each and every one of the steps in its life, from the plant that produces the leaf to the smoker who transforms it into smoke and ashes. For this reason the tobacco industry employs *escogedores* and *rezagadores,* who by touch, sight, smell, and taste can distinguish and select the leaves and tobacco, just as wine-tasters do with the fermented juice of the vine. For each product of the tobacco industry a constant selection of the tobacco used is necessary. From the time tobacco is set out in the field until it goes up in smoke an innumerable series of selections and eliminations are required. In the color field alone the nomenclature of smoking tobacco in Cuba is as finely shaded as that employed by anthropologists to describe the human race. The color of the different types of cigar, like that of women, cannot be simply reduced to blondes and brunettes. Just as a Cuban distinguishes among women every shade from jet-black to golden white, with a long intervening series of intermediary and mixed pigmentations, and classifies them according to color, attractiveness, and social position, so he knows the different types of tobacco: *claros, colorado-claros, colorados, colorado-maduros, maduros, ligeros, secos, medios-tiempos, finos, amarillos, manchados, quebrados, sentidos, broncos, puntillas,* and many others down to *botes y colas,* these last "from the wrong side of the railroad track," used only in the proletarian mass of cut tobacco. There are selectors who can distinguish seventy or eighty different shades of tobacco, with the technical exactitude of the most painstaking anthropologist. It is not to be wondered at that there are "tobaccologists" as bold and self-seeking as certain proponents of racial theories today who, for the sake of defending the tobacco interests of their own countries, have created varieties, blends, names, and brands as absurd and artificial as the imaginary races invented by the race theorists of the present. And the races of tobacco, as well as its mixtures and adulterations, are so on the increase at present that outside of Cuba there are hybrid cigars, of unconfessable ancestry, some not even of tobacco; and the Havana cigar of good family has always to be on the alert for innumerable and hateful bastards who would usurp the legitimacy of his good name.

As the taste of all refined sugars is the same, they always have to be taken with something that will give them flavor. No one, except a child with a sweet tooth, would think of eating sugar by itself. When people are starving, they will take it dissolved in water: the Cuban revolutionists, the *mambises,* drank *canchánchara* sometimes in the everglades, and the slaves drank cane juice as it ran from the press, just as today poor Cubans buy a glass of cane juice for a penny to fill their bellies and quiet their hunger. When one chews pieces of peeled cane and sucks out the juice, there is a mixture of flavors in it, and the same is true of cane syrup and raw sugar. From the time the Arabs with their alchemy brought *"alçucar,"* as it was still called in the royal decrees having to do with America, into our Western civilization, it has been used in syrups, frosting, icings, cakes, candy, always with other flavors added to it.

Tobacco is proud; it is taken straight, for its own sake, without company or disguise. Its ambition is to be pure, or to be so considered. Sugar by itself surfeits and cloys, and for this reason it needs company and uses a disguise or a chaperon. It must have some other substance to lend it a seductive flavor. And it, in turn, repays the favor by covering up the flatness, insipidness, or bitterness of other ingredients with its own sweetness. A miscegenation of flavors.

This basic contrast between sugar and tobacco is emphasized even more throughout the whole process of their agricultural, industrial, and commercial development by the amorphism of the one and the polymorphism of the other.

Sugar is common, unpretentious, undifferentiated. Tobacco is always distinguished, all class, form, and dignity. Sugar is always a formless mass whether as cane, juice, or syrup, and then as sugar, whether in loaf, lump, grain, or powder, the same in the sack as in the sugar-bowl, or when it is absorbed in syrup, compote, preserves, candy, ice cream, cake, or other forms of pastry. Tobacco may be good or bad, but it always strives for individuality.

Sometimes, even when an attempt is made to bring about a similarity and even a confusion between different types, the indomitable individualism of tobacco thwarts the effort and turns the tables on the designing manufacturers. When during the past century cigars were manufactured in Seville with Virginia filler and Havana wrapper for the Spanish market, to the detriment of the Cuban product, the critical smokers could detect at a glance the difference between the two by the fact that the wrapper of the Havana cigars was rolled from right to left, and that of the Peninsular article from left to right. It almost amounted to saying that the Cubans were leftists, and the inhabitants of Seville rightists. Perhaps the distinction still holds good.

The best smoker looks for the best cigar, the best cigar for the best wrapper, the best wrapper for the best leaf, the best leaf for the best cultivation, the best cultivation for the best seed, the best seed for the best field. This is why tobacco-raising is such a meticulous affair, in contrast to cane, which demands little attention. The tobacco-grower has to tend his tobacco not by fields, not even by plants, but leaf by leaf. The good cultivation of good tobacco does not consist in having the plant give more leaves, but the best possible. In tobacco quality is the goal; in sugar, quantity. The ideal of the tobacco man, grower or manufacturer, is distinction, for his product to be in a class by itself, *the best*. For both sugar-grower and refiner the aim is *the most;* the most cane, the most juice, the most bagasse, the most evaporating-pans, the most centrifugals, the highest crystallization, the most sacks, and the most indifference as to quality for the sake of coming as close as possible in the refineries to a symbolic hundred per cent chemical purity where all difference of class and origin is obliterated, and where the mother beet and the mother cane are forgotten in the equal whiteness of their off-

CASA DE CALDERAS DEL INCENIO ARMONIA

BOILING-ROOM OF THE ARMONÍA SUGAR MILL

INCENIO ACANA.

ACANA SUGAR MILL

INGENIO MANACA

MANACA SUGAR MILL

INGENIO BUENA VISTA

BUENA VISTA SUGAR MILL

spring because of the equal chemical and economic standing of all the sugars of the world, which, if they are pure, sweeten, nourish, and are worth the same.

The consumer of sugar neither knows nor asks where the product he uses comes from; he neither selects it nor tries it out. The smoker seeks one specific tobacco, this one or the other. The person with a sweet tooth just asks for sugar, without article, pronoun, or adjective to give it a local habitation and a name. When, in the process of refining, sugar has achieved a high degree of saccharose and of chemical purity it is impossible to distinguish one from the other even in the best-equipped laboratory. All sugars are alike; all tobaccos are different.

Sugar is, strictly speaking, a single product. To be sure, cane always yielded, in addition to crystallized saccharose, alcohol, brandy, or rum. But this was a by-product and not sugar, just as the nicotine extracted from tobacco is not tobacco. All through the Antilles alcohol was distilled from molasses, and liquors were made from it: rum in Cuba, eau-de-vie in the French West Indies, rum in Jamaica, bitters in Trinidad, curaçao, and so on. Alcohol was always the cargo for the slaver's return trip, for with it slaves were bought, local chieftains bribed, and the African tribes corrupted and weakened. Out of this strongly flavored, caramel-colored alcohol manufactured in the West Indies for the slave-runners, mixed with the sugar with which they were provisioned along with jerked beef, codfish, and other foods designed to withstand the long crossings, and the lemons that the sailing ships always carried to ward off epidemics of scurvy, a mixed drink indigenous to the slave-trading boats came into being. It is known today as a Daiquirí, its name coming from the place where the United States soldiers first made its acquaintance. But rum was never a prime factor in the social economy of Cuba any more than were the heartwoods, hides, shellfish, and other secondary products. It is also true that the sugar-refiners in olden times manufactured different kinds of sugar, such as muscovado, loaf sugar, brown sugar, white sugar, and so forth. But these sugars were all the same product of the cane, refined to varying degrees, within the same mill, varying only in crystallization or purity.

Tobacco, on the other hand, from its first appearance in history as an article of trade, came in different forms, which were prepared in different ways. In the industrial field, tobacco has yielded six typical products. The first was that which we Cubans call tobacco by antonomasia and by adhering strictly to history, for this was the name given it by the Cuban Indians. Tobacco, strictly speaking, consists, as in the days of the Indians, of a variable number of dried tobacco leaves, called *tripa,* rolled up and enveloped in another leaf called *capa,* all forming a cylindrical roll about half an inch thick and four to eight inches long, pointed at both ends. It was in this form that the Spaniards first made the acquaintance of tobacco, and they gave it the popular name *cigarro.*

In addition to tobacco, or cigar, there were and are other products of the same plant: namely, *andullo,* or plug tobacco; twist, for chewing or pipe smoking; *picadura,* or fine-cut tobacco to be smoked in a pipe or rolled in husk or paper; *cigarrillos,* which are not little cigars or *tabaquitos,* but leaf or cut tobacco rolled in paper; and snuff, or powdered tobacco. These products of the tobacco industry do not represent successive phases in the same process of manufacture. They are all different products, and in their fabrication the tobacco from the start is handled differently, depending on the article desired. Tobacco is also exported in bulk to be made up abroad, to the detriment of Cuba's commercial standing. In this case tobacco occupies the status of a semi-raw material, like the raw sugar that is bought up by foreign refiners for the benefit of the country where it is processed rather than Cuba. These last years have seen the development of a new industrial tobacco product, the stripped leaf, in the preparation of which cheap Cuban labor is employed, and which is then exported and sold to foreign factories, which save the difference in the salaries they would have to pay and then, as though they were tobacco refineries, utilize this high-grade Cuban product, depriving our country of the profits of its final elaboration.

In any case the cultivation, processing, and manufacture of tobacco is all care, selection, attention to detail, emphasis on variety; this extends from the different botanical varieties to the innumerable commercial forms to satisfy the individual taste of the consumer. In the production of sugar the emphasis is on indifference to selection, lumping all the cane together, milling, grinding, mixing with an eye to uniformity. It passes from the botanical mass to the chemical product with the sole aim of satisfying the largest and most general tastes of the human palate.

The consumption of tobacco—that is to say, smoking—is a personal, individualized act. The consumption of sugar has no specific name; it is the humdrum satisfaction of an appetite. For this reason there is a word for smoker; there is no such word as "sugarer."

The cultivation of tobacco demands the most delicate attention at every stage; it cannot be allowed, as can sugar cane, to follow its own natural impulses. "The planter who babies his tobacco most is the one who gets the best crop," a grower said to the naturalist Miguel Rodríguez Ferrer. And Martí extolled the unflagging devotion of the tobacco-grower who spends his time caring for each tobacco plant "with his protecting hands, against the excessive heat of the sun, the treacherous cricket, the rough pruner, the rotting dampness." It is as though tobacco demanded the solicitous, pampering care of its cultivator, while cane grows by itself, as it pleases, allowing its raiser months of idleness. Who taught the tobacco-growers of Cuba these painstaking operations? Could it have been the Indian *behiques,* who gradually discovered the mysteries of the plant's cultivation, its medicinal powers, and its charms? Tobacco is one of the most difficult crops in the world to cultivate, and its technique is the most highly developed in the whole field of Cuba's agriculture. The grower has always made it his business to improve the plant with an eye to greater returns in quality, pleasure, and profit.

Tobacco is planted every year. The same cane root will produce several harvests; if the land is new and rich, as many as fifteen. Humboldt said in 1804 that there were fields on a plantation in Matanzas that were still producing forty-five years after they were set out.

The planting of tobacco is a complicated operation; the carefully selected seeds are first sown in a seed-bed, and when the young plants are ready, they are transplanted to the field where they are to be raised and cut. Cane is not planted from seed, but from cuttings of the stalk. For this reason the selection of Cuban tobacco has been an uninterrupted process, facilitated by the great quantity of seed yielded by each plant, and practiced each year by the planter who, empirically, selects in each crop the seed of the best plants for the next year's seed-bed. Foreign growers come to Cuba for quality seed. Sugar cane, on the other hand, is practically invariable, inasmuch as new plants are raised from cuttings, and the seed is not used to produce new varieties. For this reason, to develop greater resistance to disease and a greater yield of saccharose, planters have recourse to the expedient of importing foreign varieties such as Cristalina, Tahiti, Cinta, or Natal cane, P.O.J. 2725, etc.

In the cultivation of tobacco, for a long time watering and chemical fertilizers have been considered indispensable, and this means more work for the planter. The watering must be in the form of a fine spray so as not to wash away the seeds. And the fertilizer must be carefully studied and gauged by an expert tobacco dietitian. In the canefields of Cuba irrigation is exceptional and of recent origin, and fertilizer, when it is used, is of common variety and carelessly applied.

In the cultivation of tobacco the object is to obtain and make use only of the leaves, seeking a variety of classes and colors. The value of each leaf of tobacco lies in its size, its aroma, its consistency, its texture, and its color. To obtain the best qualities there is a complete technique, which is more empirical than scientific and is adapted to the needs of each grower. There is no need to explain it here, but I might mention one of the most curious practices, which consists in covering whole fields with a shedlike structure of palm fronds or with huge canopies, like mosquito-nets, of cheese-cloth. In these canopied fields less sun penetrates and the color of the leaves is lighter. It is as though the plants of these fields avoided the sun, exposing themselves to it only veiled, like lovely ladies, who, solicitous of the whiteness of their skin, protect themselves with hats, veils, and parasols. This is how on the best plantations of Vueltabajo, which can afford such luxuries, those light leaves are produced that are at a premium in certain foreign markets, especially Germany, so obsessed at the moment with the Nordic type and its blondness. Perhaps this is a dangerous moment to observe that this practice, though it originated in Cuba, is undoubtedly of Jewish and Marxist origin, for it was introduced in the Cuban tobacco fields by that great grower Don Luis Marx.

The stems and even the ribs of the tobacco are discarded; it is stripped and only the pulp of the leaves is of value. The stems were used only for snuff; but the truth of the matter is that there was considerable deceit in this, for the trimmings and sweepings of the tobacco factories were ground in with the mixture, and not a little filth, even the cigar butts and cigarette stubs dropped on the floor by the workers. The present-day industrial system makes use of tobacco stems for compounds, tars, oils, and other chemical derivatives, but the essential and noble part of tobacco is the leaf. In the cultivation of cane the objective is the heaviest stalks, which yield the most juice; the leaves are stripped off and left on the field. These leaves and the bagasse are now utilized for cattle feed, fertilizer, fuel, cellulose, and the other wonders of modern industry; but these are all by-products, younger sons, while sugar is the real heir of the cane.

The cane harvest consists in the cutting of the stalks; that of tobacco, in the cutting of the leaves. The moment of the tobacco harvesting must be carefully chosen, not too early and not too late, before the leaf starts to yellow, for in that case it will have lost much of its weight, its quality, its taste, and its aroma. And when the rainy season is over, so the tobacco will not start growing again. And in the dark of the moon, according to the old planters, so the leaves will not get holes in them, as happens if they are cut in the light of the moon. And the leaves must be cut as they ripen, by hand, with a tool as sharp as a scalpel, leaving on the plant the leaves that are still green. First the top leaves are cut, which are those the sun has ripened first; then the others, which were shaded by the upper ones. The plant goes on producing leaves. From the stump of the stalk new shoots develop quickly and yield a new crop of leaves, more pointed, which are known as *capaduras*. Still other shoots spring up, which are called *mamones,* or suckers.

If the leaves are to be used for filler, they are cut in pairs joined by a piece of the stem. But if the leaves are for wrappers, they must be cut with greater care, one by one, and then strung together on thread, two by two. Each tobacco plant produces four, five, six, seven, or more of these pairs of leaves, for they are all cut from the stalk, leaving only a short stump without any leaves. The leaves are carefully placed on the ground, as in a mother's lap, before going out in man's company and falling into his hands to be consumed in his fire. The cutting of the leaves cannot be done at any time of the day, but must be at noon, when the sun is hottest, so that when they are cut and spread on the ground, face down, it may wither them with its last kiss and the heat dry them a little and remove the moisture. On the other hand, it used to be the custom in olden times to work certain nights in the fields to pick off the tobacco-worms or to gain time with the planting and harvesting. José Aixalá, with his romantic imagination and his recollection of bygone days, tells us how they used to work with blazing torches and how on those nights the tobacco fields of Vuelta Abajo looked like a diabolical version of the "March of the Torches" of Meyerbeer. This may have been an Indian survival, for by the light of their *cuabas,* or torches of resin wood, they performed an agricultural rite in honor of the tobacco gods who dwelt in the shadows of the underworld. After completing the harvest with this night or day rite, of shadow or light, the planter soberly gathers up all the leaves from the ground, one by one, and putting them over his left arm, carries the armload from the field to the storeroom.

The season for the cutting of the cane is not so exact nor the hour so important. At times it waits for months, and at times over a year, and it is always done by the light of the sun of working hours. And the cutting is not with a fine knife, so the incision will be small and clean, but with a machete, as though one were furiously cutting down an enemy. Nor is it leaf by leaf, but slashing the stalk completely off, right down to the ground, felling it with one single blow.

Sometimes the cane catches fire in the fields and burns madly, but even after this happens, it can still be cut and ground, for the heat does not make it lose its juice right away. Tobacco never burns in the field, but only when it is dry, in the storerooms or in the smoker's mouth. When it burns, there is nothing left that can be utilized, only fine ash that disappears with the least puff.

The special requirements of tobacco cultivation have made it necessary (a most imperative economic need) for tobacco to be grown in small plots, like vegetable gardens, and not on great acreage like the canefields of the sugar plantations. Each of these tobacco fields was called, and should still be called a

tabacal; but the preferred name is *vega,* which was applied to
the river bottom lands, which were the best for the growing of
tobacco because of their fertility, the ease with which they could
be watered, and their sheltered location. *Veguerío* now is used
to designate all the vegas of a region.

Each vega is a unit in itself, where the complete agricultural
cycle of tobacco begins and ends. It is in no way connected with
the subsequent operations of the tobacco industry. The vega is
independent, unlike the canefields and the colony that springs
up about them, which is dependent upon the industrial processing and the marketing of sugar through its final stage. The
vega is not the slave of a mechanical installation whose voracity
it must feed as is the sugar plantation of the tentacled structure
of the mill. In the tobacco industry there are no centrals.

With the cutting of the cane the work of the sugar-raiser is
finished. It is loaded, weighed, and all his work is turned into
a figure that tells him the number of arrobas he has delivered
and a voucher for the money due him. But the work of the
tobacco-raiser is not over with the cutting of the leaves from the
plant. On the contrary, it is now redoubled, and requires great
skill. The drying of the leaves involves delicate and patient handling. And once more weather conditions become a factor. The
enmatulado, for instance, must be done at daybreak and under
the right conditions, for the wrong temperature could spoil the
whole crop. The tobacco leaves then go through three stages,
in the *cujes,* in the *pilones,* and in the *tercios,* and in all of them
the tobacco is cured or fermented more or less, depending on
the quality of the leaf and the amount of sap in it, and the help
given it. On the care and success of the curing much of the
value of the product depends, its aroma, taste, appearance, flexibility, combustibility.

The grower visits his treasure every day, first to touch the
leaves and gauge their degree of dryness, then to smell them
and judge the progress of curing by their scent. If the tobacco
gets too dry it may crumble to powder in handling. It is here
that the painstaking skill of the grower comes into play to keep
the leaves at the right degree of flexibility. Thus they wait for
the *pilón,* which consists in piling up the tobacco leaves, one by
one and one on top of another, in prescribed formation and

with many precautions to beautify and make uniform the color of the leaf, get rid of the excess of resin, attenuate its bitterness, and soften it so it will be more flexible and silky.

Then comes the stripping, which is removing from each leaf the stem left on it. After this comes the tedious process of selection, in which, after pulling out all the threads on which they have been strung together, those leaves to be used for filler must be separated from the wrappers, which are like fillets of tobacco. These are delicate operations, generally performed by women, *abridoras, rezagadoras* or *apartadoras,* and *repasadoras.* And finally the leaves must be reclassified, according to whether they are fillers or wrappers, in hands, hanks, bunches, and bales, prepared and ready for the market. There is no exaggeration in saying that in the cultivation and harvesting of tobacco in Cuba the human contribution is the most important element because of the great variety of special skills, physical and mental, involved in achieving the best results, as though it were a question of a work of art, the miracle of an ever changing and harmonious symphony of smells, tastes, and stimuli.

After the selection of each tobacco crop, or, to be more exact, after the selection of each leaf of each plant of each field, tobacco emerges from its agricultural cycle into that of industry, business, pleasure; it goes forth, as Martí wrote, "to employ workmen, to enrich merchants, to amuse the idle, while away sorrow, accompany lonely thoughts."

In the olden days the selection of leaf tobacco was made by the planter himself. But even before the war of independence (1895) a division of labor was introduced and the selection was made in the towns nearest the fields, where it was easier and cheaper to find adequate space and workers. In this, too, the production of tobacco differs from that of sugar. The only centers of population with which the latter is concerned is the *batey,* the mill yard around the central, and the port of embarkation alongside the warehouses. The tobacco industry, on the contrary, gives life to selected rural centers. Guanajay, Pinar del Río, Consolación, and other towns in the Pinar del Río section, and Artemisa, Alquízar, San Antonio de los Baños, Santiago de las Vegas, and Bejucal have been selected locations for Vueltabajo; Camajuaní, Remedios, and others for Vueltarriba.

In the canefields and in the mills there is no selection. All canes go to the conveyor belt and the grinder together, and all the juice is mixed together in the same syrup, the same evaporating-dishes, the same filters, the same centrifugals, and the same sacks.

Whereas the steps in the harvesting and selection of tobacco are slow and studied, those connected with sugar cane always demand haste. The cane must be ground as soon as it is cut or else the yield of juice shrinks, ferments, and spoils. This characteristic of the sugar cane is responsible for social and historical consequences of incalculable importance. The workmen who do the cutting cannot be the same as those who, later on, carry out the grinding and boiling of the syrup. With tobacco, as with wheat, the agro-industrial operations are consecutive over a whole cycle. The same farmers can carry out the different phases of the work one after another. This is not the case with sugar cane. The offhand manner in which it can be treated in the field is transformed into the most breathless haste once it has been mutilated to steal its juice, and not a moment can be lost. Cut cane begins to ferment and rot in a few days. The operations of cutting, hauling, grinding, clarification, filtration, evaporation, and crystallization must theoretically be carried out one after the other, but without interruption; nearly all of them are going on at the same time in the mill. While one field of cane is being cut, others are being converted into sacks of sugar. And all at top speed. From the time the machete fells the cane until the receptacle of the sugar is closed, there is only a short lapse, a few hours. The grinding season of a plantation lasts months because of the volume of cane, but the conversion of each stalk into sugar is always quick. For this reason the milling season requires the simultaneous co-operation of many workers for a short time. The rapidity with which the cane must be ground after cutting and milled in an unavoidably brief space of time gave rise to the need for having on hand plenty of cheap, stable, and available labor for work that is irregular and seasonal. The intermittent concentration of cheap and abundant labor is a fundamental factor in the economy of Cuban sugar production. And as there was not sufficient labor available in Cuba, for centuries it was necessary to go outside the country to find it in the amount, cheapness, ignorance, and permanence necessary. The result has been that this urgent agricultural-chemical nature of the sugar industry has been the fundamental factor in all the demogenic and social evolution of Cuba. It was due principally to these conditions governing the production of sugar that slave-trading and slavery endured there to such a late date.

It was not the existence of latifundia that was responsible for the large Negro population of Cuba, as has been erroneously supposed, but the lack of native labor, of Indians and white men, and the difficulty of bringing in from other parts of the world except Africa workers who would be equally cheap, permanent, and submissive. The latifundium in Cuba was only a consequence of stock-raising first, and then of cane cultivation, and of other concomitant factors, just as was the influx of Negro population. Both have been the almost parallel effects of the same basic causes, sugar being mainly responsible, and the Africanoid population is not a direct result of latifundism. There was an abundance of Negroes in Cuba even when there was no shortage of land, and the great sugar plantations did not constitute a primordial economic factor.

The transportation of sugar is always in bulk and in the largest possible weight and volume. As cane it is carried from the field in carts, in freight cars, or on the conveyor belt of the mill; as juice or syrup it runs through a complicated web of pipes, tubes, filters, clarifiers, and evaporating-pans; then, in the form of raw sugar, it is packed in sacks of 13 arrobas (325 pounds), which tax the strength of the muscular loaders and stevedores who handle them on the trucks, on the wharfs, and in the ship's hold. The moving of tobacco is always a careful job and done in small lots. From the fields it is carried on foot or horseback, and the work in the selecting-room and factories is done by hand, in *matules, mancuerdas, gavillas, cujes, tareas, ruedas, mazos, cajetillas,* and *cigarros.* The *tercio,* which is the heaviest load of leaf tobacco, never weighs as much as a sack of sugar.

While sugar production is all in bulk, tobacco, from the pairs of leaves threaded together and the *matules* of the vegas down to the final form in which it is presented to the consumer in the store, goes through a long series of handling, preparation, and packaging. And all of these are subject to no fixed measure and vary according to the locality and the qualities of the product.

The tobacco leaves reach the selecting-room in bales known as *matules;* if the bale is of wrappers, it contains 420 leaves, but if it is filler its content is not determined by the number of leaves, but by weight, which is in the neighborhood of three pounds. From the selecting-room the tobacco is sorted into *gavillas, manojos,* and *tercios;* but for these there is no standard measure either.

If the *gavilla* is made up of filler, the leaves are not counted, but go only by weight; if it contains wrappers, there is a set number; but even so, the number may vary from 35 to 60, depending on the class of tobacco and the care of the sorters. The *manojo* consists of four *gavillas* tied together by the *manojeadores* with palm fiber or *seibon,* or an East Indian plant that is not necessary for this tying in Cuba and that has been foisted upon us by commercial interests abroad. And with the *manojos* the *enterciadores* make up the tercios.

In Cuba leaf tobacco is packed in tercios. The tercio is a tightly packed bale of tobacco wrapped in *yaguas* (flexible strips of the trunk of the royal palm) and tied with *majagua* fiber rope in a way that can be easily untied. The old custom of Arabian and Andalusian muleteers was to divide an animal's load into two parts known as *tercios,* because the two bundles were slung across (*terciar*) the animal's back, and this method of carrying a load, together with its denomination, came to us from Spain and was applied in Cuba to the mule teams that carried the tobacco from the fields to the warehouses and thence to the city. It is from this that the bundle and its name comes to us, and not from the three *enterciadores,* or packers, who usually carry out the operation, as popular etymology has supposed. This word is so old that it was employed in the sixteenth century by the learned priest Juan de Castellanos in one of his elegies (Elegy XIII, Second Canto).

But the tercio does not contain a fixed amount of tobacco, either in the number or size of the leaves, or in volume, weight, or quality. José Comallonga in his *Lecturas Agrícolas* says that "a tercio weighs in the neighborhood of a quintal and a half" (that is, in Cuban weight, about 150 pounds), "this weight varying according to the quality of the tobacco in the bale." Moreover, the weight is not always the measure of the tercio. If it contains filler, there will be no leaf count because the *ga-*

villas that comprise it, and the *manojos* of which these are made up, have not been assembled on the basis of leaves; the only measure of this kind of tercio is its weight. But even this is not invariable. As a rule it contains 120 pounds of tobacco net weight, 125 pounds if it is Semivullta leaf, and as much as 150 pounds if it is Remedios tobacco. But if the tercio contains the precious leaf produced in the favored vegas of Vueltabajo, then the weight is no longer the unit of measure, and the *manojos* that make up the tercio are counted, as though it were wrapper leaf. If the tercio is wrapper, then it generally contains 80 *manojos* or 320 *gavillas,* but the number of leaves varies, depending on the quality. Therefore the tercio is not an exact unit of measurement for leaf tobacco. It is the expression of an approximation of volume, quantity, variety, and weight; but it does not give a precise indication of any of these factors. And if, at the same time, one takes into account the great variability in quality, it may be said that in the leaf-tobacco trade there is no such thing as measure.

In contrast, everything in sugar production is almost always measured by universal standards: acreage of the canefields, weight for cane and sugar, pressure for the mills, capacity of filters and drying-pans, sucrose content of syrups and molasses, heat for the driers and evaporators, degree of viscosity in the crystallizers, of light for the polarization, of shrinkage in shipping, algebraic proportions for the extractions, returns, and economy of each step in the agro-industrial process according to the figures of an involved system of accounting. In addition there is the timing of each phase to make sure that the necessary work is carried out punctually and to calculate the time involved to know its cost, careful bookkeeping to know the average returns and the percentage due the farmers who bring in their cane to be milled on the sugar produced, and unremitting attention to the prices quoted on the national and foreign stock exchanges.

The bell that put all social life on an exact schedule as it pealed the four canonical hours in the convents of the Middle Ages, and that later marked the watches on the ships sailing to the Indies, introduced the chronometry of work into America, signaling the beginning and end of work for the slaves on the sugar plantations. The bell that rang for the slaves in the mill yard was broken on the La Demajagua plantation on the 10th of October 1868, ringing the tocsin of liberty; but it was replaced by the steam whistle that now calls the workers with its ear-piercing note, like the whistle of an inhuman steel overseer.

Nothing of this sort happens with tobacco. The tercio, rather than an exact cubic measure or commercial weight, is a comfortable-sized bundle for the handling or storing of leaf tobacco. And, aside from the tercio, the production of tobacco is carried on without any other measure than the simple numeration of its different units. In its cultivation only the number of plants, leaves, mancuerdas, cujes, matules, gavillas, manojos, and tercios are counted. Its manufacturer counts the number of cigars, mazos, ruedas, and millares. And time in the tobacco fields is counted only by the sun. Tobacco is not cultivated or harvested in a day's work marked by a bell, but by seasons and the moon, rainy and dry seasons, sunshine and showers, the pleasure of the heavenly bodies; and its only rhythm is that of the constellations. In the tobacco factories time does not count either, for the cigar workers often do piece work. And, finally, tobacco is smoked to "kill time," to do away with it, or so it will pass without measure and without feeling it, which is the same thing. Likewise the quality values of tobacco are not susceptible of measurement. The colors, the aroma, the taste, the freshness, the ripeness, the way it is made, the shape, everything that determines the individuality of each cigar is appreciated empirically by the senses alone, without the help of precision instruments, except in the very rare case of some special shape made by a poor workman, a *bonchero,* who has to use *cartabón* and *cepo.* (A)

This lack of measuring in the case of tobacco leads to a curious consequence in mercantile ethics. As leaf tobacco has no stipulated measure, the shrewdness of the growers and buyers who deal in it must likewise be immeasurable. In these transactions in leaf tobacco there is no market price fixed by a foreign stock exchange, nor uniformity of varieties, or of chemical composition or of volume or weight. For this reason the buying and selling of tobacco requires highly expert personal negotiations between men who understand the complexity of the merchandise they are dealing with, in addition to general market conditions.

There is no type of trading that affords such a margin for subtle fraud as that in tobacco. For this reason the good repute of a dealer, based on past dealings, is almost a *sine qua non* for staying in business and doing well at it. Uprightness and reliability are the accepted thing in the tobacco business, not out of virtue or principle, but because experience has proved, even to those least burdened with scruples, that honesty is the best policy. In the same way the development of trade and credit since the sixteenth·century has made it necessary to glorify as a cardinal virtue the honoring of agreements and the prompt payment of loans and interest when due. This was the underlying motive of Puritan morality as typified in the Protestant middle class, and the same thing may be said to have taken place in the agricultural aspect of the tobacco industry. José Aixalá, in his interesting *"Recuerdos Tabacaleros del Tiempo Viejo"* (*Horizontes,* Havana, August 1936), has left us this typical picture of the old Vueltabajo: "I still remember seeing Negroes meet the train running from Villanueva to Batabanó on Thursday, with wheelbarrows loaded with sacks of gold doubloons, which were unloaded at Punta de Cartas, Bailén, and Cortés and then carried on muleback through the tobacco country, leaving at each vega payment for its crop."

But when tobacco begins to circulate commercially on a large scale, it becomes frankly knavish and seizes upon every opportunity to get the highest returns on the lowest investment. As soon as tobacco hits the European market, it loses its virtue. Especially is this true of Havana tobacco, whether in cigars or as cut tobacco, which, because of its exquisite quality, its reputation, and the high price it brings on the foreign market, has always been the victim of every manner of adulteration. The sugar industry, on the other hand, is laced with fraud at every step, from the scales on which the cane is first weighed to the shrinkage in weight and polarization that the buyer falsely claims from the warehouse at the port of embarkation, and to the refiner who bleaches and markets the final product abroad.

The sugar trade always involved contracts, promissory notes, foreign drafts, and complicated litigation in the Cuban courts. Dealings in tobacco have always followed the system of "cash on the barrel head," the money passing directly from buyer to seller, and of credits extended by a simple country storekeeper.

A deal in sugar had to be written out on paper; with tobacco a man's word was enough. Nevertheless, sugar boasts of its orderly methods, and accuses tobacco of carelessness. But it has already been observed that the one is a conservative and the other a liberal, and each has its own prejudices, likes, and dislikes.

One might say that working with sugar was a trade, and with tobacco an art. In the first, machines and brute strength are the essentials; in the latter, it is always a matter of the individual skill of the workman. From the utilitarian point of view, tobacco is all leaves, cane all stalk. And this fundamental difference between a pliable leaf and a tough stalk gives rise to the greatest divergences between the two industries, with striking social and economic repercussions. In the harvesting of tobacco all that is needed is a little knife to cut the stem of the leaf; even without a knife, with the hands alone, it can gently be pulled from the plant. For the harvesting of cane a long, sharp machete is required, which hacks down the stalk, trims off the leaves, and cuts it into lengths.

All the operations in the preparation of tobacco are carried out without machinery, using only the complex apparatus of the human body, which is the tobacco central. In leaves cut by hand and one by one, the vega yields its harvest to the grower, and from his hands the leaf passes into other hands, and from hand to hand it reaches the warehouse and the factory, where still other hands work it up into cigars or cigarettes that will be consumed in another hand, that of the smoker. Everything having to do with tobacco is hand work—its cultivation, harvesting, manufacture, sale, even its consumption.

In the tobacco industry delicate hands are required, of woman or man, to manipulate the leaves and cut them with a slender-bladed knife. In the sugar industry human hands alone cannot crush the tough, woody canes, which must be ground and shredded in mills before they yield the treasure of their juice. For tobacco, deft and gentle hands; for sugar, powerful and complicated machinery.

In reality, the enjoyment of tobacco, like that of sugar, does not come until after a series of complicated operations beginning with the harvesting of the plants and ending in the consumer's mouth. Like sugar, tobacco has to go through a series of physical and biochemical phases, such as cutting, sweating, drying, curing, pressing, and combustion; like sugar it, too, needs the heat of fire to yield its substance, and it gives off a refuse of waste and ash which are its bagasse and settlings. But the operations involving sugar consist of machete strokes, shredding, crushing, boiling, dizzy rotation, and continual filtration and separation, while those having to do with tobacco are all delicate and caressing touches, "as though each plant were a delicate lady," to use José Martí's phrase.

This explains why the tobacco worker is better-mannered and more intelligent. And why he is generally a more individualized and prepossessing person; this holds true of the grower as of the selector, and of the man who works and shapes the tobacco. In the tobacco business each individual has a reputation of his own, like a brand, and a standing that has a price value. Just as there are planters and selectors famous for their special ability, so there are workers who are gifted in the manufacture of certain types of cigars and receive higher salaries for their work. It is said that part of their skill depends on the length of their fingers and their sense of touch; it is generally believed that the colored tobacco worker of Cuba excels those of other countries. Possibly this is a racial distinction of a somatic character; but whether it is an inherited trait or merely individual, the result of habit and training, lacks scientific proof. A French journalist wrote that Havana cigars owed their excellence to the fact that they were rolled by beautiful mulattoes on their bare thighs. This smacks of lewdness, and probably has its origin in the custom of the old tobacco-growers who rolled their *monteras,* flattening out the leaves with their hands across their right leg, as is still done in certain selection operations.

The great secret lies in the individual ability of the worker. A man who was once a tobacco worker and is now a professor has said of the members of his erstwhile profession: "This work demands not only apprenticeship and practice but aptitude and a natural gift. Whoever lacks this and taste will never be anything more than a maker of plug tobacco. The good cigar-maker is an artist. The variety in the form, size, and workmanship creates different classes of cigar-makers, from those known as master cigar-makers to those who make *brevas* or *londres.*

INGENIO EL PROGRESO

EL PROGRESO SUGAR MILL

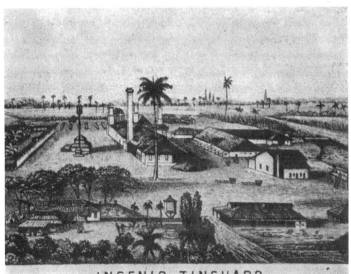

INGENIO TINGUARO

TINGUARO SUGAR MILL

CASA DE CALDERAS DEL INGENIO ASUNCION.

BOILING-ROOM OF THE ASUNCIÓN SUGAR MILL

CASA DE CALDERA DEL INGENIO Sᵗᵃ ROSA.

*BOILING-ROOM OF THE SANTA ROSA
SUGAR MILL*

The lowest level is represented by the *bonchero* who works with a mold" (G. M. Jorge: *El Tabaquero Cubano*.) If there had been cigar-makers in the Middle Ages, they would have formed a guild and a secret society like the Masons, by reason of the diabolical fame of the leaves they worked with and the strict system of apprenticeship and rank. The master cigar worker had to serve an apprenticeship of three or four years; even today he represents only two per cent of his profession. The good cigar worker is a "master." At every step tobacco is the work of a tobacco worker, while sugar is the product of a whole mill.

For this reason, also, the division of labor in a sugar factory is different from that in a tobacco workshop. In the sugar central many workmen are needed to carry on the different industrial operations necessary from the time the cane is put on the conveyor belt to the grinder until the sugar comes out of the centrifugals into the sacks in which it is packed. In the making of sugar certain workmen tend to the fires, others to the grinding, others to the syrup, others to the chemical clarification, others to the filtering, others to the evaporating, and so on, through the successive steps involved in the sugar cycle. Each workman attends to one single job. No workman of the mill can by himself make sugar out of the cane; but each cigar worker can make a cigar out of a leaf of tobacco unassisted. The same tobacco worker can take charge of making each cigar from beginning to end, from cutting with his own knife the leaf for the filler to twisting the wrapper to its final turn. And any smoker can do it, too, as the tobacco-planter does with the leaf of his own harvest for his own smoking. He can cut the leaf, dry it, shred it, stuff it into a pipe or roll it in a cigarette paper, and light it and smoke it, as he pleases. At the mill many work together in joint and successive operations, which in their totality produce sugar in great quantities. In the tobacco industry many workers are occupied in individual but identical tasks, all of which when added together produce many cigars. The manufacture of sugar is a collective job; that of tobacco is by individual efforts.

The consumption of tobacco is likewise individual, and this is borne in mind during its processing; but this is not the case with sugar. One cigar or one cigarette for each smoke. No smoker smokes two cigarettes at the same time. Even when made up in the big factories, tobacco is put out for the exclusive use of each smoker; in the singular, by portion and form, thinking of the potential consumer. Not so sugar, which is packed and consumed in bulk. And even though now certain refined sugar is made up in cubes, equivalent to the lumps into which the old sugarloaves used to be broken, these are not individual portions, but rather small doses of granulated sugar, like spoonfuls without a spoon, of which the consumer uses one or several, without selection, according to his needs.

Sugar comes into the world without a last name, like a slave. It may take on that of its owner, of the plantation or mill, but in its economic life it never departs from its typical equalitarian lack of caste. Nor does it have a first name, either in the field, where it is merely cane, or in the grinding mill, where it is just juice, or in the evaporating-pan, where it is only syrup. And when in the dizzying whirl of the turbines it begins to be sugar and receives this name, that is as far as it goes. It is like calling it woman, but just woman, without family name, baptismal name, name of hate or of love. Sugar dies as it is born and lives, anonymously; as though it were ashamed of having no name, thrown into a liquid or a dough where it melts and disappears as if predestined to suicide in the waters of a lake or in the maelstrom of society.

From birth tobacco was, and was called, tobacco. This was the name the Spaniards gave it, using the Indian word; so it is called in the world today, and so it will be known always. It is tobacco in the plant, in the leaf, while it is being manufactured, and when it is going up in smoke and ashes. Moreover, tobacco always had a family name, that of its place of origin, which is the vega; that of its epoch, which is the harvest; that of its school, which is the selection; that of its gang, which is the tercio; that of its regiment, which is the factory; that of its distinguished service, which is its commercial label. And it has its citizenship, which, if it is Cuban, it proudly proclaims.

There was a time when tobacco was largely sold in packages of fine-cut. Then a tobacco pouch full of cut tobacco was like a sugar-bowl full of sugar, from which the consumer took what he wanted. But this has fallen almost completely into disuse, and the smoker prefers his cut tobacco served up to him in individual, specially prepared portions—like sugar in lump—known as cigarettes. Snuff too was kept in bulk, like sugar, in special little receptacles, but this Indian custom has gone out of fashion never to return.

Tobacco is born a gentleman, and in its economic development at each step it acquires titles and honors by reason of its color, its aroma, its taste, its powers of combustion, until it achieves the aristocratic distinction of shape, *"vitola,"* [1] brand, and band. All tobacco strives for elegance of form and figure, blue-blooded lineage, nobility of manners, and pride of birth. And if it is Havana tobacco, it can display crowns, scepters, the emblems of royalty, and even an emperor's title.

Tobacco proudly wears until the moment of its death the band of its brand; only in the sacrificial fire does it burn its individuality and convert it to ashes as it ascends to glory. There

---

[1] The *vitola* is the figure of the cigar. It is not merely, as the Academy defines it, "the measure of size by which cigars are differentiated." It is not so much an expression of size as of shape.

Originally this word, taken from sailors' lingo, like so many others in the Spanish language of America, meant the model used in shipyards for measuring the different parts required in ship construction by naval architecture. It also meant the templet to calibrate bullets in a munitions plant; but the *vitola* of a cigar is more than all this. It is form, figure, mold, but it is more than just a geometrical arrangement. The *vitola* includes a social attitude. It is the form of the cigar, but in making his selection the smoker looks for the one that goes with his own make-up. It is the shape of the cigar, it is also its advertisement, and becomes that of the person who is smoking it, like a gesture or extension of his personality. There is always something a bit presumptuous about a cigar. The *vitola* of the cigar is its calibrating gauge, but in the smoker it becomes a mark of positive individuality,

a manner that is in itself a measure.

The *vitola* of the cigar is an outward manifestation of the *vitola* of the smoker. *Vitola* figuratively is transformed from a matter of shape and size measured in inches to a matter having human connotations; one speaks of *vitola* referring to the gesture, aspect, or appearance of a person. "Tell me what *vitola* you smoke and I will tell you who you are." In the impressive *vitola* of the Havana cigar smoked ostentatiously by Winston Churchill, who has known Havana this many a long year, there is a gesture of empire. If the cigar always demands its *vitola,* which is its appearance, every smoker, even the humblest, takes on a certain air of importance as he smokes a cigar.

And what a variety of them there are! In the collection of *vitolas* of a factory in Havana, which are reproduced in carved wooden models, there are over 996 different types of cigar. What a field for the psychologist attempting to translate them into corresponding traits of human character!

are smokers who smoke their fine cigar to the end without removing the band, which bears witness to the quality of its
brand, just as the critical drinker derives greater pleasure from
an aged wine if it comes from an old, unopened bottle, coquet-
tishly covered with dust and bearing a gold label indicating the
unquestionable aristocracy of its vintage and its origin.

In its desire for individuality, tobacco adorns itself, sometimes with a band bearing the portrait of the consumer for
whom it is intended—Napoleon III, Lord Byron, Bismarck,
the Prince of Wales, or some fashionable personage or fop who
satisfies his desire for smoking and adulation at the same time.
The latter aspiration is far the more subtly poisonous and produces more harmful effects.

Tobacco is a seeker after art; sugar avoids it. In Havana the
trade in boxed cigars has created a tradition of art in its designs,
colors, and embossing that is recognized the world over because
of its anachronisms, its style, its pompousness, and its inspiration in folkways. In Cuba boys make collections of cigar bands
and boxes just as they do of stamps, postcards, and pictures of
celebrities. Here even the makers of matches to light the cigars
have sought artistic effects in the adorning of their matchboxes.

At times, to heighten its inborn rank with outward signs of
pomp and circumstance, tobacco employs for its setting the
beautiful native woods of Cuba and has itself packed in luxury
boxes, like a majestic ceremonial carriage, to make a more fitting entrance into the palaces of royalty or of the pleasure-lov-
ing aristocracy. And even that tobacco made into shapeless
twist or shredded into fine-cut, often seeks the graceful pipe
adorned with paintings or set with medallions carved of amber,
ebony, and ivory, or molded from the finest clay or porcelain.
Even high-quality cigars are often smoked with great show, set
in a mouthpiece or a pipe. Not here in Cuba, where this fashion
would seem a profanation of the most typical and authocthonus
custom left to us, but in European countries when such baroque
display was tolerated and the smoking of a Havana cigar was a
mark of prosperity, like those canes of tortoise shell with heads
and ferrules of embossed gold.

This was not tobacco's only act of display. When, over a century ago, it was the fashion to take pinches of snuff, sniffing it
up the nostrils, tobacco had to have recourse to art to atone for

what was undeniably a revolting habit. Snuff-taking was considered a vice, disgusting but fashionable, brought in from foreign parts, like an exotic drug of rare virtues and even with claims to elegance and distinction.

The taking of powdered tobacco was for the çourtiers, highborn gentlemen, and clergy of olden times in Europe a habit of great social distinction, like having little black slaves, monkeys, parrots, and macaws around them. The dandies adorned their snuffboxes with miniatures, cameos, and precious stones, turning them into exquisite jewels, to match the lace handkerchiefs with which they wiped away the sticky traces of that revolting vice. The English statesman Petersham had a different snuffbox for every day of the year. Following the same criterion of aristocracy as the courtiers of the Grand Monarch, the Bantu Negroes of South Africa decorate their tiny snuffboxes that hold only enough for one time by covering them with beads of many colors, a labor of infinite patience. The showcases in the museums of art and customs now display the luxurious snuffboxes of the grandees of Europe and of Africa as the sacramental vessels of this diabolical cult. The same thing happens, too, with the collection of pipes, cigar-holders, cigar cases, tobacco pouches, cigarette tongs, matchboxes, lighters, ash-trays, and other smoking accessories from all the nations of the world. We can recall those old goblets or hand braziers of bronze, copper, or brass, and in America often of silver, that our grandfathers had on their tables and in which, half covered by ash, a few live coals smoldered at which they lighted their cigars.

Tobacco displays its presumptuous individualism not only in its trappings but in the art and luxury with which it enriches them to emphasize its distinction. Sugar seeks no art, either in its presentation or in its consumption. Yesterday it used to be packed into rough hogsheads or wooden boxes without style or character; today it travels about the world in sackcloth, like a hermit's robe. No designs, no colors aside from the simplicity of shape and color of bonbons and caramels. Sugar takes on the form of its container: hogshead, box, sack, or sugar-bowl. With tobacco, on the contrary, it is the holder that adapts itself more and more to its contents, at times to the point of being designed to hold a single cigar. There are boxes made to hold only one cigar, like coffins, which house a single corpse, and tinfoil wrappings for each cigar, which are like the armor of a knight of old. Or those very modern cellophane boxes in which each cigar displays its own particular shape and color and at the same time conceals its shortcomings of quality and workmanship, like a dancer whose gauzy veils reveal her voluptuous curves and luscious cinnamon skin, yet at the same time conceal her less pleasing features.

Sugar is to be found in the cradle, in the kitchen, and on the table; tobacco in the drawing-room, the bedroom, and the study. With tobacco one works and dreams; sugar is repose and satisfaction. Sugar is the capable matron, tobacco the dreaming youth. Sugar is an investment, tobacco an amusement; sugar enters the body as nourishment, tobacco enters the spirit as a cathartic. The former contributes to the good and the useful; the latter seeks beauty and personality.

Tobacco is a magic gift of the savage world; sugar is a scientific gift of civilization.

Tobacco was taken to the rest of the world from America; sugar was brought to America. Tobacco is a native plant, which the Europeans who came with Columbus discovered, in Cuba, to be exact, at the beginning of November of the year 1492. Sugar cane is a foreign plant of remote origin that was brought to Europe from the Orient, thence to the Canary Islands, and it was from there that Columbus brought it to the Antilles in 1493. The discovery of tobacco in Cuba was a surprise, but the introduction of sugar was planned.

It has been said, though I do not know on what authority the statement is based, that when Columbus returned from his second voyage he took tobacco seed with him to Andalusia, as did the Catalonian friar Ramón Pané later, and planted it, but without success. It would seem that it was Dr. Francisco Hernández de Toledo who gave a scientific account of tobacco half a century later in a report to King Philip II, who had sent him to Mexico to study the flora of that country. The cultivation of tobacco was spread less through a desire for gain than through the spontaneous and subversive propaganda of temptation. Tobacco was the delight of the people before it became that of the upper classes. Its appeal was natural and traditional rather than studied and commercial. It was the sailors who spread its use through the ports of Europe in the forms in which they used it aboard ship, either for chewing or for pipe smoking. The courtiers of Europe made its acquaintance later through travelers returning from America.

Centuries and even millenniums went by before sugar left Asiatic India, passing into Arabia and Egypt, then traveling along the islands and shoreline of the Mediterranean to the Atlantic Ocean and the Indies of America. A few decades after a handful of adventurers discovered it in Cuba, tobacco had already been carried not only through America, where the Indians used it before the arrival of the Spaniards, but through Europe, Africa, and Asia, to the distant confines of Muscovy, the heart of darkest Africa, and Japan. In 1605 the Sultan Murad had to place severe penalties on the cultivation of tobacco in Turkey, and the Japanese Emperor ordered the acreage that had been given over to its cultivation reduced. Even today many nations still lack sugar, but hardly any lack tobacco or some substitute for it, however unworthy. Tobacco is today the most universal plant, more so than either corn or wheat. Today the world lives and dreams in a haze of blue smoke spirals that evoke the old Cuban gods. In the spread of this habit of smoking the island of Cuba has played a large part, not only because tobacco and its rites were native to it, but because of the incomparable excellence of its product, which is universally recognized by all discerning smokers, and because Havana happened to be the port of the West Indies most frequented by sailors in bygone days. Even today to speak of a Havana cigar is to refer to the best cigar in the world. And that is why, as a general thing, in lands remote from the Antilles the geographical name of Havana is better known than that of Cuba.

The economy of sugar was from the start capitalistic in contrast to that of tobacco. From the earliest days of the economic exploitation of these West Indies this was perfectly evident to Columbus and his successors who settled the islands. Aside from the fertility of the land and the favorable climate, the efficient production of sugar always required large acreage for plantations, pastures, timberland, and reserves of land—in a word, extensions that verged upon the latifundium. As the historian Oviedo said, "an ample supply of water" and accessible "forests for the hot and continuous fires" and, in addition, "a large and well-constructed building for making the sugar and another in which to store it." And, besides, a great number of "wagons for hauling the cane to the mill and to fetch wood, and an uninterrupted supply of workers to wash the sugarloaves and tend and water the cane." Even all this was not enough, for there was the investment in the required number of those automotive machines known as slaves, on which Oviedo commented: "At least from eighty to one hundred Negroes must be on hand all the time, or even a hundred and twenty and more to keep things running smoothly," and, besides, other people, "overseers and skilled workmen to make the sugar." And to feed all this crew still another and larger investment. According to Oviedo, "a good herd of cattle, from one thousand to three thousand head, close by the mill to feed the workers" was necessary. For this reason he concludes logically enough that "the man who owns a plantation free from mortgage and well equipped has a property of great value." Sugar is not made from patches of cane but from plantations of it; cane is not cultivated by the plant but in mass. The industry was not developed for private or domestic consumption, nor even for that of the locality, but for large-scale production and foreign exportation. (See Part II, Chapter viii.)

Tobacco is born complete. It is nature's gift to man, and his work with tobacco is merely that of selection. Sugar does not spring full-fledged; it is a gift man makes to himself through the creative effort of his labor. Sugar is the fruit of man's ingenuity and the mill's engines. Ingenuity where sugar is concerned consists in the human and mechanical power of creation. In the case of tobacco it is rather in the personal selection of that which has been naturally created.

Of the tobacco leaves, the invention and gift of nature, the knowing countryman selects the best, and, with the simple effort of his hands to roll them into shape, he can smoke the best cigar that can be made. Just with the hands, without tools, machinery, or capital, one can enjoy the finest tobacco in the world; but one cannot get sugar that way, not even the poorest grade.

There can be no manufacture of sugar without machinery, without milling apparatus to grind the cane and get out its sweet juice, from which saccharose is obtained. The mill may be an Indian *cunyaya*—a pump-handle device resting against the branch of a tree, which as it moves up and down presses the cane against the trunk—or a simple two-cylinder roller moved by animal or human power, or a titanic system of mills, wheels, cogs, pumps, evaporating-pans, boilers, and ovens, powered by water, steam, or electricity; but it is always a machine, fundamentally a lever that squeezes. Sugar is made by man and power. Tobacco is the voluntary offering of nature.

It is possible for the *guajiro* living on his small farm to make a little sugar squeezing the juice out of the cane by the pressure of the *cunyaya*, that simple device with its single lever which the Indians used, simpler even than the *cibucán* with which they pressed the yucca. Probably it was with the Indian *cunyaya* that the first juice was squeezed out in America, from the cane planted in Hispaniola by Christopher Columbus. But it was impossible to develop production on a commercial scale with so rudimentary an instrument. The first settlers in Hispaniola devised and set up grinding mills operated by water or horse power.

To be sure, these mills which were known in Europe before the discovery of America were all of wood, including the rollers. The maximum of juice that could be extracted from the cane was thirty-five per cent, and the sugar yield was only six per cent. But in the manufacture of sugar the grinder was always as essential as the evaporating-dishes and the other vessels for the filtration of the settlings and the clarification of the syrup.

For centuries sugar was manufactured in these *cachimbos*. (See Part II, Chapter ix.) In the year 1827 Cuba had over one thousand centrals. The limited capacity of the mills was the cause of the small scale of their operations. At this time the average size of the numerous plantations in Matanzas, for example, was only about 167 acres of cane, and some 750 in wood and pasture land. For a good central 1,000 acres was enough.

In 1820 the steam engine was introduced into Cuba and marked the beginning of an industrial revolution. The steam engine changed everything on the central. The process of the penetration of the steam engine into the sugar industry was slow; half a century went by from the time it was first employed in the grinding mill in 1820 until 1878, when it was applied to the last step of the process—that is, in the separation centrifugals. By the end of the nineteenth century everything about the central was mechanical, nothing was done by hand. Everything about the organism was new. The framework continued the same, but the organs, the joints, the viscera had been adapted to new functions and new dimensions. For as a result of the introduction of steam not only was completely new machinery installed, but everything grew in size. The increased potential of energy called for enlarged grinding capacity of the mills, and this, in turn, made it necessary for all the other apparatus in the sugar-milling process to expand. But only in the last third of the nineteenth century did the Cuban sugar central begin that intense growth which has brought it to its present-day dimensions.

The Cuban sugar mill, despite the complete transformation of its machinery brought about by the steam engine, grew slowly in productive capacity, both in machinery and in acreage. As late as 1880 the size of the centrals was not extremely large. At that time the centrals of Matanzas Province, for example, averaged some 1,650 acres all together, of which only about 770 were planted to cane. This delay in the growth in size of the centrals, despite the possibilities afforded by the introduction of steam-powered machinery, was not due so much to the revolutions and wars that harassed the colony and laid much of its land waste for years as to the economic difficulties that impeded the development of transportation by steam—that is, the railroads. Railways were first introduced into Cuba in the year 1837, before Spain had them, by a company of wealthy Creoles. But it was after the ten-year revolution that steel rails were invented and that they became cheap enough so they could be used on a large scale on the centrals, not only on lines from the mill yard to the canefields, but to link up the mills and the cane-growing zones with each other and with the ports where sugar was stored and shipped. From this time on, the railway lines reached out steadily toward the sugar cane and wrapped themselves about it like the tentacles of a great iron spider. The centrals began to grow in size, giving way to the great latifundium. By 1890 there was a central in Cuba, the Constancia, that produced a yield as high as 135,000 sacks of sugar, at that time the largest in the world.

The machine won a complete victory in the sugar-manufacturing process. Hand labor has almost completely disappeared. The mechanization has been so thorough that it has brought about a transformation in the industrial, territorial, judicial, political, and social structure of the sugar economy of Cuba through an interlinked chain of phenomena which have not been fully appreciated by Cuban sociologists.

In the twentieth century the sugar production of Cuba reached the peak of its historical process of industrialization, even though it has not yet passed through all the phases necessary for its perfect evolutionary integration. Mechanization, which reached Cuba in the nineteenth century with the steam engine, began to triumph in that century and created the central; but it is in this twentieth century that the machine has given rise to the typical present-day organization, the *supercentral*. This type of mill has been the logical outgrowth of mechanization, and from it have streamed a whole series of derivations that because of their complicated interlocking structure and the relation of cause and effect have not been clearly understood or properly analyzed. It is sufficient to point out here that the principal characteristics typical of the Cuban sugar industry today, and the same holds true in a greater or lesser degree of the other islands of the Antilles, and happens to a certain extent in other similar industries, are the following: mechanization, latifundism, sharecropping, wage-fixing, super-capitalism, absentee landlordism, foreign ownership, corporate control, and imperialism.

Mechanization is the factor that has made possible and necessary the increased size of the centrals. Prior to this the central's radius of activity was the distance suitable for animal-drawn haulage. Now, with railroads, the limits of extension of a central are measured by the cost of transportation. It is a known fact that cane cut in Santo Domingo is milled in Puerto Rico and transported to the mill in ships. The mill and the railroad have developed simultaneously and their growth has made necessary planting on a larger scale, which explains the need for vast areas for cane plantations. This phenomenon also gave rise to the occupation of virgin lands in the provinces of Camagüey and Oriente and the consequent shifting of the agricultural center of Cuba. These Cyclopean machines and those great tentacles of railways that have turned the centrals into monstrous iron octopuses have created the demand for more and more land to feed the insatiable voracity of the mills with canefields, pasture land, and woodland.

On the heels of the mechanization came the great latifundism—that is, the use of a great extension of land by a single private owner. Latifundism was the economic basis of feudalism, and it has often reproduced this state. The struggle of the modern age has always been, particularly since the eighteenth century, to give man freedom and sever him from his bondage to the land, and for the freedom of the land, liberating it from the monopolistic tyranny of man. Today this process is on the way to being repeated in the Antilles, and one day we shall see agrarian laws enacted to disentail the lands held in the grasp of mortmain. The agrarian latifundism today is a fatal consequence of the present universal system of the concentration of capital. Every day industry needs more and more means of production, and the land is the most important of them all.

The central is now more than a mere plantation; there are no longer any real planters in Cuba. The modern central is not a simple agricultural enterprise, nor even a factory whose production is based on the raw materials at hand. Today it is a complicated "system of land, machinery, transportation, technicians, workers, capital, and people to produce sugar." It is a complete social organism, as live and complex as a city or municipality, or a baronial keep with its surrounding fief of vassals, tenants, and serfs. The latifundium is only the territorial base, the visible expression of this. The central is vertebrated by an economic and legal structure that combines masses of land, masses of machinery, masses of men, and masses of money, all in proportion to the integral scope of the huge organism for sugar production.

Today the sugar latifundium is so constituted that it is not necessary for the tracts of land or farms that constitute it to be contiguous. It is generally made up of a nuclear center around the mill yard, a sort of town, and of outlying lands, adjacent or distant, linked by railroads and under the same general control, all forming a complete empire with subject colonies covered with canefields and forests, with houses and villages. And all this huge feudal territory is practically outside the jurisdiction of public law; the norms of private property hold sway there. The owner's power is as complete over this immense estate as though it were just a small plantation or farm. Everything there is private—ownership, industry, mill, houses, stores, police, railroad, port. Until the year 1886 the workers, too, were chattels like other property.

The sugar latifundium was the cause of important agro-social developments, such as the monopolizing of land that is not cultivated but lies fallow; the scarcity of garden produce or fruits that would complement the basic crop, which is sugar—the reason for the latifundium's existence—because the effort required for this can be turned to more profitable use from the economic standpoint; the depreciation in value of land that it does not need within the zone monopolized by the central, and so on.

Within the territorial scope of the central, economic liberty suffers serious restrictions. There is not a small holding of land nor a dwelling that does not belong to the owner of the central, nor a fruit orchard or vegetable patch or store or shop that does not form part of the owner's domain. The small Cuban land-owner, independent and prosperous, the backbone of a strong rural middle class, is gradually disappearing. The farmer is becoming a member of the proletariat, just another laborer, without roots in the soil, shifted from one district to another. The whole life of the central is permeated by this provisional quality of dependence, which is a characteristic of colonial populations whose members have lost their stake in their country.

The economic organization of the latifundium in Cuba has been blamed for consequences that are not properly attributable to it, such as the importation of cheap labor, especially colored. First Negro slaves were brought into the country, then laborers from Haiti and Jamaica. But this immigration, which lowers the wage level of the whole Cuban proletariat and the living standard of Cuban society and upsets its racial balance, thus retarding the fusion of its component elements into a national whole, is not the result of the latifundium system. The use of colored slaves or laborers has never been nor is it a social phenomenon due to latifundism or to the monopolizing of the land. Both these economic developments are essentially identical: with the concentration of the ownership of land comes the concentration of laborers, and both depend directly upon the concentration of capital resulting from industry, especially when the process of mechanization demands more land for the plantations upon whose crop it depends, more labor to harvest it, and, in an endless progression, more machines and more and more money. The land and the laborer, like the machine itself, are only means of production, which, as a rule, are simultaneously augmented, but often the increment of one is followed by that of the others. When there was an abundance of land and before the machines had reached their full development, sugar-planting used large numbers of Negro slaves brought in from Africa; at this time the latifundium had not yet come into being. Later, as the machines grew in power, they demanded more and more cane plantations, and these, in turn, more and more labor, which was supplied by white immigration and the natural growth of population. But as the speed of the development of the sugar industry outpaced that of the population, and great centrals were established on vast tracts of virgin land, everything had to be brought in: machines, plantations, and— population. It was the swift occupation of large and new sections of Camagüey and Oriente that, aside from other secondary economic considerations such as the scale of wages, brought about a revival of "traffic in Negroes," who were now hired on terms of miserable peonage instead of being bought outright, as under the earlier system of slavery. In Puerto Rico the latifundium developed after its great demographic expansion, and as it has a dense and poverty-stricken white population, it has not been necessary to bring in cheap labor from the other islands.

In the tobacco industry the process is exactly the reverse. It was an industry without machinery. In the beginning it used very few manual devices, and these of the simplest, to twist the tobacco or grind it to powder or shred it. The largest of these apparatuses was a simple wheel. At the Quinta de los Molinos in Havana one can still see the insignificant stream of water that turned the little mills, from which it derived its name, that were formerly used to make the snuff that was exported. In addition to the preparation of snuff and cut tobacco, it was in the manufacture of cigarettes that the machine began to be used; but for hundreds of years these were made by hand at home.

Machinery did not come into the life of tobacco with the invention of the steam engine, but years after the Jamaica engines had been invented for the sugar mills and were introduced into Cuba.

There is always a stationary quality to sugar. Where the canefields are planted, there they stay and last for years, around the mill installation, which is permanent and immovable. The canefields are vast plantings and the central is a great plant. Tobacco is a volatile thing. The seeds are planted in a seed-bed, then transplanted to another spot; sometimes even from one vega to another, and tobacco's cycle ends with the year's harvest. Nothing is left in the field, and it has to be planted all over again.

The rental arrangements for tobacco lands are usually for a brief period; the crop-sharing may be on an annual basis. In the case of sugar they are of lengthy duration, depending on how long the root stock continues to produce cane before it turns into worthless stubble.

Without a large investment of money in lasting plantations and powerful machinery it is impossible to set up a central or produce any form of sugar, unless one excepts the honey produced by the communistic bees in their hives. Tobacco's economical arrangements could be limited to a small patch of fertile land and a pair of skillful hands to twist the leaf into cigars or shred it for pipe smoking. For the widespread distribution of sugar great advances had to be made first in the secrets of chemistry, in machinery, in maritime shipping capacity, in tropical colonization, in the securing of slave labor, and, above all, in the accumulation of capital and in banking organization. In the case of tobacco all that was required was for a few sailors and traders to scatter about the world a few handfuls of seed, which are so small they will fit anywhere, even in a cabin-boy's duffel-bag.

The social consequences deriving from tobacco and sugar in Cuba and originating in the different conditions under which the two crops are produced can be easily grasped. The contrast between the vegas where tobacco is grown and the sugar plantation, particularly if it is a modern central, is striking. Tobacco gave origin to a special type of agricultural life. There is not the great human agglomeration in the tobacco region that is to be found around the sugar plants. This is due to the fact that tobacco requires no machinery; it needs no mills, nor elaborate physical and chemical equipment, nor railway transport systems. The vega is a geographical term; the central is a term of mechanics.

In the production of tobacco intelligence is the prime factor; we have already observed that tobacco is liberal, not to say revolutionary. In the production of sugar it is a question of power; sugar is conservative, if not reactionary.

I repeat, the production of sugar was always a capitalistic venture because of its great territorial and industrial scope and the size of its long-term investments. Tobacco, child of the savage Indian and the virgin earth, is a free being, bowing its neck to no mechanical yoke, unlike sugar, which is ground to bits by the mill. This has occasioned profound economic and social consequences.

In the first place, tobacco was raised on the land best suited for the purpose, without being bound to a great indispensable industrial plant that was stationary and remained "planted" even after it had impoverished all the land about it. This gave rise to the central, which even in olden times was at least a village, and today is a city. The vega was never anything but a rural holding, like a garden. The vega was small; it was never the site of latifundia, but belonged to small property-owners. The central required a plantation; in the vega a small farm was enough. The owners of a central are known as *hacendados* and live in the city; those of the vegas remained *monteros, sitieros,* or *guajiros* and never left their rural homes.

The cultivation of tobacco demands a yearly cycle of steady work by persons who are skilled and specialized in this activity. Tobacco is often smoked to kill time, but in the tobacco industry there is no such thing as "dead time," as is the case with sugar. This, together with the circumstance that the vega was a small holding, has developed in the *veguero* a strong attachment to his land, as in the rancher of old, and made it possible for him to carry on his tasks with the help of members of his family. Only when this is not feasible does he hire workers, but in small groups, never in gangs or by the hundred, as happens with sugar cane. The vega, I repeat, is merely a topographical denomination; the *colonia* is a term having complex political and social connotations.

For these same reasons, while during slavery Negroes were employed as sugar-plantation hands, the cultivation of the vegas was based on free, white labor. Thus tobacco and sugar each have racial connections. Tobacco is an inheritance received from the Indian, which was immediately used and esteemed by the Negro, but cultivated and commercialized by the white man. The Indians at the time of the discovery raised tobacco in their gardens, considering it "a very holy thing," in the words of Oviedo, distinguishing between the mild cultivated variety and the stronger wild species, according to Cobo. The whites were familiar with it, but did not develop a taste for it at once. "It is a thing for savages." The historians of the Indies did not smoke, and some abominated the habit. Benzoni tells that the smell of tobacco was so offensive to him that he would run to get away from it. When Las Casas wrote his *Apologética Historia de las Indias,* in the second quarter of the sixteenth century, he called attention to the unusual fact that he had known "an upright, married Spaniard on this island who was in the habit of using tobacco and the smoke from it, just as the Indians did, and who said that because of the great benefit he derived from it he would not give it up for anything."

It was the Negroes of Hispaniola who quickly came to esteem the qualities of tobacco and not only copied from the Indians the habit of smoking it, but were the first to cultivate it on their owners' plantations. They said it "took away their weariness," to use Oviedo's words. But the Spaniards still looked askance at it. "Negro stuff."

In Cuba the same thing probably happened; tobacco was a thing "for Indians and Negroes," and only later, as it worked its way up from the lower strata of society, did the whites develop a taste for it. But by the middle of the sixteenth century in Havana, where each year the Spanish fleets assembled and set out across the ocean in convoy, tobacco had already become an article of trade, and it was the Negroes who carried on the business. The whites realized that they were missing a good venture, and the authorities issued ordinances forbidding the Negroes to go on selling tobacco to the fleets. The Negro could no longer sell or cultivate tobacco except for his own use; the

Negro could not be a merchant. From then on, the cultivation and trade in tobacco was the economic privilege of the white man.

Sugar was mulatto from the start, for the energies of black men and white always went into its production. Even though it was Columbus who brought the first sugar cane into the Antilles from the Canary Islands, sugar was not a Spanish plant, nor even European. It was native to Asia, and from there it was carried along the Mediterranean by the Arabs and Moors. For the cultivation of the cane and the extraction of its juice the help of stout slaves and serfs was required, and in Portugal, as in Spain and Sicily in Europe, in Mauritania and Egypt in Africa, in Arabia, Mesopotamia, Persia, and India in Asia, these workers were as a rule of Negroid stock, those dark people who from prehistoric times had penetrated into that long strip of supertropical areas and gave them their permanent dark coloring, the same stock that in the Middle Ages invaded it anew with the waves of Moslems, who never felt any hostile racial prejudice toward the Negro. Sugar cane and Negro slaves arrived together in the island of Cuba, and possibly in Hispaniola, from across the sea. And since then Negro labor and sugar cane have been two factors in the same economic binomial of the social equation of our country.

For centuries the workers in the centrals were exclusively Negroes; often even the overseers were colored. This was true of the mill workers as well as of the field workers, with the exception of the technicians and the management. It was not until the abolishment of slavery, the influx of Spanish immigrants after the Ten Years' War, and the introduction of the sharecropping system that white farmers were to be found on the Cuban sugar plantations.

The nineteenth century in Cuba was marked by the change in the labor system brought about by the prohibition of the slave trade and, much later, by the abolition of slavery and the substitution for it of hired workers. The abolition was proclaimed by the Cubans fighting a war of secession against the mother country, and later by Spain in 1880–6. The cessation of the slave trade coincided with the introduction of the steam engine, which increased the productive capacity of the mills, and the abolition of slavery (1886) was simultaneous with the use of steel rails and the development of the railroads, which increased the radius of activity of the centrals. Cheap labor was an imperative need, so Spain, no longer able to smuggle in slaves or bring in more Chinese coolies or peons from Yucatán, began to export her own white laborers. As a result the proportion of Negroes in the Cuban population began to diminish. In the distribution of colored population in Cuba today the greatest density is to be found in the old sugar-growing sections, not in the tobacco-raising areas, which were settled in the main by white immigrants from the Canary Islands and peasants of old Cuban stock. Tobacco drew upon the free white population, whereas for sugar cane black slaves were imported. This also explains why there are no invasions of migrant seasonal workers in the tobacco industry, and still less of Haitians and Jamaicans, who were brought in to make the harvesting of cane cheaper.

It should be noted that this process which took place in Cuba was not paralleled in other countries, such as Virginia, for example, whose early economy was based on tobacco. In Virginia, at that time an English colony, when the settlers began to raise tobacco they depended wholly on slave labor to cultivate it— white or black slaves, but preferably black.

This was due to the fact that the growing of tobacco there did not follow the same pattern as in Cuba, where, just as the Indians had done, it was treated as a small-scale, garden product, but in Virginia it employed the system of large plantations. The reason for this was that from the start the growing of tobacco in Virginia was a business, and the product was for foreign export, with the largest possible profit. That is to say, from its beginnings it was a capitalistic enterprise. For this reason, in the Anglo-American colonies there were never small growers nor any concern with the distinctive qualities of the leaf. There capitalism was in control of tobacco production from the first moment, and its objective was quantity rather than quality. The organization of the processes of curing and grading followed the same lines. Moreover, tobacco there was never rolled or made into cigars, which were unknown, but only into plugs for chewing, into twist for pipe smoking, and later into snuff for inhaling. For this mass industrial production the fine skill of the cigar worker was not necessary. As a result, in England

and her American colonies tobacco was "a thing for Negroes and Indians," and for centuries the trademark of the tobacco business was a Negro Indian—that is, a figure with Indian features and attire, but with the skin of an African slave.

This also explains why tobacco, instead of encouraging the small farm or holding in Virginia, gave rise to a growing appetite for land, and the planters who cultivated tobacco with slave labor kept pushing westward in search of new lands, thus increasing the territory and pushing back the frontiers. It is evident that the large land grants there were not responsible for slavery; rather it was the existence of slavery and the possibility of securing larger numbers of slaves that determined the creation of latifundia for the cultivation of tobacco. The existence of slavery, which in the last analysis is a form of capitalism, whose ownership of the means of production in this case took the form of large numbers of slaves, was the moving force behind the drive to secure correspondingly large landholdings and made the cultivation of tobacco, which in Cuba was intensive, individual, free, and middle-class, a large-scale capitalistic slave enterprise in Virginia.

It must also be set down that the union between sugar and the Negro had nothing to do with the latter's race or pigmentation; it was due solely to the fact that for centuries Negroes were the most numerous, available, and strongest slaves, and cane was cultivated by them throughout America. When there were no Negroes, or even together with them, slaves of other races were to be found on the plantations—Berbers, Moors, mulattoes. The alliance was not between the canefield and the Negro, but between the canefield and the slave. Sugar spelled slavery; tobacco, liberty. And if on the tobacco plantations of Virginia along with the black slaves there were white ones, purchased in England with bales of tobacco, on the sugar plantations of the British West Indies there were also black and white slaves, Irish condemned to slavery by Cromwell, and even Englishmen who had been sold for 1,550 pounds of sugar a head; or, as we would put it today, the price of an Englishman was five sacks of sugar. This did not happen in Cuba. There may have been an occasional white slave there, more probably a white female slave, in the early days of the colony, but not afterwards; and although it is true that there were near-white mulattoes who were still slaves, the whiteness of the skin was always the sign of emancipation in Cuba.

The seasonal nature of the work involved in sugar, in both the fields and the mill, is likewise very characteristic and of great social consequence. The cutting is not continuous, and whereas it used to last almost half a year, it is now almost never longer than a hundred days, and even less since legal restrictions have been placed upon it. All the rest of the year is "dead time." When it is finished, the workers who came to Cuba for the harvest in swallow-like migrations leave the country, taking their savings with them, and the native proletariat goes through a long stretch of unemployment and constant insecurity. A large part of the working class of Cuba has to live all year on the wages earned during two or three months, and the whole lower class suffers from this seasonal work system, being reduced to a state of poverty, with an inadequate, vitamin-deficient diet consisting principally of rice, beans, and tubers, which leave it undernourished and the ready prey of hook-worm, tuberculosis, anemia, malaria, and other diseases. This does not occur to the same degree with the tobacco workers, for both the agricultural and the industrial activities require steadier work; but even so, unfortunately for the country, they are also coming to suffer from undernourishment.

The unflagging devotion of the tobacco-grower to his field, his constant concern with weather and climatic conditions, the painstaking manual care the plant requires, have prevented the development of the vegas into great plantations, with great capital investments and submission to foreign control. González del Valle writes that "there is not one known case of an American or other foreigner who has grown rich cultivating tobacco in Cuba; as a matter of fact, foreigners who have tried it have lost most if not all of their capital." There are foreign landowners, but they are not the growers, with the exception of a few Spaniards who became quickly naturalized because of their easy adaptability to Cuban ways. Tobacco has always been more Cuban than sugar. It has been pointed out that tobacco is native to the New World, while sugar was brought in from the Old.

Foreign predominance in the sugar industry was always great, and now it is almost exclusive. Tobacco has always been more Cuban because of its origin, its character, and its economy. The reason is obvious. Sugar has always required a large capital investment; today it amounts to a veritable fortune. A century ago a well-balanced central could be set up with a hundred thousand pesos; today the industrial plant alone is worth a million. Moreover, ever since the centrals were first established in America, all their equipment, with the exception of the land, has had to be brought in from abroad. Machinery, workers, capital, all have to be imported, and this makes necessary an even larger outlay. If the sugar industry was capitalistic in its beginnings, with the improvement in mechanical techniques and the introduction of the steam engine more elaborate mills were required, more canefields, more land, more slaves, greater investments and reserves—in a word, more and more capital. The entire history of sugar in Cuba, from the first day, has been the struggle originated by the introduction of foreign capital and its overwhelming influence on the island's economy. And it was not Spanish capital, but foreign: that of the Genoese, the Germans, the Flemings, the English, the Yankees, from the days of the Emperor Charles V and his bankers, the Fuggers, to our own "good-neighbor" days and the Wall Street financiers.

Even in the palmy days of the Cuban landowning aristocracy, which sometimes unexpectedly acquired fabulous fortunes and titles of nobility through their centrals, the sugar-planters always suffered a certain amount of foreign overlordship. The sugar they produced was not consumed in our country and had to be shipped raw to foreign markets, where it became the booty of the refiners, without whose intervention it could not enter the world market. The sugar-planter needed the underwriter, and he, in turn, the rich banker. As early as the middle of the sixteenth century the sugar-planters were requesting loans of the brokers of Seville and of the kings, not only to continue with their enterprises, but even to set them up. This was another factor that contributed to sugar's foreignness. Its capitalistic character obliged it to seek abroad the creditors and bankers not to be found here or who, when they existed, were merely agents of the brokers of Cádiz or the English refiners, who supplied machinery and financial support but who through their loans at usurious rates could dictate their own terms and prices from London and Liverpool, and later from New York. When María de las Mercedes, the Countess of Merlin, wrote her *Viaje a la Habana,* well along in the nineteenth century, she was amazed at the fact that the rate of interest charged the Cuban planters by foreign loan-brokers was thirty per cent a year, or two and a half per cent a month.

By the end of the Ten Years' War, when through the progress in metallurgical techniques the great mills and the networks of railways were introduced in the centrals, the capital required for a venture of this sort was enormous, beyond the possibilities of any one person. This brought about three economic-social developments: the revival of the sharecropping system of cultivation, the anonymous stockholders' corporations, and the direct control of foreign capital over the management and ordering of the centrals. And finally, as a result of the financial depression after the first World War, industrial and mercantile capitalism was replaced by the supercapitalism of banks and financial companies, which today constitute the foreign plutocracy that controls the economic life of Cuba. One of the effects of this has been the greater dependence of the tenant farmer, who, according to Maxwell, received his fairest share of returns in Cuba, his gradual disappearance, and, finally, the complete proletarianization of the workers in the central, from the fields to the mill, where an executive proconsul holds sway as the representative of a distant and imperial power. The "foreignness" of the sugar industry in Cuba is even greater than that of Puerto Rico, which is actually under the sovereignty of the United States.

The foreign control of the central is not only external but internal as well. To use the language in vogue today, it has a vertical structure. There are not merely the decisions of policy taken by the sugar companies in the United States, from that radiating center of moneyed power known as Wall Street, but the legal ownership of the central is also foreign. The bank that underwrites the cutting of the cane is foreign, the consumers' market is foreign, the administrative staff set up in Cuba, the machinery that is installed, the capital that is invested, the very land of Cuba held by foreign ownership and enfeoffed to the central, all are foreign, as are, logically enough, the profits that flow out of the country to enrich others. The process does not end here; in some of the supercentrals even the workers are foreigners, who have been brought into Cuba, under a new form of slavery, from Haiti and Jamaica, or by immigration, from Spanish villages.

This foreignness is further aggravated by absentee landlordism. There were already absentee owners a century ago, who lived at ease in Havana, leaving the mill in the hands of a manager. But since 1882, when a North American, Atkins, bought the Soledad central, becoming the first Yankee planter of Cuba, absentee landlordism has been on the increase and has become more permanent, more distant, more foreign, and, in consequence, more deleterious in its social effects on the country.

Before, this absentee landlordism was periodically attenuated by inheritance, through which, upon the death of the planter, this accumulated wealth returned to society through his children and heirs. This is not so any longer, for the planter, if this name can be given to the organization that in the eyes of the law is the owner of the central, is born outside the country and dies a foreigner, and even has no heirs if it is a corporation. The great wealth of capital needed for these supercentrals could not be raised in Cuba, and the tendency toward productive capitalism could not be held in check from within. And so the sugar industry became increasingly denaturalized and passed into anonymous, corporative, distant, dehumanized, all-powerful hands, with little or no sense of responsibility.

By 1850 the trade of Cuba with the United States was greater than that with the mother country, Spain, and the United States assumed for all time its natural place, given geographical conditions, as the principal consumer of Cuba's production as well as its economic center. In 1851 the Consul General of Cuba in the United States wrote officially that Cuba was an economic colony of the United States, even though politically it was still governed by Spain. From then on sugar for North American consumption was king in Cuba, and its tariffs played a greater part in our political life than all the constitutions, as though the whole country were one huge mill, and Cuba merely the symbolic name of a great central controlled by a foreign stockholders' corporation.

Even today the most pressing problem confronting the Cuban Treasury Department is that of being able to collect its revenues by levying them directly against the sources of wealth and their earnings, making no exception of foreign holdings, instead of continuing the indirect taxes that fall so burdensomely upon the Cuban people and fleece it. Cuba will never be really independent until it can free itself from the coils of the serpent of colonial economy that fattens on its soil but strangles its inhabitants and winds itself about the palm tree of our republican coat of arms, converting it into the sign of the Yankee dollar.

This has not been so with tobacco, either in the field or in the workroom. The tobacco-grower was a simple countryman who required no machinery beyond a few tools and who could supply his own needs from the limited resources of the local general store. Whereas the sugar-planter acquired wealth, titles of nobility, government posts, refinement, and, at times, a desire for progress, the *veguero* was always a small, rustic, rule-of-thumb farmer. While the planter gave Cuba railroads before they were introduced into Spain, and Havana had its flourishing theater presenting plays and operas as good as those of Madrid, the *veguero* still rode his horse through the woods and found his entertainment in cock-fights, songs, and country dancing.

The personal element always predominated in tobacco-growing, and there was a patriarchal, intimate quality about its work. Sugar was an anonymous industry, the mass labor of slaves or gangs of hired workmen, under the supervision of capital's overseers. Tobacco has created a middle class, a free bourgeoisie; sugar has created two extremes, slaves and masters, the proletariat and the rich. "There is no middle class in Havana, only masters and slaves," the Countess of Merlin wrote about her own country a century ago. Then she goes on to say: "The *guajiro* prefers to live on little for the sake of having his freedom." On the sugar plantations there existed the overlordship and the serfdom of underling and master; on the vegas there was the free industry of the humble peasant. The old colonial aristocracy of Cuba was almost always made up of rich planters on whom a title had been conferred because of their wealth in mills and slaves. Sugar titles rested on black foundations. The sprightly archpriest long ago observed (op. cit., stanza 491):

Suppose a man's an utter fool, a farmer or a boor,
With money he becomes a sage, a knight with prestige sure;
In fact, the greater grows his wealth the greater his allure,
While he not even owns himself who is in money poor.[1]

It is easy to see how the social organization involved in sugar production (mill and plantation) had, in addition to its capitalistic character, certain feudal and baronial features. Another clergyman, Juan de Castellanos, who was also a poet, put it very well in one of the thousands of verses that make up his famous *Elegías*: "A plantation is a great estate." But he also said, referring to the plantations: "Each of these is a domain."

The tenant farmer, who made his appearance in the sugar set-up in the role of an intermediate class, never had anything but a walk-on part, important only for what he stood for and said as he entered and left the stage.

The lack of need for machinery and the limited amount of land cultivated made money in large quantities unnecessary in the tobacco industry years ago. The historian Pezuela, in the second half of the nineteenth century, said of tobacco when the mechanization of the sugar industry was becoming general: "In addition to its recognized superiority over the other products of the island, the future of tobacco is assured because of the fortunate circumstance that it does not demand an accumulated fortune for its exploitation. Without this, nobody can think of starting a central or even a coffee plantation. With a small amount of capital and the will to work, a man can acquire and work a vega." Even though this circumstance did not wholly eliminate the cruel practices of usury by which tobacco-growers were victimized, it did prevent the effects of capitalistic control and concentration at the time and to the extent that this has taken place in the Cuban sugar industry.

It was not until the middle of the nineteenth century that machinery on a large scale penetrated the tobacco industry. The twisting of tobacco into rolls or grinding of it into snuff or cutting it was done by machines as primitive and simple as the spinning-wheel, and all these operations were done by hand, too. It was after this that machines were introduced for cigarette-making, cutting, rolling, and packing. Today the machine for mak-

[1] Translation by E. K. Kane.

ing cigarettes is a perfect precision instrument. As early as 1853 a cigarette factory powered by steam was set up in Havana, by Don Luis Susini, which turned out as many as 2,580,000 cigarettes a day. At the same time the railroad, which is the great machine for overland transportation, brought the vegas closer to Havana, facilitating the sale and purchase of the leaf without the need of agents and middlemen, even though they were not completely eliminated. Today there are machines for rolling cigars, invented and used in the United States; but the Cuban proletariat refuses to accept them.

The dependence of the Cuban sugar industry on banking and foreign influences has also been a factor in the difference between its relations with the government and its tax programs and those of tobacco.

Cuban tobacco always had a more difficult time than did sugar with the Treasury Department and its burdensome systems of monopolies, government stores, tariffs, and restrictions of every sort. Sugar, once an article of luxury, is today a necessity; tobacco, which was a religious and medicinal necessity, has become, paradoxically speaking, "an everyday luxury." This explains, to a certain extent, the merciless attitude of the Treasury Department toward tobacco, which has taken the form of restrictions on its cultivation, its industry, and its commerce and of a great variety of taxes. During centuries the cultivation of tobacco in Cuba motivated many royal and governmental edicts of a contradictory nature: they were prohibitory, restrictive, permissive, but rarely encouraging.

By the sixteenth century, when the English, French, and Dutch engaged in smuggling came to America in search of tobacco, Philip II began the legal restrictions on its planting and sale. In 1606 the cultivation of tobacco in Cuba and the other islands and lands of the Spanish Main was forbidden for ten years. In 1614 this infamous ban was lifted, but the entire crop had to be sent to Seville, and disobedience of this order was punishable by death.

As the consumption of Cuban tobacco increased, so did the interest of the exchequer, and it became the object of government restrictions and monopolies, from that created by the royal edict of April 11, 1717 to the more sweeping one of 1740, which gave Martín Aróstegui a monopoly not only on the tobacco trade but on all the island's commerce. Not for nothing were the first armed insurrections of Cuba those of 1717, 1718, and 1723, which were incited by the outraged tobacco-growers against this brutal system of abusive privileges. These trade monopolies lasted for a century, and they were as corrupt as they were corruptive. But even since the establishment of freedom of trade in 1817 the fiscal difficulties originating in domestic and foreign taxes on tobacco have been a constant thorn in the flesh to both growers and dealers.

With sugar, on the contrary, everything was favor and privilege. The sixteenth century was not yet half run and Cuba was already receiving money from the royal coffers to assist in the establishment of cane plantations, without any strings attached and with free grants of land, which then abounded and which the crown wished to see settled. In 1517, barely five years after the conquest of the island, the planters of Cuba obtained from the King the first moratorium for their debts. In 1518, under a royal edict of December 9, the royal exchequer undertook to act as land bank for all who wanted to start a sugar plantation in Hispaniola, offering them "aid from the royal treasury" and canceling their debts. And the privileges did not end here. (See Part II, Chapter x.)

All the colonial governments favored the sugar-planters. They received loans of money, grants of land; the forests were cleared for them, experts in the manufacture of sugar brought in, duties were suspended, sales tax forgotten, smuggling winked at, a moratorium declared on debts, railroads built, loans made, treaties drawn up, monopolies ignored, religion weakened, heretics tolerated, civil liberties curtailed, the people tyrannized, and independence delayed. And to work the mills and plantations thousands and thousands of miserable wretches were killed or enslaved: Negroes from Africa, Indians from Yucatán, Mongolians from China. For the profit of the sugar plantation whole communities were dragged from their homes, blood flowed like the syrup from the cane, and all races suffered the lash, the stocks, and the prison cell.

Even today Cuba's national economy is governed by the sugar industry, which enjoys constant protection, even though the centrals are no longer Cuban, in exchange for special tariffs on imports, which are not Cuban either.

By the end of the nineteenth century capitalism was beginning to invade the tobacco industry to an ever greater degree, introducing changes in all branches of its cultivation, manufacture, and trade. Even in the ownership of the land, for capitalism has been getting control of the vegas. In the last fifteen years the number of landowning tobacco-growers has dropped from 11,200 to some 3,000. The landowner is disappearing from the vegas, and the *guajiro* is joining the ranks of the proletariat, becoming undernourished, poverty-stricken, preyed upon by intestinal and social parasites. The economic system of tobacco is gradually approaching that typical of the sugar industry, and both are being strangled by heartless foreign and native tentacles.

Tobacco goes out and comes in; sugar comes in and goes out —and stays out. The whole process of tobacco's development in Cuba, by reason of its native origin, its superior quality, and other collateral factors, is one of economic centripetalism. This is a product for foreign consumption, and its production is carried on with a view to its exportation to markets in other countries, but the profits return here and are spent here. The sugar industry, on the other hand, because of its exotic origin, its European antecedents, and the foreign capital invested in it, is economically centrifugal. It came to the country from abroad; it is the trader in it for foreign consumption who attempts to establish himself in Cuba and encourage its cheap production here; but those in control are not Cubans and the profits are reaped far from here. And for this sugar has exercised an almost tyrannical pressure throughout our history, introducing a constant note of oppression and force, without contributing toward the creation of robust institutions such as education, government, and civic responsibility. It was sugar that gave us slavery, that was responsible for the conquest of Havana by the English in 1762, that dictated their leaving in 1763, that caused the slave trade to flourish, that evaded the restrictions laid upon it, that robbed Cuba of its liberties throughout the nineteenth century, that brought about and maintains its colonial status and economic backwardness. As far as its primary dirigents are concerned, sugar in Cuba has always been an exogenous force, from without to within, to get what it could from the country, an oppressive, weakening force, whereas tobacco has been an en-

dogenous force, from the country outward, bringing back returns, an expansive, integrating force. The economic parabola
of sugar is a curve that cuts through Cuba, but has its beginning and end outside it; the parabolic curve of tobacco begins
in Cuba, cuts across other countries, and returns to the place of
its origin. For this reason the economy of tobacco has always
been more Cuban, and, more specifically, of Havana, with its
principal control in the capital of the island, while that of sugar
has never been controlled by Cubans, but by absentee and almost always unknown foreigners.

Tobacco has always been under the control of home government, economically and politically; whatever party has been in
power in Cuba has been in control, for better or for worse, of
tobacco. Sugar, on the contrary, has been under foreign control
superimposed on the island's government. The history of Cuba,
from the days of the conquest to the present moment, has been
essentially dominated by foreign controls over sugar, and the
greater the value of our production, the greater the domination.
During the centuries of the colonial period this power which
was and is the controlling force in the economy of the Antilles
was not, properly speaking, located in Madrid, inasmuch as ever
since the sixteenth century the Spanish crown was only the legal
machinery that, in exchange for the comfortable, well-paid, par-
asitical upkeep of its dynastic, aristocratic, military, clerical, and
administrative bureaucracies maintained order among the peoples of the Peninsula and America and exploited their inhabitants under systems of feudal absolutism, leaving the economic
initiative and control in the hands of the commercial, industrial,
and financial capitalism of the more astute centers of Europe—
Genoa, Augsburg, Flanders, London, and, in the nineteenth
century, New York. By the same token we sons of free Cuba
have sometimes asked ourselves whether our officials and politicians are serving the interests of our people or those of some
anonymous sugar corporation, playing the part of deputized
guards of the great Cuban sugar mill at the orders of foreign
owners.

It is apparent from the foregoing that since the beginning of
the sugar industry in the sixteenth century the whole history
of Cuba has developed around this foreign domination which
has always placed its own interests above those of the country.

For this reason Cuban tobacco has had to bear the weight of export taxes levied against it for the benefit of the island's exchequer, whereas foreign-controlled sugar has always successfully evaded, until the present moment, which is exceptional, the payment of export duties to the Cuban Treasury Department even in those times, which today seem fabulous, when the returns on the capital invested in land, mill, and plantations was better than one hundred per cent. In the history of Cuba sugar represents Spanish absolutism; tobacco, the native liberators. Tobacco was more strongly on the side of national independence. Sugar has always stood for foreign intervention. But today, unfortunately, this capitalism, which is not Cuban by birth or by inclination, is reducing everything to the same common denominator.

It throws light on these serious political contrasts to observe those that exist in the commercial field.

The trade in tobacco and sugar came about and developed in very different fashion.

Tobacco is characterized by its individuality, sugar by its amorphousness. Five factors had a decisive influence on the history and commercial vicissitudes of tobacco: namely, (1) the fact that tobacco is an article of pleasure and vice, a luxury article, like sparkling wines; (2) that Havana cigars, like champagne, are something unique, that cannot be surpassed or substituted; (3) that, notwithstanding, the use of tobacco is subject to the influences of caprice, fashion, and the degeneration of taste; (4) that, in spite of its nonessential and frivolous character, its use is as widespread as though it were an article of primary importance; and (5) that it is a product against which taxes can be readily assessed. On the basis of these special conditions, which do not apply to sugar, a whole structure of appetites, fables, vices, anathemas, profits, enterprises, restrictions, duties, taxes, frauds, fashions, dreams, and invectives has grown up about it.

From the beginning sugar represented a planned economy, tobacco free enterprise. Sugar came about through the application of scientific alchemy; tobacco's origins are to be found in folklore.

When Christopher Columbus brought to these cisatlantic Indies the first cuttings of sugar cane, it was in keeping with a carefully thought-out economic plan. The object was to plant them, grind the cane, and extract the sugar to sell at a handsome profit. When the Admiral discovered tobacco and took it to the Catholic kings he had no idea of profiting by it, or planting it, or manufacturing it, or selling it on the other side of the ocean. Columbus had been dead a long time when people began to consider the possibility of trading in tobacco and making a business of it. Sugar was an enterprise that had always received serious thought as a means of acquiring wealth, and a lifetime undertaking; tobacco-growing was a casual, venturesome undertaking, like a whimscial device to while away an hour.

It was Columbus who exported the first tobacco and imported the first sugar cane. Tobacco left Cuba with him on his return from his first voyage; sugar cane came in with the Admiral on his second voyage. But tobacco was only an exotic novelty, like the Indians and their gold trinkets, their native skirts, the seats used in their ceremonies, their shell belts and breastplates, the totems and masks of their mysteries, pineapples, cassavas, corn, sweet potatoes, prickly pears, parrots, macaws and hammocks; whereas sugar cane was brought in as a tried and established source of wealth, along with wheat, vegetables, fruit trees, horses, cattle, swine, and barnyard fowl.

Tobacco was unknown in Europe before the early part of the sixteenth century; there was no taste for it, there was no economic interest in developing such a taste, and it never occurred to anyone that factories could be set up to turn out cigars by the thousand or shops where they could be profitably sold. Sugar, on the other hand, was already a prized delicacy, in a class with pepper, cinnamon, and other spices, and the demand for it was very great. The European market had already experienced crises and fluctuations in the output of sugar and its price, and the difficulty lay not in stimulating its consumption, but in finding fertile lands where it could be grown in quantities that would make it more accessible in price. The taste for tobacco developed after Columbus's time; sugar antedated him.

As soon as the Spaniards began building up the colonial economy of the West Indies, on their return voyages to Europe the ships carried sugar as well as gold, pearls, lignum-vitæ, hides, and cassia pods. But there is no record of cargoes of tobacco.

Sugar was shipped by the boatload, tobacco in an occasional bale. From the beginning of the sixteenth century the demand for sugar in Europe far exceeded the supply America could offer. (See Part II, Chapter xi.)

But if there was a scarcity of sugar here, there was a plentiful supply of tobacco growing around the Indians' huts and in the Negroes' garden patches, where they raised it for their own use. There was as yet, however, no thought of organizing its cultivation. Sugar was the great business for dealers and shippers; tobacco was a little side line for sailors, who carried it in pigs' bladders.

There are those who have believed that the evolution in the use of tobacco was from snuff to the pipe, from the pipe to the cigar, and from the cigar to the cigarette. But even though this morphologic scale is very interesting and this pattern may have been followed in certain countries of Europe, I hardly feel that it can be accepted as universal. Spain, for example, must have first made the acquaintance of the cigar, which was discovered in 1492, and perhaps chewing tobacco and snuff. The pipe must have crossed the Atlantic several lustrums later, when its use was learned on the mainland. It is impossible to know whether snuff or smoking tobacco was first used by the American Indians, or whether the pipe, representing a technical step forward, was preceded by the cigar, either that made of tobacco leaves rolled in a leaf of the same plant or that made of tobacco filler wrapped in the leaf of some other plant.

It seems likely that the typical cigar, consisting of tobacco twisted and wrapped in a tobacco leaf, to be lighted at one end and the smoke inhaled at the other, was carried back to Spain by the caravels in which Columbus discovered America. But this manner of using tobacco spread more slowly than the plug or powdered form for medicinal purposes. The Spaniards chewed this stimulating leaf of the natives of Cuba just as they did later the coca of the Indians of the continent. Smoking tobacco was more complicated. It required a pipe or special skill in twisting the leaves into *mosquetes,* as Father Bartolomé de las Casas called them. Besides, fire to light the tobacco in this form was needed, and the paraphernalia to start it.

The use of tobacco gradually spread among the sailors making the American run, who carried it in powder to snuff it and, especially, in plug for chewing and twist for smoking it in a pipe, which was the way it could best be smoked on a sea voyage. And these seafaring men introduced the custom in their native lands. By the middle of the sixteenth century the magic plant of the Taino medicine-men had made its way into the habits of the better-class Spaniards, who took it in the form of "powder" and "smoke."

The invention of snuff is still generally attributed to the Grand Prior of France, who happened to inhale the powder and liked its effects; for this reason, it is said, tobacco was also known as the *herbe du Grand Prieur*. But it is an established fact that powdered tobacco, by itself or mixed with other more toxic plants, was used by the Indian priests of America in their ceremonies before it was taken by the Catholic priests of Europe, without rites and for the pleasure it gave them. It may be that this dignitary of the French Church encouraged its use in France without inventing it, for in those days the use of snuff was a clerical vice, perhaps because they considered it less ostensible than smoking. Not for nothing did Father Bernabé Cobo, a Jesuit well versed in such matters, say that snuffing tobacco through the nose was a hypocritical invention of the Spaniards for the purpose of "taking it with dissimulation and less offense to bystanders." Brunet states that it was the priests who used snuff in Spain when it was introduced there. The clergy was so given to "taking tobacco and drinking chocolate," both products of America, that *El Diablo Cojuelo,* after all a minor devil, boasted of having triumphed over them by means of these temptations of the Indian deities. Quevedo mocks at all smokers and calls them *tabacanos* and slaves of the devil.

Although the smoking habit acquired by the discoverers of America and the seamen who followed in their wake probably introduced the use of tobacco into Europe, the spreading of the habit was due primarily to the medicinal virtues attributed to the plant, regarded as a magic panacea, rather than to its gustatory or stimulating properties. In the year 1560 a page of Catherine de Médicis was suffering from ulcers, and the French Ambassador to Portugal, Jean Nicot, sent for some tobacco plants, and when the leaves were placed over the page's ulcers the sores healed as by magic in no time at all, and thus the fame of tobacco's medicinal qualities spread beyond Portugal and Spain.

It is said that it was Sir Walter Ralegh who introduced it at the court in London. There were teachers of the art of smoking there, just as there were dancing masters, and the ability to blow rings and spirals of smoke was an accomplishment as esteemed as knowing the steps and figures of the latest dance. In 1599 a pound of good Cuban tobacco cost over a hundred and twenty dollars in England. In 1612 John Rolfe, the husband of Pocahontas, took some seeds of West Indian tobacco to Virginia, and the settlers there began to export their crop to Europe, thus competing with Spain and breaking her monopoly.

The Casa de Contratación de Indias of Seville attempted to organize the cultivation of tobacco in the colonies. The first royal edict dealing with Cuban tobacco is dated October 20, 1614. The first shipments of tobacco registered in Havana were made in 1626. (See Part II, Chapter xii.) But these were not the first cargoes of Cuban tobacco to cross the Atlantic, nor does the record state whether it was in twist, fine-cut, roll, powder, or leaf to be made up in cigarettes by the tobacco girls of Seville, or for the use of the Lutheran heretics.

By the seventeenth century the use of tobacco was a firmly established custom in Europe. Teniers in his realistic paintings has many scenes of Dutch smokers enjoying their pipes.

Tobacco seed was planted in every country, and plants bearing leaves with strange flavors and aromas sprang up in alien lands. And with these leaves it was possible to prepare tobacco for smoking everywhere.

In Cuba tobacco was cured at home for the domestic use of the settlers, who had developed a great liking for the stimulating qualities of this native contribution. But the real cigar, the *tabaco* of the Cuban Indians, in the form in which the discoverers found it and which we still have in mind when we speak of a "Havana," was exported and known abroad much later. It was not so well adapted for use by sailors on their sea voyages. And outside Cuba there were no workers who knew how to prepare it properly, twisting it, rolling it, fitting it into the wrapper with that precision and sureness of touch as characteristic of the Cuban cigar-maker as of the Savile Row tailor.

After many amusing and contradictory ups and downs, tobacco finally came to be smoked everywhere. Few people smoked cigars; most of them used pipes, chewed it, or took it in the form of snuff. In Europe, outside of Spain, the cigar was almost unknown, and its use did not become general until the middle of the nineteenth century. But even then the cigar was not popular, and in England a little book was published in the year 1840 showing the trade-mark of a tobacconist which consisted of three hands joined to a single arm, the first holding between its thumb and forefinger a pinch of snuff, the second a pipe, and the third a plug of chewing tobacco, with these lines underneath:

> *We three are engaged in one cause*:
> *I snuff, I smoke, and I chaws.*

For centuries it was only in Spain that the typical *tabaco* of the Indians of Cuba was used to a considerable extent. An attempt was made on the part of the educated classes to introduce the name *tubano,* referring to its tubular shape, but this cultivated term made little headway; people preferred the popular name of *cigarro,* which had been given it because of its resemblance in shape, size, and color to certain cicadas found in Andalusia. Then later it came to be known as *puro,* to distinguish it from the cigarette, that wasted, poor, little cigar, without filler or wrapper, that mixture of scraps of tobacco wrapped in paper. It was never considered common to smoke Havana cigars, even in Spain. They were always expensive, and Spain was poor for centuries, even under the pomp and circumstance of the Habsburgs, which was the period in which the literature of roguery flourished.

It must have been in these very rogues' circles that the use of the cigarette won its popularity. It was not a product of the house of Monipodio nor was it the invention of Rinconete, or Miguel de Cervantes would have told us about it. Neither was it in Turkey, as some have maintained, that the cigarette was invented. It is a known fact that in Spain in the seventeenth century it had become a practice to make cigarettes of shredded tobacco wrapped in paper, which were known as *papeletas, papeletes, papelotes,* and *papelillos.* Some poverty-stricken emigrant back from the Antilles who recalled the cigars the Indians smoked wrapped in cornhusk or banana leaf probably hit upon the idea of using the wrapper he could most easily come by in everyday city life: a sheet of paper. Its paper coat identifies the cigarette as a city product.

The paper-covered cigarette seems to have originated in Seville, thanks to the ingeniousness of some guttersnipe, who, like the sage of the fable, was happy to "gather up the leaves another threw away." The cigarette was the invention of the stub-collector. Thus a symbiosis developed between rich tobacco and the poverty of the lower depths. Every cigarette seems to smack a little of fraud and contraband. And outside Spain the cigarette grew even more knavish, and took on an effeminate quality that enabled it to worm itself into the companionship of the ladies. In Turkey it was flavored and seasoned until it lost its masculine Indian vigor and sallied forth, like a eunuch, to find its fortune in the harems of the world. It was there in the Moslem lands that those adulterate mixtures known among tobacconists by the name of *harman,* a word taken from the Turkish, were developed. But the returning Spanish emigrant, the priest, the soldier, and the civil servant, who had made his fortune in America, clung to the expensive and aristocratic vice of smoking Havana cigars, which they had sent them from Cuba to their Peninsular retreat.

It was well on in the second half of the eighteenth century, after the conquest of Havana by the English in 1762, that Havana cigars in turn set out to conquer the world. It was then that Havana cigars traveled to England in the red coats of the British officials, and to North America with the Yankee officers who had been in charge of the colonial regiments that helped occupy Havana and not long afterwards, in 1776, were to win the independence of their own country. After this momentous episode in Cuba's history the taste for cigars began to spread beyond Spain. In 1788 the first factory manufacturing cigars was set up in Hamburg by H. H. Schlottmann, and by 1793 they were in wide use in all Germany. The philosopher Kant in 1798 still used the German version of the Spanish word *zigarro* in his *Anthropologie.*

In the nineteenth century it was the invasion of Spain by Napoleon's armies and Lord Wellington's troops, and later by the Hundred Thousand Sons of St. Louis, that spread the use of Cuban tobacco through the countries from which the troops proceeded. Just as it was in the snowy trenches of the Crimean

War that the use of cigarettes became generalized among the soldiers, who preserved the habit on their return from the campaign. The cigarette was sponsored by the beggar, the soldier, and later the workingman; the pipe was for the use of sailors, farmers, and shepherds. The cigar has borne the seal of the clergy, the Indian chief, the man of power, and the wealthy middle class.

In our own day, when capitalism dominates, speeds up, transforms, and puts a money value on everything, the cigarette is winning ground because of economic factors. It has won over the women, the proletariat, and a large part of the middle class. Even the mighty have taken a liking to its unpretentiousness, leaving the fine makes of cigars for special occasions of display. A good cigar is expensive, it is big, and it lasts a long time. Today there is no time to smoke it with the relaxation it demands; in the feverish haste of everyday life it would often have to be thrown away almost as soon as it was lighted, and this would be an unpardonable waste. The cigarette is small and burns fast and, when necessary, can be tossed aside without loss or regret, for it costs very little and the loss is insignificant. People prefer the cigarette for reasons of economy, because of the increasing money valuation put upon time and because of the generalization of the luxury of smoking. Even in Havana the production of cigarettes now exceeds that of cigars.

As was to be expected, from 1762 and 1776, but especially after 1825 and 1826, when tobacco could be exported without government restrictions, a great wave of trade in tobacco, both leaf and manufactured, sprang up between Cuba and the United States, England, and Germany. In 1849 the export of leaf tobacco had tripled. It has been said that from the middle of the past century the increase in exports influenced the growers in their methods of cultivation, to aim at size and number of leaves rather than fragrance and color—that is to say, quantity rather than quality. The prostituting effects of trade! One can note the effects of capitalism which tends to convert the tobacco industry into an amorphous mass production, maintaining the traditional appearances of quality and selection. The magic of money! The miracle-working powers of capitalism! There comes to mind the observations and experiences of Juan Ruiz, the minstrel priest of Hita (op. cit., verses 493 and 494): "I saw there in

Rome, the seat of Sanctity," that "money not only can buy heaven and win a man salvation," but also "makes truth of lies, and lies truth." In Rome, "the seat of Sanctity," and in Cuba, where sanctity is less abundant, these miracles and transmutations of facts and merchandise are the natural consequence of "the power that in money lies," to use the words of this same sagacious archpriest in one of his versified psychosocial analyses.

Also on account of the growing concentration of capitalism, with its imperialistic tentacles and its deals with the treasury officials and the government leaders who control them, tobacco like sugar now finds itself involved in the same network of treaties, monopolies, reciprocal trade agreements, tariffs, quotas, agricultural restrictions, price-fixing, cartels, trusts, and other legal snares that for many years in this part of the world have been choking liberalism to death, substituting for it the intervention of the state in the economic life of the country, setting up a kind of one-legged socialism, unilateral and halfway, without equitable intention or benefit for the people as a whole.

Capitalism is also establishing a parallelism between the industrial aspects of tobacco and sugar, subjecting them both to increasing foreign domination, with devastating results for Cuba. Sugar has always been under foreign economic control, and Cuba's share in its returns has always been held at a minimum, to what it made from producing the raw material; and the same thing is now being attempted with tobacco.

During the Ten Years' War the tobacco production of Cuba suffered severely. A large share of the vegas at this time were in the province of Oriente, the center of the War of Secession. Yara tobacco, which was grown in the vegas of Cauto, was famous, as was that of Mayarí. As a result of the Ten Years' War many of these tobacco fields were wiped out. It was during this period of upheaval also that a foreign operating center for our tobacco industry was set up on the neighboring islet of Cayo Hueso, known in English as Key West.

Gerardo Castellanos G. in his book *Motivos de Cayo Hueso* tells how a group of Cuban cigar-makers had established themselves in Key West around 1831—about fifty of them. Over a century ago the brothers Arnao set up a little hole-in-the-wall factory with sixteen Cuban workmen. But it was at the outbreak of the Ten Years' War that many Cuban tobacco workers fled from Havana and its surrounding districts to the rocky neighboring island, a traditional refuge for Cuban exiles. Because of political passions, then at fever pitch, two cigar-manufacturers of Havana, one a Spaniard from Valencia, Don Vicente Martínez Ibor, and the other a Cuban, Don Eduardo Hidalgo Gato, decided it would be wise to leave Cuba and set themselves up in cities of Florida where they could establish themselves in their business, employing raw materials and workmen from Cuba. Tampa, Ibor City, Key West, and even New York were havens for Cuban and Spanish tobacco workers seeking political freedom and better salaries. The continual crises in Cuban affairs increased the emigration of workers to the Florida factories, and they were the principal actors in the struggle for independence outside of Cuba. In this way capitalism set up its factories abroad and took away from Cuba its tobacco, its skilled workers, and their wages. This has been the process of de-Cubanization of tobacco in its industrial aspects.

The contrast with sugar in this respect is striking. In Cuba's economy sugar has always been a raw material. We have never been able to refine it here freely for exportation and put it on the market as a finished product for foreign consumption. There was a time when Cuba did not even refine sugar for home consumption, and our sugar, which was shipped out raw, was sent back to us refined, with an increase in price that represented the foreign refiners' profit.

Tobacco, on the other hand, was always grown and prepared in Cuban factories and shipped to the foreign markets as a manifest product of the country, with its place of origin clearly marked, a fact that added to the market value of the genuine Havana. This is not always so today. Cigars which often contain little or no Cuban tobacco are sold abroad as Havanas, and an effort is being made to reduce the industry here to its purely agricultural phase—that is to say, to the growing of the leaf and possibly stripping it. This is further complicated by the increasing importation into Cuba of cigarettes manufactured outside the country, with foreign tobacco and with flavors that are foreign, too. And this process of foreign domination has not come to a close. Mechanization and capitalism are exerting more and more pressure in the direction of keeping Cuba in the economic status of a colony, a state of affairs that has been typical of its history ever since the Genoese, Christopher Columbus, hit upon his economic plan for the Spanish West Indies down to our own times, when foreigners are intent on working out plans for us to follow. Once more the pertinent satire of that greatest poet of the Spanish Middle Ages comes to mind (op. cit., stanza 510):

> *Above all, let me tell you this, do with it what you can:*
> *Throughout the world Sir Money is a most seditious man*
> *Who makes a courtesan a slave, a slave a courtesan,*
> *And for his love all crimes are done since this old earth began.*[1]

The relations of tobacco and sugar with their workers have been very different.

Sugar has always preferred slave labor; tobacco, free men. Sugar brought in Negroes by force; tobacco encouraged the voluntary immigration of white men.

In the production of sugar, agriculture and industry are concentrated at the same spot, and the result is the creation of that complicated social-economic institution which is the *ingenio* or central, consisting of a vast cane plantation, a huge factory with all its apparatus for grinding, evaporating, crystallizing, separating, and shipping the sugar, and the urban center, village or city, which is the *batey* with its sheds, dwellings, machine-shops, stores, stables, and other services. With tobacco, on the other hand, there is a separation between its cultivation and its manufacture. The former has remained strictly rural, whereas the latter has always been urban, and mainly of Havana. For this reason the best leaf tobacco is known as *vueltabajo,* taking its name from the region where it is grown, whereas the best finished cigar in all the world is known as a Havana, the name of a great city. A cigar factory is a simple, movable street location; a central is a complicated and permanent geographical accident.

As a result of the unavoidable combination and concentration of agriculture and industry in the production of sugar, the central has always needed large masses of laborers. In olden times this was possible only by bringing in Negro slaves from Africa, for in a short time there were no natives left to be enslaved in Cuba. It took a ten years' war (1868–78) to eliminate slave labor completely from the sugar industry. The growing of tobacco, on the contrary, was on a garden scale, on small patches in the

[1] Translation by E. K. Kane.

bottom lands, where the soil was extremely fertile and the workers were, as a rule, members of the family. For this reason the Cuban tobacco-grower was, in the majority, white and free, aside from an occasional Negro slave, especially in the nineteenth century, for certain heavy work.

The manufacture of tobacco was organized in the city, and was promoted by merchants and exporters. It may be said that in tobacco's economy there were dealers before there were manufacturers; whereas in the case of the sugar of the West Indies, even though its economy was established by the commercial interests of the settlers, the first step was the setting up of the agricultural-industrial plant and then organizing the business of exportation.

In the beginning cigars were rolled by the workers in their own homes, individually, as a supplementary task to their regular work, or in little *chinchales* or workrooms, as can still be seen in the manufacture for local consumption in Havana and even in New York. The bundles of finished cigars were delivered by these individuals or small groups of cigar-makers to the export traders, who bought them, classified them according to size or shape, packed them for shipment, and sold them all under the guarantee of their own special trade-mark. Some of these trade-marks are over a hundred years old, and the manufacturers or mere exporters of labeled brands have always had several brands to suit the interests of the factory as well as the varying tastes of their foreign customers. As has been pointed out, capitalism got a hold upon tobacco's economy, as on that of many other products, through trade. At first production was limited, subject to the will of the worker, who sold his output to the dealer. It was only later, with the growth of the proletariat, that the cigar factory came into existence, with its shifts of workmen and capitalism's control of the industry.

Cigar and cigarette factories, with numbers of steady workers, and workrooms did not spring up until the nineteenth century was well advanced. For this reason the tobacco workers, like the growers, were mainly free men, even though some skilled slaves were hired out by their owners to help in harvesting the crop. The exporters or manufacturers would have liked to be able to depend on slave workers, as cheaper and easier to hold down, but inasmuch as the individual specialized, painstaking, and delicate work demanded for tobacco was incompatible with slavery, it was hoped that prison inmates could be made into tobacco workers, the "slaves of punishment," for whom the work involved in tobacco manufacture was ideally suited to their confined state. As the workshops of the penitentiaries are still known as galleys, in memory of the ships where those who had run afoul of the law were forced to work as galley slaves, so the workrooms of the tobacco factories are also known as galleys, recalling the original tobacco workrooms of the prisons. But the prisoners could not provide an adequate supply of skilled cigar workers, and the employers had to resort to the free, salaried labor market. In the manufacture of cigarettes, however, much less delicate and specialized work, the use of prison labor lasted much longer, up to our own times. It is curious to note that at the same time prison labor was employed, soldiers were made use of, too, in their enforced idleness in barracks. In 1863 in Havana cigarettes were being made for its 36 cigarette factories by 700 soldiers and 350 prisoners.

In the beginning the heavy work around the cigar and cigarette factories was done by Negroes, freed slaves, and Chinese. This is natural enough if one bears in mind that these rude tasks were not to the liking of the whites owing to medieval prejudices, which were then very deep-rooted and have not even yet been wholly extirpated. The tobacco-growers were predominantly white; the tobacco workers mostly colored. But the manufacturers of cigars were as a rule Spaniards who had settled in Cuba, mainly Catalonians, Asturians, and Galicians.

As time went on and the white population grew, as the slave trade disappeared and the proletariat of whites and native half-breeds increased, the tobacco factories employed workmen of every race and origin.

There were few foreigners in the tobacco business; nearly all were Cubans and Spaniards. This was not so with sugar, which brought in hordes of Negroes and Chinese to work in the fields, French chemists and Anglo-Saxon engineers, not to mention the Spaniards who were formerly the masters of the country. Gaspar Manuel Jorge is of the belief that "in proportion to the volume of production, more Cubans live off tobacco than off sugar" (*"El Tabaquero Cubano," Lyceum,* Vol. I, p. 76).

Whereas sensual sugar requires the rude strength of men for

its preparation, which is hard work, virile tobacco calls for
delicate hands, those of women or those having a woman's soft
touch, for its gentle handling. In olden times the cigars smoked
by the Cuban tobacco-grower were rolled "by his wife, his
daughter or his sweetheart," to quote the Countess of Merlin.
And on the farms, in the selecting-rooms, stripping-rooms, and
factories, women workers are employed. The woman stripper is
a popular figure in Cuba, and in Spain Carmen was a cigarette-
maker. It was at the end of the Ten Years' War, in 1877, that a
woman went to work in a Havana factory for the first time; it
was in the cigarette factory La Africana. Before that women had
wrapped and packed cigarettes by hand at home. From then
on, women came to form part of the factory proletariat. This
chronological coincidence is very significant. As slavery, which
was abolished in 1880, was giving its last gasps, industrial greed,
unable to depend on slave labor any longer, but unwilling to
pay the salaries of free men, created the feminine proletariat,
which is cheaper.

Women never worked in sugar, with the exception of a few
Negress slaves who were strong enough to plant and cut cane,
or some who were forced to it by hunger or the higher wages it
offered. In certain regions of Africa that supplied slaves to the
dealers in this merchandise it was an age-old custom for the
women to plant and cultivate the crops with a *coa,* or sharp-
pointed stick; anthropologists, moreover, have noted certain
traces of sexual dimorphism in some Negro races.

The workers in both tobacco and sugar have had their conflicts and difficulties with owners and employers. Contrary to the
general idea, strikes sprang up in these West Indies almost as
soon as the whites of Europe began to take possession of them.
In 1503, before Cuba had been colonized, Governor Ovando of
Hispaniola complained of the insurrections of Indians and Negroes who refused to do forced labor for the white man's exclusive benefit. There was an uprising of the workers on the
plantation of Columbus's son, Diego, in 1522. And in Cuba,
which was conquered in 1512, there were outbreaks of protest
against the system of slavery from the moment it was established. In 1538 the slaves sacked Havana, aiding and abetting
the French pirates who had descended on the city.

As may be easily deduced, there has always been a conflict of interests between the sugar and tobacco workers and their owners and employers. In these last decades, since the process of mechanization and the growing power and concentration of capital have tended to synthesize and unify labor problems in all fields of production, the demands of the workers in these two industries have approached each other more closely than in past epochs, when their industrial set-ups differed.

In both industries the problem of contracts, wages, hours, accident compensation, vacations, retirement, working and sanitary conditions has been discussed. But, nevertheless, the differences in the history of the labor conflicts of sugar and tobacco are striking, owing to the different systems under which the two industries are carried on.

Sugar was the product of latifundium and fief, which created serfs; tobacco, of the small farm and town, which were the abode of free men.

Alvaro Reynoso tells how on the early Cuban plantations the slaves lived in *bohíos,* those rustic huts which the Taino Indians used as dwellings; but as uprisings became more frequent, and more Negroes ran off to the hills, the slaves were housed in barrack-like quarters that resembled a jail. Some of these huge sugar-plantation prisons are still standing, with their single door and high, barred windows, into which the slaves, men, women, and children, black overseers, and even the semi-slave Chinese and their foremen were shut up when the day's work was done. There were centrals that had watchtowers and blockhouses in the mill yard, and private hired troops to defend them against workers' uprisings. In the vegas it was the government troops who occasionally burned the defenseless cabin of the poor tobacco-grower in the name of law and order.

The rural tobacco-raiser fought against the taxes, monopolies, and unjust restrictions placed upon his product. Not so the sugar-grower. As for the sugar worker, he had to fight in the centrals and warehouses and on the docks to have a limit set upon the load he had to carry, the maximum now being the 325-pound sack. The tobacco worker never was confronted by this problem.

Because of the individual nature of his work and his product, the cigar-maker always was entitled to his own "smokes"—that is, a certain number of the cigars he made for his personal use. This privilege came to acquire a tangible economic value. The cigar-maker could sell his smokes to a passing customer, and the manufacturer came to regard this as a part of the worker's wages, paid in kind. The attempt to treat this privilege as a part of the worker's wages gave rise at times to acrimonious disputes and strikes. Nothing of the sort happens in the case of sugar, aside from the stalk of cane the cutters or carters occasionally chew. There is no privilege of cane-sucking on the centrals analogous to the cigar factories' smokes. And if the mill worker wants sugar to sweeten his coffee, he has to buy it in the store just like anything else he needs.

The history of labor in Cuba until the last third of the nineteenth century was, with few exceptions, a record of rural slavery. Contrary to the general opinion that the Negro accepted his state of subjection passively, there are the frequently recurring episodes of uprisings and flight to the hills by runaway Negroes, and even of the collective suicide of bands of desperate slaves. The Mandingas were known for their tendency toward group suicide; in this way they freed themselves from their labors and had the last laugh on the master with a strike for which there was no settlement, and their successful escape to the other world. Despite the heap of alien earth that covered their bodies, the poor creatures believed that they would be resurrected in body and soul back in their native African villages. And the masters, aware of this belief, mutilated their bodies, even after death, cutting off vital organs so that when they came back to life, it would be without head or limbs, and through fear of this, as terrifying and as mythical as the torments of hell, they discouraged the living from following the others' example.

The strikes on the plantations and in the coffee groves were the Negro slaves' rebellions. Some of the great slave revolts were presented by the authorities as real social revolutions planned to secure liberation from the work of the sugar plantations. When slavery was abolished, the proletariat of the country, which took the slaves' place, was, as a rule, quiet both before and after the last war of independence, which had a different social significance from that of 1868, which brought about the abolition of slavery. Not only the *guajiro* on the plantations and the cane-cutters, but the men who operated the machinery came from the same essentially rustic background as the proletarian sugar workers; they were more disunited, having read less and being less prepared for permanent, directed collective organization. Only now, in the twentieth century, when the centrals are cities with hundreds of workers living around the plant, does the mass of the mill operatives, less rustic than they used to be, begin to show signs of class consciousness, attempts at organization, and the determination to have its rights.

The tobacco-grower, who was white, free, and as a rule attached to the land he owned or rented, even though he stayed in the hills, was not an outlaw, nor did he set up rebel colonies, but he did take part in uprisings, such as those which took place in the vegas near Havana during the eighteenth century, which were ruthlessly put down by the military forces. But these fierce revolts were not protests against slavery nor complaints about salary on the part of the worker in the productive phase of the industry, but the result of abuses committed during the most advanced phase, the commercial. And if they caused repercussions among the growers, it was because the agricultural labors, as they were carried on, received their compensation at the moment of the sale of the harvested crop. Nothing corresponding took place in the case of sugar; there were no strikes on the part of the tenant farmers when they began to appear in the middle of the nineteenth century; they sold on the open market when they could, or else accepted the centrals' terms.

In the city, the cigar worker, who formerly did not work by the hour but did piece work, discussed the price of his wares by units, layers, or thousands, and not by days, hours, and shifts as did the sugar worker.

The nineteenth century did not go by without struggles and reverses for the tobacco industry. The fact that tobacco is a so-called luxury article, which is at the same time comparable to a necessity because of the scope of the demand for it, makes it extremely vulnerable to heavy excise, export, and import taxes in all countries. These last often far exceed the original value of the tobacco, especially on the manufactured article. This increasing scale of customs duties has had its repercussions in Cuba, on more than one occasion unexpectedly, giving rise to difficult situations and always affecting the market and upsetting the industry. One domestic result of these barriers to exportation has been frequent unemployment among the Cuban cigar workers, not to mention sharp struggles between them and their employers over wage rates. For instance, in 1856 there were a number of cigar workers idle in Havana, owing to the fact that in 1855, in addition to a large amount of leaf tobacco, 356,582,500 cigars were exported, the greatest volume of export trade Cuba had ever known. This happened because the United States market wanted to stock up before the sharp increase in customs duties effective March 3, 1857. The emigration of Cuban cigar workers to Key West during the second half of the last century may be considered in large part a defensive measure taken to protect themselves against persecution in Havana because of their liberal, antislavery, or secessionist ideas, which were frowned upon by the factory-owner, who, besides his economic resources, had the backing of the rifles of the Spanish volunteers. Key West and Tampa were "civilian camps" of the national revolution, to use the phrase of Castellanos. Martí called Key West "a Creole stronghold, where from all the sufferings and anxieties of life arose all the sublimities of hope." There "Martí visited the factories, presided at the meetings, and by his eloquence infused the tobacco workers with his own fire" (G. M. Jorge). And the emigrant cigar workers "openly contributed ten per cent of their weekly earnings toward the revolution." It was a holy tithe laid upon the altar of country. For this reason, according to Tesifonte Gallego, Captain General Salamanca plotted to "destroy the tobacco workers' centers of Key West and Tampa to wipe out the rebels' organization."

In the nineteenth century, too, there were great strikes of tobacco workers. Even today it may be said that the relation between employer and worker in the tobacco industry is one of the most controversial in the whole field of Cuban labor. This is undoubtedly due to the fact that there was little infiltration of slave labor in the industry; that it involved hand work, at which the Cubans were adept; that the constant fluctuation of prices on the diverse and distant foreign markets fell outside the market quotations and the workers' knowledge. Above all, it was due to a proletarian class consciousness that developed among the tobacco workers earlier than in other groups.

In all this one highly influential factor was a custom that is typical of the Cuban tobacco workrooms, where the work is hand labor in contrast to the highly industrialized nature of other occupations, such as sugar. All the operations connected with sugar, from the conveyor belt to the mill, through the filters and evaporating-pans, the centrifugals and the packing, are done standing and moving from one place to another in the midst of an infernal racket. The mill workers find it almost impossible to talk to one another or to hear.

Reading is impossible in the sugar mill, for the noise in the boiling-vat rooms is so great that it drowns out the sound of the human voice. One no longer hears the rhythmic work songs with which in olden days the slaves accompanied their tasks in the grinding-rooms, the furnaces, the packing-house, and the refuse dumps. Today the mill is a mechanical monster that produces as it moves a deafening symphony of wheels, presses, piston rods, cogwheels, plungers, pistons, valves, hydroextractors, dumpcarts, with safety valves that give off a noise like the roar of a wild animal and ear-piercing whistles like enraged sirens.

In the case of tobacco, on the contrary, there is silence in the workroom if the chatter of conversation stops. The preparation of tobacco is carried on by workers, each seated at his own table, side by side, like students at their desks in school. This has made it possible to introduce into the tobacco workrooms a custom adopted from the refectories of convents and prison dining-halls: that of reading aloud so all the workers can listen while performing their tasks.

Sugar is produced to the orchestration of noisy machines; tobacco is worked up in silence or to the accompaniment of the spoken word. Sugar calls for choral harmony; tobacco for a solo melody. The sugar worker's tasks are active, heavy, deafening, and monotonous; the cigar worker sits down to his labors and can enjoy the pleasure and advantages of talking and listening.

It is said that the custom of reading to the cigar workers was introduced in the latter half of the nineteenth century in the two galleys of prison cigar-makers that had been set up in the Arsenal of Havana, and from there it spread to the other tobacco workrooms. The Reverend Manuel Deulofeu says that the custom was first permanently established, on the initiative of the workers, in the factory that existed in the town of Bejucal in the year 1864. And he recalls the name of the first reader,

Antonio Leal, in the workroom of the Viñas factory. It seems that it was in Bejucal, too, in the factory of Facundo Acosta, that the reader first read to the workers from a platform. It is difficult to know exactly where the custom originated and was instituted, but it is certain that it was not by accident or imitation, but done with a definite plan of social propaganda. A campaign was carried on to establish the custom, presenting it as an imitation of the instructive and democratic "public reading-rooms" then in vogue in the United States. These tobacco workroom readings were championed by the workers' weekly, *La Aurora* of Havana, in 1865, almost from its founding; and it was defended by the liberals in an editorial in *El Siglo* (January 25, 1866) of Havana, entitled "Readings in the Tobacco Workrooms," in which it was recalled that public readings were a common thing in other countries and that the public paid to hear them. It was thus that the eminent novelist Charles Dickens had toured the United States, reading from his own works, and there were many readers who had no literary gift of their own who read from the works of others. "Imagine paying to hear someone talk, to hear someone read," observed the writer of *El Siglo* pessimistically. But his lack of faith was ungrounded, and there were readings in the factories every day, and the workers paid to hear them. In Havana the custom was introduced in the factories in 1865, sponsored by Nicolás Azcárate, the Cuban liberal. The El Figaro factory was the first to allow reading in its workrooms, and was followed the next year by the factory of Don Jaime Partagás. But there is no doubt that it had been recommended many years before by the Spanish traveler Salas y Quiroga in his observations on the coffee groves of Cuba; he had suggested its introduction during the coffee grading, but it had never been put into effect.

This reading aloud in the tobacco workrooms became an instrument of local propaganda. The first reading in one of the factories of Havana was from a book entitled *The Struggles of the Century*. It was symbolic. The reading-table of each tobacco workroom became, according to Martí, "an advanced pulpit of liberty." When in 1896 Cuba rose in revolt against the Bourbon despotism and fought for its independence, an official order was issued silencing the pulpits of the tobacco factories. A number of these readers became leaders of the people's party, even though some wound up as deserters to the cause, and even paid traitors.

As a result of this advanced political consciousness on the part of the Cuban cigar workers, in conjunction with other causes, a strange phenomenon took place: two contradictory and parallel migrations, the emigration of Cuban tobacco workers abroad, coinciding with the immigration of foreign laborers to work in the sugar industry of the country. At the same time that Cuban workmen had to leave their country to be able to work, foreigners were coming into it to work and make a living.

The custom of reading also explains the fact that the tobacco workers in Cuba were the first Cuban workmen to form associations to protect class interests. In 1865 a cigar worker, Saturnino Martínez, founded the weekly *La Aurora.* That same year cigar workers founded the Workmen's Mutual Aid Society of Havana, the Brotherhood of Santiago de las Vegas, and the Workmen's Society of San Antonio de los Baños. In 1878 and '79 the Workers' Guild and the Workers' Center were established in Havana, and in 1885 the Workers' Circle. In 1878 the tobacco selectors former their association, whose statutes were drawn up by the aforementioned politician Nicolás Azcárate. In 1892 the cigar workers organized and held the first workers' convention, not without arousing opposition. It was the Cuban cigar workers who most courageously and unflaggingly supported José Martí's revolutionary efforts on behalf of Cuban independence. From Key West, rolled in a cigar made by Fernando Figueredo, a great citizen, general, and cigar worker, the order for the revolution for national independence reached Havana in 1895. To the poet's mind the single star of the Cuban flag evokes the tobacco flower; it is a five-pointed star, with a white corolla whose petals are pink-edged.

Even today the Society of Selectors is one of the oldest and most firmly established among the Cuban workers' organizations. In our own day education has become much more general among Cuban workingmen; there was a time when the cigar workers were the "brain trust" of the Cuban proletariat. The workroom readers were "graduates of the factory," in the words of Martí. As he told the cigar workers of Tampa in his famous revolutionary speech of November 26, 1891, they worked "with the table at which they studied alongside that at which they

earned their bread." And he spoke of "those factories which are
like academies where reading and thinking are continuous, and
those lyceums where the hand that folds the tobacco leaf by day
picks up the textbook by night." They worked with tobacco
leaves and book leaves. This was the cigar worker. In Cuba he
is still the "enlightened" workman whose "intellectual veneer
makes him feel superior in this respect to the other workers.
This, he feels, entitles him to talk about everything and pass
opinion on everything" (G. M. Jorge). He is given to argument
and controversy. There are those who believe that because of his
intellectual tendencies and his romantic tradition "he does not
completely grasp the new theories of class struggle," or if he
understands them he is unwilling to submit to the discipline
necessary to put them into effect. But there is no doubt that the
tobacco worker is a nonconformist who thinks and insists upon
a new design for living.

At the present time mechanization, which years ago took possession of cigarette-manufacturing and is now trying to eliminate the cigar-maker, has had an influence on the typical custom
of reading aloud. In 1923 a radio loudspeaker was installed in
the Cabañas y Carvajal factory. The reader used earphones and
rebroadcast to the workroom the radio news. In 1936 the reader
and the radio still existed side by side; now the machine is tri-
umphing over the reader by means of the radio, which trans-
mits readings to the workers over the air. But this is no longer
the typical reading of news and selections chosen by the workers of each galley, like one more selection of their art. Now the
reader's platform is deserted, and is occupied only on rare occasions; new books and controversial material are no longer
read there. The fellow workman is silent and is replaced by
anonymous speakers. And over the air there comes into the
workroom a deplorable mixture of the lowest-grade mental
vulgarities, interlarded with music, like the poorest-grade cigars, which are wrapped in showy tinfoil because of its resemblance to silver.

Cuba had two parallel sources of pride, the synthesis of this
strange contrast I have outlined, that of being the country that
produced sugar in the greatest quantity and tobacco of the fin-

est quality. The first is disappearing; nobody can take away the second.

We have seen the fundamental differences that existed between them from the beginning until machines and capitalism gradually ironed out these differences, dehumanized their economy, and made their problems more and more similar.

But it should be observed at the same time that although there are differences between sugar and tobacco, there have never been any conflicts between them. Cane sugar has had and has a bitter struggle with beet sugar; a world-wide war has been going on between them for over a century, the "War of the Sugars," like the Wars of the Roses. The authentic tobacco of Cuba has had and has its fierce struggle with foreign tobacco, especially with that which usurps its name, the fight between the Havana and the *mabinga*. There has been a global conflict going on between tobaccos for centuries, just as there has been and will be between men. But there was never any enmity between sugar and tobacco.

Therefore it would be impossible for the rhymesters of Cuba to write a "Controversy between Don Tobacco and Doña Sugar," as the roguish archpriest would have liked. Just a bit of friendly bickering, which should end, like the fairy tales, in marrying and living happy ever after. The marriage of tobacco and sugar, and the birth of alcohol, conceived of the Unholy Ghost, the devil, who is the father of tobacco, in the sweet womb of wanton sugar. The Cuban Trinity: tobacco, sugar, and alcohol.

It may be that one day the bards of Cuba will sing of how alcohol inherited its virtues from sugar and its mischievous qualities from tobacco; how from sugar, which is mass, it received its force, and from tobacco, which is distinction, its power of inspiration; and how alcohol, the offspring of such parents, is fire, force, spirit, intoxication, thought, and action.

And with this laud of alcohol the counterpoint comes to an end.

II

THE ETHNOGRAPHY *and* TRANSCULTURATION *of*
HAVANA TOBACCO *and the Beginnings*
of SUGAR *in America*

I

On Cuban Counterpoint

THE preceding essay is of a schematic nature. It makes no attempt to exhaust the subject, nor does it claim that the economic, social, and historical contrasts pointed out between the two great products of Cuban industry are all as absolute and clear-cut as they would sometimes appear. The historic evolution of economic-social phenomena is extremely complex, and the variety of factors that determine them cause them to vary greatly in the course of their development; at times there are similarities that make them appear identical; at times the differences make them seem completely opposed. Nevertheless, fundamentally the contrasts I have pointed out do exist.

The ideas outlined in this work and the facts upon which they are based could be substantiated by full and systematic documentation in the form of notes; but in view of the nature of the work I have preferred to add some supplementary chapters. They deal with a basic theme of their own, but bear upon certain fundamental aspects of "Cuban Counterpoint" and will be of interest to readers who care to go deeper into the subject.

2

On the Social Phenomenon of "Transculturation" and Its Importance in Cuba

With the reader's permission, especially if he happens to be interested in ethnographic and sociological questions, I am going to take the liberty of employing for the first time the term *transculturation,* fully aware of the fact that it is a neologism. And I venture to suggest that it might be adopted in sociological terminology, to a great extent at least, as a substitute for the term *acculturation,* whose use is now spreading.

Acculturation is used to describe the process of transition from one culture to another, and its manifold social repercussions. But *transculturation* is a more fitting term.

I have chosen the word *transculturation* to express the highly varied phenomena that have come about in Cuba as a result of the extremely complex transmutations of culture that have taken place here, and without a knowledge of which it is impossible to understand the evolution of the Cuban folk, either in the economic or in the institutional, legal, ethical, religious, artistic, linguistic, psychological, sexual, or other aspects of its life.

The real history of Cuba is the history of its intermeshed transculturations. First came the transculturation of the paleolithic Indian to the neolithic, and the disappearance of the latter because of his inability to adjust himself to the culture brought in by the Spaniards. Then the transculturation of an unbroken stream of white immigrants. They were Spaniards, but representatives of different cultures and themselves torn loose, to use the phrase of the time, from the Iberian Peninsula groups and transplanted to a New World, where everything was new to them, nature and people, and where they had to readjust themselves to a new syncretism of cultures. At the same time there was going on the transculturation of a steady human stream of African Negroes coming from all the coastal regions of Africa along the Atlantic, from Senegal, Guinea, the Congo, and Angola and as far away as Mozambique on the opposite shore of that continent. All of them snatched from their original social groups, their own cultures destroyed and crushed under the weight of the cultures in existence here, like sugar cane ground in the rollers of the mill. And still other immigrant cultures of the most varying origins arrived, either in sporadic waves or a continuous flow, always exerting an influence and being influenced in turn: Indians from the mainland, Jews, Portuguese, Anglo-Saxons, French, North Americans, even yellow Mongoloids from Macao, Canton, and other regions of the sometime Celestial Kingdom. And each of them torn from his native moorings, faced with the problem of disadjustment and readjustment, of deculturation and acculturation—in a word, of transculturation.

Among all peoples historical evolution has always meant a

vital change from one culture to another at tempos varying from gradual to sudden. But in Cuba the cultures that have influenced the formation of its folk have been so many and so diverse in their spatial position and their structural composition that this vast blend of races and cultures overshadows in importance every other historical phenomenon. Even economic phenomena, the most basic factors of social existence, in Cuba are almost always conditioned by the different cultures. In Cuba the terms Ciboney, Taino, Spaniard, Jew, English, French, Anglo-American, Negro, Yucatec, Chinese, and Creole do not mean merely the different elements that go into the make-up of the Cuban nation, as expressed by their different indications of origin. Each of these has come to mean in addition the synthetic and historic appellation of one of the various economies and cultures that have existed in Cuba successively and even simultaneously, at times giving rise to the most terrible clashes. We have only to recall that described by Bartolomé de las Casas as the "destruction of the Indies."

The whole gamut of culture run by Europe in a span of more than four millenniums took place in Cuba in less than four centuries. In Europe the change was step by step; here it was by leaps and bounds. First there was the culture of the Ciboneys and the Guanajabibes, the paleolithic culture, our stone age. Or, to be more exact, our age of stone and wood, of unpolished stone and rough wood, and of sea shells and fish bones, which were like stones and thorns of the sea.

After this came the culture of the Taino Indians, which was neolithic. This was the age of polished stone and carved wood. With the Tainos came agriculture, a sedentary as opposed to a nomadic existence, abundance, tribal chieftains, or caciques, and priests. They entered as conquerers and imposed the first transculturation. The Ciboneys became serfs, *naborías,* or fled to the hills and jungles, to the *cibaos* and *caonaos.* Then came a hurricane of culture: Europe. There arrived together, and in mass, iron, gunpowder, the horse, the wheel, the sail, the compass, money, wages, writing, the printing-press, books, the master, the King, the Church, the banker. . . . A revolutionary upheaval shook the Indian peoples of Cuba, tearing up their institutions by the roots and destroying their lives. At one bound the bridge between the drowsing stone ages and the

wide-awake Renaissance was spanned. In a single day various
of the intervening ages were crossed in Cuba; one might say
thousands of "culture-years," if such measurement were admissible in the chronology of peoples. If the Indies of America
were a New World for the Europeans, Europe was a far newer
world for the people of America. They were two worlds that
discovered each other and collided head-on. The impact of the
two on each other was terrible. One of them perished, as though
struck by lightning. It was a transculturation that failed as far
as the natives were concerned, and was profound and cruel for
the new arrivals. The aboriginal human basis of society was destroyed in Cuba, and it was necessary to bring in a complete
new population, both masters and servants. This is one of the
strange social features of Cuba, that since the sixteenth century
all its classes, races, and cultures, coming in by will or by force,
have all been exogenous and have all been torn from their
places of origin, suffering the shock of this first uprooting and
a harsh transplanting.

With the white men came the culture of Spain, and together with the Castilians, Andalusians, Portuguese, Galicians,
Basques, and Catalonians. It could be called a crosscut of the
Iberian culture of the white Pyrenean subrace. And in the first
waves of immigration came Genoese, Florentines, Jews, Levan-
tines, and Berbers—that is to say, representatives of the Mediterranean culture, an age-old mixture of peoples, cultures, and
pigmentation, from the ruddy Normans to the sub-Sahara Negroes. Some of the white men brought with them a feudal economy, conquerors in search of loot and peoples to subjugate and
make serfs of; while others, white too, were urged on by mercantile and even industrial capitalism, which was already in its
early stages of development. And so various types of economy
came in, confused with each other and in a state of transition,
to set themselves up over other types, different and intermin-
gled too, but primitive and impossible of adaptation to the
needs of the white men at that close of the Middle Ages. The
mere fact of having crossed the sea had changed their outlook;
they left their native lands ragged and penniless and arrived as
lords and masters; from the lowly in their own country they
became converted into the mighty in that of others. And all of
them, warriors, friars, merchants, peasants, came in search of

adventure, cutting their links with an old society to graft them-selves on another, new in climate, in people, in food, customs, and hazards. All came with their ambitions fixed on the goal of riches and power to be achieved here, and with the idea of returning to their native land to enjoy the fruits of their labors in their declining years. That is to say, the undertak-ing was to be bold, swift, and temporary, a parabolic curve whose beginning and end lay in a foreign land, and whose intersection through this country was only for the purpose of betterment.

There was no more important human factor in the evolution of Cuba than these continuous, radical, contrasting geographic transmigrations, economic and social, of the first settlers, this perennial transitory nature of their objectives, and their un-stable life in the land where they were living, in perpetual dis-harmony with the society from which they drew their living. Men, economies, cultures, ambitions were all foreigners here, provisional, changing, "birds of passage" over the country, at its cost, against its wishes, and without its approval.

With the whites came the Negroes, first from Spain, at that time full of slaves from Guinea and the Congo, and then di-rectly from all the Dark Continent. They brought with them their diverse cultures, some as primitive as that of the Ciboneys, others in a state of advanced barbarism like that of the Tainos, and others more economically and socially developed, like the Mandingas, Yolofes (Wolofs), Hausas, Dahomeyans, and Yorubas, with agriculture, slaves, money, markets, trade, and centralized governments ruling territories and populations as large as Cuba; intermediate cultures between the Taino and the Aztec, with metals, but as yet without writing.

The Negroes brought with their bodies their souls, but not their institutions nor their implements. They were of different regions, races, languages, cultures, classes, ages, sexes, thrown promiscuously into the slave ships, and socially equalized by the same system of slavery. They arrived deracinated, wounded, shattered, like the cane of the fields, and like it they were ground and crushed to extract the juice of their labor. No other human element has had to suffer such a profound and repeated change of surroundings, cultures, class, and conscience. They were transferred from their own to another more advanced culture, like that of the Indians; but the Indians suffered their fate in their native land, believing that when they died they passed over to the invisible regions of their own Cuban world. The fate of the Negroes was far more cruel; they crossed the ocean in agony, believing that even after death they would have to re-cross it to be resurrected in Africa with their lost ancestors. The Negroes were torn from another continent, as were the whites; but not of their own will or choice, and forced to leave their free and easy tribal ways to eat the bitter bread of slavery, whereas the white man, who may have set out from his native land in despair, arrived in the Indies in a frenzy of hope, con-verted into master and authority. The Indians and the Span-iards had the support and comfort of their families, their kin-folk, their leaders, and their places of worship in their suffer-ings; the Negroes found none of this. They, the most uprooted of all, were herded together like animals in a pen, always in a state of impotent rage, always filled with a longing for flight, freedom, change, and always having to adopt a defensive atti-tude of submission, pretense, and acculturation to a new world. Under these conditions of mutilation and social amputation, thousands and thousands of human beings were brought to Cuba year after year and century after century from continents beyond the sea. To a greater or lesser degree whites and Ne-groes were in the same state of dissociation in Cuba. All, those above and those below, living together in the same atmosphere of terror and oppression, the oppressed in terror of punishment, the oppressor in terror of reprisals, all beside justice, beside ad-justment, beside themselves. And all in the painful process of transculturation.

After the Negroes began the influx of Jews, French, Anglo-Saxons, Chinese, and peoples from the four quarters of the globe. They were all coming to a new world, all on the way to a more or less rapid process of transculturation.

I am of the opinion that the word *transculturation* better ex-presses the different phases of the process of transition from one culture to another because this does not consist merely in ac-quiring another culture, which is what the English word *ac-culturation* really implies, but the process also necessarily in-volves the loss or uprooting of a previous culture, which could be defined as a deculturation. In addition it carries the idea of the consequent creation of new cultural phenomena, which could be called neoculturation. In the end, as the school of Malinowski's followers maintains, the result of every union of cultures is similar to that of the reproductive process between individuals: the offspring always has something of both parents but is always different from each of them.

These questions of sociological nomenclature are not to be disregarded in the interests of a better understanding of social phenomena, especially in Cuba, whose history, more than that of any other country of America, is an intense, complex, unbroken process of transculturation of human groups, all in a state of transition. The concept of transculturation is fundamental and indispensable for an understanding of the history of Cuba, and, for analogous reasons, of that of America in general. But this is not the moment to go into this theme at length, which will be considered in another work in progress dealing with the effects on Cuba of the transculturations of Indians, whites, Negroes, and Mongols.

When the proposed neologism, *transculturation,* was submitted to the unimpeachable authority of Bronislaw Malinowski, the great figure in contemporary ethnography and sociology, it met with his instant approbation. Under his eminent sponsorship, I have no qualms about putting the term into circulation.

3

Concerning Tobacco Seed

There is something marvelous even about the unusual number of seeds produced by tobacco. This was one of the reasons for its rapid spread in all lands, once the Spaniards found the plant in America and succumbed to its temptation.

The seeds of tobacco are incredibly numerous and very tiny. There are from 300,000 to 400,000 to an ounce. One ounce of seed could theoretically produce 300,000 plants. Each tobacco plant can yield as many as a million seeds, according to William George Freeman. Each of these little seeds in turn could produce another million. If there were not factors that impede the germination of so many seeds, in a few generations the whole surface of the globe would be covered with tobacco plants.

4

Concerning the Low Nicotine Content of Cuban Tobacco

I feel that this is the proper occasion to refute a generally accepted error concerning the tobacco of Cuba, which is said to have an excessively high nicotine content, when exactly the opposite is the case. About the middle of the nineteenth century Schloesing discovered that the tobacco of Cuba was lowest in nicotine. (T. Schloesing: *"Investigaciones acerca del tabaco,"* in *Documentos relativos al cultivo del tabaco,* collected by Alvaro Reynoso, Havana, 1888, Vol. I.)

5

On How Tobacco was Discovered in Cuba by the Europeans

The theory has been advanced that tobacco was known to the old Eurasian world before Christopher Columbus made his first visit to our cisatlantic world. Some believed that the forerunners of tobacco were to be found in certain stupor-inducing plants used by the ancients, the ancients of both the Old World and the New, or in the burning of certain plants as incense to honor the gods, who are as susceptible as all rulers to the fumes of human flattery. There are others who, knowing that tobacco has been smoked in Oriental lands for centuries, have assumed from this that the plant was of native origin there, even though unknown to Europeans. The bibliography on the subject is large but not convincing. It is no longer possible to doubt that tobacco, along with syphilis, was America's most eagerly accepted gift to mankind. Nobody any longer attempts to uphold the non-American origin of tobacco. Quite properly Jerome E. Brooks calls it "heretical" in his learned introduction to the extensive work entitled: *Tobacco, Its History Illustrated by the Books, Manuscripts and Engravings in the Library of George Arents, Jr.* (edition of 700 copies, New York, 1937, 4 vols., $75 a volume).

Brooks's sociological acumen (op. cit., p. 91) is responsible for one of the strongest arguments in proof of the American origin of tobacco, based on the observation that tobacco, which constitutes one of the most primitive and typical elements of the mythologies and religious rituals of numerous peoples of America, is not to be found in any cosmogonal myth or sacramental rite of any people except in America. Certain folkloric allusions to tobacco in connection with the Biblical tradition that have been collected in Abyssinia were undoubtedly grafted on the old trunk only a few centuries ago by a process of simple cultural syncretism in an attempt to Christianize the use of heathenish tobacco and make it suitable for adoption by certain ecclesiastical fanaticisms of the Old World. In view of the archaic nature of religious rites among all peoples (the use of dead languages, cryptic words and signs, indecipherable formulas, hieratic gestures of symbolic and almost always forgotten meaning) this argument in favor of the American origin of tobacco carries great weight, especially since it is supported by conclusive documents of written history.

Tobacco is a genuinely native plant of America. It existed from pre-Columbian times in both Americas, and if it was not indigenous to these Antilles it was brought here from the neighboring continent. The absence of wild tobacco in Cuba would perhaps permit the hypothesis that the plant does not form part of the island's abundant flora, but was brought to Cuba from over the sea. If this was the case, it would be necessary to assume from the Cuban Indians' mode of using it, which was different from that of the North American Indians, that tobacco belongs to the culture of the Arawak or Taino Indians, who some time before the arrival of the Spaniards had invaded Cuba and settled its eastern region, where they had subdued the more primitive and savage Ciboneys. In any case, tobacco is Indian and it is on record that it was in Cuba that the Western civilization of the white man first discovered it, although there are writers who deny this.

Tobacco is mentioned in the first book that was written about the discovery, the Diary of Columbus, which he wrote aboard the very ships that were making the discovery as he turned the geographical pages of the new truth he was finding. After him there is not one historian of the Indies who does not refer to this surprising custom of the natives. The book that gave origin to the name of America also contains the first printed reference to the use of tobacco in this part of the world, the famous *Cosmographiæ* of Martin Waldseemüller, which describes Amerigo Vespucci's explorations.

In this same sixteenth century an English writer affirmed that Columbus first saw tobacco used by the natives of the island of San Salvador, the first American territory he discovered, and that these Indians smoked it in a kind of funnel made of palm leaf (perhaps he meant yagua), into which they put the dried leaves of tobacco and then lighted them and inhaled the smoke (Lobel in *Nova Stirpium Adversaria,* appendix to his *History of Plants,* 1576). It may well have been that Columbus found tobacco on the island of Guanahaní, or San Salvador; but the description of this palm-leaf funnel and the method of smoking the tobacco is untenable. Undoubtedly the English botanist was writing from hearsay, and he was confused by what he had heard of certain Indians smoking tobacco leaves rolled or put into the leaf of another plant, as we shall see farther on. What probably led him most astray were the accounts he read in the book *La Cosmographie universelle* by A. Thevet, which had been published a year previous. This writer was in Brazil about the year 1555, and on his return to Europe he described the *petum,* or cigar, and the way the Indians of that country smoked it.

The Cuban historian Antonio Bachiller y Morales, in his book *Cuba Primitiva* (Havana, 1883, p. 269), referring to the word *Exuma* says: "Island of the Bahama group where Admiral Columbus saw the tobacco which is today known throughout the world." Bachiller refers to an episode that took place on October 15, 1492, which I shall speak of shortly. After Bachiller, a North American, J. B. Thacher (*Christopher Columbus, His*

Life, His Work, His Remains, New York, 1903, Vol. I, p. 560, n. 4) also states that these leaves Columbus saw on the little island of Exuma were "undoubtedly" tobacco.

The idea has been upheld that tobacco was discovered by Christopher Columbus on that same island of Guanahaní, or San Salvador, where he first set foot on the New World. This is based on the interpretation of a passage in the Admiral's Diary that reads:

"Monday, October 15. . . . halfway across the gulf from these two islands . . . I came upon a man alone in a canoe who was going from the island of Santa María to Fernandina, and he was carrying a little bread of the kind they use, about the size of a fist, and a gourd of water, and a piece of red earth crumbled to dust and then pressed into a cake, and some dry leaves, which must be something highly esteemed among them, for they brought me some as a present in San Salvador. . . ."

To be sure, that offering of leaves by the Indians of San Salvador was a ritual peace offering from the astonished natives of this side of the Atlantic to the unexpected natives from the opposite coast. It was a typically Indian gesture, and we could even classify it as Taino or Arawak, for even today, according to the Marquis of Wavris *Mœurs et coutumes des Indiens sauvages de l'Amérique du Sud,* p. 377, when two of them meet they greet each other with an exchange of a handful of coca, the magic leaf, and the same thing happens with the exchange of tobacco.

The theory of the finding of tobacco on Guanahaní is very plausible, but there is hardly sufficient proof to consider it conclusive. Tobacco was not the only dried plant those Indians used, nor does Columbus say in his Diary that the leaves were twisted or rolled up, a detail that would have identified them as smoking tobacco. Moreover, this paragraph of the Admiral's calls attention to the fact that this Indian, alone in his canoe (which Columbus referred to as an *almadía,* as the Indian word *canoa* was still unknown to him), carried no other provisions than "a little bread of the kind they use, about the size of a fist, and a gourd of water," in addition to "a piece of red earth crumbled to dust and then pressed into a cake, and some dry leaves. . . ." This must be interpreted in the light of its various and complex elements. These are: (1) a journey by canoe among the keys, which might well be a long one; (2) a great scarcity of supplies; (3) some highly prized leaves; and (4) a little reddish earth "crumbled to dust and then pressed into a cake." All this would seem to indicate that the "highly esteemed" leaves might be tobacco (although there is no assurance of this) and that these leaves and the cake of reddish earth were the Indian's way of keeping up his strength for a long time with very little food. The Indians in many parts of South America often carried no other grub on their trips than some powdered seeds or the leaves of certain stimulant plants such as *Erythroxylon coca, Banisteria caapi,* and others that will be mentioned later on. "The Indians can travel great distances without other nourishment." See what E. A. Wallace (*Timehri,* Demerara, December 1887, p. 317) has to say about the Guahibo Indians of the Orinoco. These kneaded powders that the Indian carried must have belonged to one of these groups of stimulants, and by their nature might form part of that cultural group which today is known as the "tobacco complex," as will be better understood farther on (see Part II, Chapter vi).

Besides, it must be admitted that if these highly prized dried leaves were really tobacco, the Admiral saw it, but did not discover it, for discovering is not merely seeing but "coming to see," finding that which was hidden, concealed, or overlooked, acquiring knowledge of something that was unknown. And Don Christopher did not know what tobacco was, nor its properties, nor that it was used for smoking until the night of Monday, the 5th day of November of 1492, or the morning of the next day when Luis de Torres and Rodrigo de Xerez showed it to him. They had discovered it sometime between the 2nd and the 5th day of that month when they went to explore the interior of Cuba at Columbus's orders.

Prior to this day neither Columbus nor the Spaniards had "come to see" what tobacco was. I copy herewith the text of Father Bartolomé de las Casas, glossing the Diary of Admiral Christopher Columbus in which he describes the discovery of tobacco on the island of Cuba.

In Chapter xlv of the *History of the Indies* by that famous friar, referring to the events of Friday, the 2nd of November of

1492, when the Admiral was at the river and port of Mares in the island of Cuba, which was probably the present-day Manatí or Sabanalamar, we read the following:

"With that idea he had that this land was the mainland and the kingdom of the Great Khan or bordered upon it, in order to get some information and learn about it he decided to send two Spaniards, one by the name of Rodrigo de Xerez, who lived in Ayamonte, and the other one Luis de Torres, who had lived with the Governor of Murcia, and had been a Jew and knew Hebrew and Chaldee, and they even say Arabic.

"With them he sent two Indians, one he had brought with him from Guanahaní, the other from the settlement at the mouth of the river. He gave them from the articles of barter he had brought along strings of beads and other things to exchange for food if they should need it, and set them a limit of six days in which to return. He gave them samples of spices so they would recognize them if they found any along the way. He gave them instructions as to how they should inquire for the King of that land, and what they should say to him in the name of the King and Queen of Castile, how they had sent out the Admiral to present their letters to him and a gift from them, and to receive news of his realm and have friendship with him and offer him their favor and help for whenever and on whatever occasion he should wish to avail himself of it, and to receive knowledge of certain provinces and ports and rivers the Admiral knew about, and how far distant they were."

The Admiral spent Saturday, the 3rd of November 1492, and Sunday, the 4th, conversing with the Indians and hunting along the banks of the port of Mares.

In the next chapter of Las Casas's work, Chapter xlvi, he says:

"The night of Monday the two Indians returned that had gone inland with them, a good twelve leagues, where they found a settlement of as many fifty houses, in which they said there were a thousand inhabitants, because it seemed to them that many lived in the same house; and this is clear proof that they are humble, gentle, and peaceful people."

And then he continues:

"These two Spaniards met many people on the way going back and forth to their villages, men and women, and the men always carried a firebrand in their hand and certain plants to take their smokes, which are some dried plants put into a certain leaf, dried, too, in the shape of a *mosquete* or squib made of paper, like those boys make on the day of the Holy Ghost, and they light it at one end and at the other they suck or chew or draw in with their breath that smoke with which their flesh is benumbed and, so to speak, it intoxicates them, and in this way they say they do not feel fatigue. These *mosquetes,* or whatever we shall call them, they call *tabacos.* I have known Spaniards on this island of Hispaniola who learned to use them and who when they were reproved for it and told that this was a vice, answered that it was beyond their power to give them up; I do not know what pleasure or good they found in them."

From all this it seems evident that Rodrigo de Xerez and the Jew Luis de Torres discovered tobacco one day, the 2nd, the 3rd, the 4th, or the 5th of November of 1492, in the lands lying beside the port of Mares or Manatí in the island of Cuba.

With perfect justification J. Alden Mason (*Use of Tobacco in Mexico and South America,* Chicago, 1924, p. 3) wrote that the "first contact" of Europeans with tobacco took place on the island of San Salvador; but that the "use of tobacco" was discovered in "what is still its principal realm, in Cuba."

The interpretation put on the text in question of the Admiral's Diary by the chronicler Herrera shows a complete lack of understanding. He does not mention the *tabaco,* and alludes only to the fact that Columbus's emissaries found "many people on the way and each one carried a firebrand in his hand to light a fire and perfume themselves with certain herbs they carried with them" (*Década I,* Chapter iv).

Besides, the same thing has happened with tobacco as with so many other discoveries in these Indies and with the discovery of America itself: they were in reality "discoveries" only for the white men of Europe, for they had been really discovered by the copper-colored natives of this land. Tobacco, like quinine and coca, corn, the tomato, the potato, the papaya, peppers, the yucca, the sweet potato, chocolate, peanuts, the cashew nut, the alligator pear, the pineapple, and other natural and cultivated products that today make up perhaps the major part of the vegetable food of the world was a "discovery" made by the intelligence of the native Americans centuries and millenniums before it "came to be seen" by the European natives.

6

On Tobacco Among the Indians of the Antilles

It has seemed to me of interest to give in abridged form the texts of some of the chroniclers and historians of the Indies referring to tobacco. Reading and comparing them will help to clear up some of the doubtful problems; but they will have to be accompanied by certain commentaries and references to comparative ethnography, without which the interpretation of some of the texts would be difficult.

Let us begin with Father Ramón Pané, generally called Pane. On various occasions in his famous *Relación* he refers to the use of the *tabaco* among the Indians of Hispaniola, but he does not employ that word. Chapter xi, entitled, "What happened to the four brothers when they were fleeing from Yaya," contains the first myth connected with tobacco.

Yaya was a man whose son Yayael wanted to kill him (is this the Freudian Œdipus myth?); but the father, suspecting his designs, killed him first and put his bones into a gourd full of water, where they turned into fish. One day when Yaya was out in his field, four brothers entered his house, the sons by a single birth of a woman named Itiba Yahuvava (the four points of the compass?). Yaya came back unexpectedly and the quadruplets dropped the gourd full of the little fish that had come from the bones of Yayael, and so much water spilled out that it covered the earth. Thus the sea filled with fish was created. The four brothers fled to the house of the patriarch Basamanaco (or Ayamanaco or Bayamanicoel). Father Pané's version is as follows:

"As soon as they reached the door of Basamanaco and saw that he had *cazabí* in his hand, they said to him: '*Ayacavo Guarocoel*,' which means: 'We know our grandfather.' Whereupon Dimivan Caracaracol, seeing his brothers before him, went into his house to see if he could find some *cazabí*, which is the bread they eat in that country.

Caracaracol went into the house of Ayamanaco and asked him for *cazabí,* the aforementioned bread. The latter put his hand to his nose and blew against the other's back a gobbet of mucus full of cohoba, which he had ordered prepared that day.

"Cohoba is a kind of powder they take sometimes to physic themselves and for other purposes that will be explained farther on. They take it through a reed, half an arm's span in length. They put one end in the nose, and the other in this powder, and breathe it in through the nose, and it physics them powerfully.

"So he gave him this gobbet of mucus instead of the bread he was making, and was very angry because he had asked him for bread. Caracaracol went back to his brothers and told them what had happened to him with Bayamanicoel, and how he had blown a gobbet of mucus against his back, which hurt him very much. Then his brothers looked at his back and saw that it was badly swollen. The swelling grew until he was on the verge of death, and for that reason they tried to cut it off, but they were unable to; taking a stone ax, they split it open and out came a living female turtle. Then they built a house and carried the turtle to it. I have not learned more about this, and what I have written is of little value."

(Transcript of the *Historia del Almirante Don Cristóbal Colón,* by his son Don Hernando. Madrid, edition of 1932, Vol. II, p. 51.)

In another Spanish edition of the *Historia del Almirante Don Cristóbal Colón,* that published in Madrid in 1892 (Vol. I, p. 292), instead of "gobbet of mucus full of cohoba" it says: "He put his hand to his nose and threw against his back a gourd that was full of *cogioba.*" The word *mucus* seems closer to reality, for it refers, as we would say today, to "nasal mucus full of powdered tobacco." The Italian editions say: *"Si mife mano al nafo e gli gittó un guanguaio dalle spalle pieno di Cogioba che aveva fatto far quel di."* According to the text of Father Pané, which was copied by Peter Martyr d'Anghiera (*Década I,* Book IX, Chapter v), this dirty mucus full of tobacco from the nose of Bayamanicoel (or Basamanaco or Ayamanaco) engendered in the back of the first-born of Dimivan Caracaracol (the name probably means "the Mangy One" or "the Syphilitic") a woman by whom the four quadruplet brothers had sons and daughters. It is a variation on the myth of Eve, who came out of her husband's side, mixed with a myth similar to that of the Flood, with the patriarch Noah and his sons populating the world anew. In this Indo-Antillean myth the erotic, genesial wickedness does not come from a serpent but from the powders of cohoba, which penetrate the flesh of Caracaracol to engender a female or a turtle, which in primitive folklore is also a genesial symbol, like the serpent of the Bible.

Father Ramón Pané was "a poor hermit of the Order of St. Jerome," and one of the first friars who came out to Hispaniola to convert the Indians. Too much credence cannot be given to his words. In the first place, Father Pané was ignorant, a "simple" man, who did not speak Spanish too well, as he was of Catalonian origin, nor the language of the Indians either, according to Father Las Casas (*Apologética,* Chapters cxx, clxvii, and clxxviii). Also because of the confusion and inconsistencies that accumulate around myths, even the Christian, but especially among peoples still close to the first phase of religious evolution, when everything floats upon that conceptual magma which ethnologists today describe by the word *mana,* taken from the Polynesians, and which in the more vulgar folklore in Cuba is preserved in the terms of African origin *cocorícamo, merequetén, timba, sumba,* etc. (See Fernando Ortiz: *El Cocorícamo y demás conceptos teoplásmicos del folklore afro-cubano.*) Moreover, the original manuscript of Father Pané's work has been lost, and in references to it and excerpts and translations from it there is considerable variation in the Indian terms employed.

Even if we did not possess abundant and reliable information regarding the sacred nature of tobacco among the Taino Indians, due, no doubt, to its stimulant and toxic qualities, as with wine, opium, and other narcotic agents employed in the different religions, this myth would give us an understanding of the supernatural importance ascribed to the powders of cohoba.

There are those who have seen in the use of tobacco by the Indians for religious purposes, especially in smoking, a kind of propitiatory offering to the gods, an adulation of the Almighty designed to win his favor and placate his wrath, or a magic rite, like the burning of incense and the liturgies of numerous other religions. Others have interpreted it as a way of driving out the evil spirits, like a fumigation to get rid of noisome insects that are a menace to the health. The three interpretations are sound, and could perhaps be joined to a new one that follows the same lines and is of basic importance. The tobacco smoke came to represent the visible form of the spirit or supernatural power, of the mysterious, powerful, and fecundating mana. Smoke was the very subtle and fleeting materialization of this power of tobacco which showed itself in its stimulant and narcotic effects and in the medicinal and genesial properties attributed to it by magic.

Among the Indians of America, where tobacco originated, and among the African Negroes, who adopted it with amazing rapidity, tobacco was a thing for men, masculine in its spirit, and could not be planted by women. It was used in the form of smoke, infusion, powder, for chewing, etc., in the fertility rites of women, animals, and fields. It was employed in the nuptial ceremonies, for censing the bride with the smoke; in agricultural rites, impregnating the seeds in tobacco smoke and blowing it over the planted fields. Sick people and children were also fumigated with it or made to inhale it. In some tribes even nursing babies smoke.

Among peoples as given to phallic cults, erotic ceremonies, and sexual propitiations as the Cuban Indians, the *tabaco* or cigar may, in its form, have been a Priapic symbol, and the smoke and the shredded tobacco may have been a figuration of the seminal potency that penetrates, fecundates, and animates life in all its forms.

As for the term *cohoba* used by Father Pané, it is also employed by Las Casas, López de Gómara, and other chroniclers, as we shall see. It is possible that the word was pronounced with the *h* aspirated, giving the word a sound similar to *cojoba,* for which reason in two earlier Spanish editions of Pané's *Relación* (those of 1747 and 1892) it is written *cogioba*. But this form would seem to be influenced by the Italian spelling in the oldest text that has been preserved of Father Pané's work (which is included in the Venice edition of 1665 of the work of Ferdinand Columbus referred to above), from which all subsequent reproductions have been taken. It is also possible that in the Indian tongue the word had the sound of *cooba* "with the first syllable long," as they said in those days, and that the aspirated *h* was influenced by the Spanish word *cohobar,* which, like the word *al-cohol,* comes from the Arabic vocabulary of alchemy, and is still to be found in Spain and Spanish dictionaries with the meaning: "to distill the same substance several times"; and

from which, in turn, that term of picaresque slang *coba* probably came, which means the adroit use of "the gift of gab" for
purposes of deceit.

Father Pané deals with cohoba again in Chapter xv, in connection with the medical practices of the Buhuitihu, whom
Oviedo calls Buhití, and we today *behiques*. These are those
persons among the Indians who exercise conjointly the functions of priest, doctor, and soothsayer. This is what the author
of the first study on the religion of America has to say:

"When a person is sick, they take the *buhuitihu,* who is the doctor,
to him. He must follow the same diet as the patient, and assume the
aspect of a sick person, which is done in the way you shall now see.
The doctor must purge himself, the same as the sick person, and to
do this he takes a certain powder known as *cohoba,* which he sniffs
up through his nose, and this intoxicates them so that they do not
know what they are doing, and they say many senseless things, and
claim to have talked with the *cemíes,* and that these have revealed to
them the cause of the sickness."

And Father Pané returns to the cohoba in Chapter xix of his
work on the consecration of the *cemíes.* He says:

"How they make and house the cemíes of wood or stone.—Those
of wood are made in this manner: when a man is walking along a
road and it seems to him that he sees a tree that moves to its very
roots, that man stops and asks it fearfully who it is. The tree replies:
'Bring a *buhuitihu* here; he will tell you who I am.' The Indian goes
to the doctor and tells him what he has seen. The magician or wizard
then goes to see the tree the other has told him of, sits down beside it,
and goes through the ceremony of the *cohoba,* which we have described in the story of the four brothers. When this is done he arises
and tells him all his titles, as though he were a great lord, and says to
him: 'Tell me who you are, what you are doing here, what you want
of me, and why you have sent for me; tell me if you want me to cut
you down, or if you want to come with me, and how you want me
to take you; I will build you a house with land around it.' Then that
tree, or *cemí,* converted into an idol or devil, answers him, telling
him what it wants him to do. The wizard cuts it down and does with
it as he has been ordered, he builds it a house with land, and many
times a year he performs the *cohoba,* which is a way of praying to it,
rendering homage to it, to learn certain bad or good things from the
cemí, and also to ask it for riches.

"When they want to find out if they are going to be victorious against their enemies, they go into a house where only the leading men of the tribe may enter; the owner is the first one who begins the rite of the *cohoba* and he plays an instrument. While he is doing the *cohoba,* none of the others speaks until he has finished. When he concludes his speech, he remains for some time with his head bowed and his arms on his knees; then he raises his head and, looking at the sky, speaks. Then they all answer at the same time in a loud voice, and when all have spoken to give thanks, he relates the vision he had while he was drunk with the *cohoba* that he took through the nose and that went to his head. He tells them that he spoke with the *cemíes* and that the Indians will be victorious, that their enemies will flee, that there will be great loss of lives, wars, famines, or other similar things, depending on what he, who was drunk, wants to say. Imagine the state they get themselves into, for they say that they have seen the houses with their foundations pointing upwards, and men walking with their feet in the air. This *cohoba* is done not only to the *cemíes* of stone and wood, but also to the corpses of the dead, as we have observed above."

In this account of Father Pané the cohoba is a powder that is sniffed up the nose through a reed "half an arm's span in length" or "a handbreadth in length," which Oviedo and Las Casas also refer to, and, before them, Christopher Columbus himself. He speaks of this in his Diary, in these words which his son Ferdinand transcribed from the text in his *Historia del Almirante Don Cristóbal Colón* (edition published in Madrid, 1932, Vol. II, p. 28):

"Beginning with those dealing with matters of religion, I shall copy here the words of the Admiral, just as he set them down: 'I have not been able to find evidence of idolatry or any other sect among them, although all their kings, who are many, in Hispaniola as well as in the other islands and on the mainland, have their own house, apart from the village, in which there are only some wooden images carved in relief which they call *cemíes.* In these houses no activity is carried on except rites in honor of the *cemíes,* with certain ceremonies and prayers that they perform there as we do in our churches.

"In these houses there is a well-built table, round in form, like a platter, on which there are certain powders that they put on the heads of the aforesaid *cemíes* with certain ceremonies; then, through a two-forked reed they put into their nose, they snuff up these powders. None of our men understand the words they say. When they take these powders they go out of their head, raving like drunken men."

Abbot Peter Martyr d'Anghiera must have seen these writings of Father Ramón Pané and Christopher Columbus, of whom he was an admirer and so close a friend that he received a letter from him in 1494 from the settlement of Isabela in Hispaniola. Peter Martyr wrote as follows about the cohoba:

"When the caciques wanted to consult the *cemí* about the outcome of a war, or the crops or their health, they would go into the hut consecrated to the god and snuff up through their nose the cohoba, which is the name they give to the plant, and this so perturbs their mind that they immediately go into a frenzy. They talk incoherently and scream that not only the house but the things in it are whirling about and that the men are walking with their head down and their feet in the opposite direction. So great is the power of these powders of cohoba that as soon as they take it, it immediately affects every sense and feeling. When this temporary madness has passed, with their head hanging low and their hands on their knees, they sit as though dazed for a time, until finally they lift their heads, as though they were awakening from a deep, heavy sleep, and, raising their eyes heavenward, they tell everything the *cemí* has spoken to them." (From *Orbe Novo,* Paris, 1857, p. 94.)

A more precise description than those of the foregoing texts is another by Father Bartolomé de las Casas, who was familiar with the Admiral's Diary, had lived longer than Columbus in the Antilles, and was better acquainted with the Indians and their customs. This is Las Casas's account:

"We have already told how on this island they have certain statues, although they are not numerous. Through these it was believed that the devil spoke with the priests who are known as *behiques,* and also to the leading men and kings when they prepared themselves for it, so these were their oracles. From this came other sacrifices and ceremonies they carried out to please him, which he must have shown them how to perform. This was done in the following manner: they had ready certain powders of certain herbs dried and finely ground, the color of cinnamon or of powdered henna; in a word, of a brownish color. They put these on a round plate, not flat but somewhat hollowed out or deep, made of wood, and so beautiful, smooth, and finely fashioned that it would not have been much more handsome if it were of gold or silver; it was almost black and shiny like jet.

"They also had an instrument of the same wood and material, and equally polished and beautiful; in construction that instrument was the size of a small flute, all hollow as is a flute, and at a point about two thirds of its length it opened into two hollow tubes as when we spread apart the two middle fingers, leaving out the thumb, when we shake hands. With one of those two tubes in each nostril and the bottom of the flute in the powders that were in the plate, they would snuff up with their breath into their nose the amount of powder they wanted to take, and when they had taken it, they went out of their minds almost as if they had drunk strong wine, and they became drunk or almost drunk.

"These powders and these ceremonies or acts are called *cohoba,* the middle syllable long, in their language; when under its influence, they spoke in a kind of jargon, or all confused as Germans do, I cannot say what words or things. With this they were in a fit state to talk with the statues and oracles, which is to say with the enemy of mankind. In this way secrets were revealed to them, and they prophesied or read the future, and heard or learned whether some good, adversity, or harm was going to come to them. This was when the priest alone prepared to talk and listen to the statue; but when all the headmen of the village gathered to perform that sacrifice, or whatever it was (they call it *cohoba*), with the permission of the *behiques* or priests, or the chieftains, it was a sight to behold. They were in the habit, when they held their councils to decide important things, such as whether they should begin one of their wars, or do some other thing they considered of moment, to perform the cohoba, and in that way to become drunk, or almost so. And this fashion of seeking counsel, full of wine and drunk or almost so, was not peculiar to them, for according to Herodotus in Book I and Strabo at the end of Book XV, the Persians, when they wanted to reach a decision about weighty matters and of great importance, employed it, because they never did this except while they were eating and drinking and replete with wine, and that council and the decisions reached in it they said were more dependable than those arrived at in a sober and frugal state.

"I have watched them celebrate their cohoba several times, and it was a great sight to see them take it and hear them talk. The first who began was the King, and while he was engaged in it, all the others kept quiet. When he had done his cohoba (which is snuffing up those powders through the nose, as has been described, sitting on a low stool, but very well constructed, which they call *duhos,* the first syllable long), he would sit for a while with his head canted to one side and his arms on his knees, and then he would raise his face to the sky and begin to speak certain words, which must have been his prayer to the true God, or whatever god they had; then all the others answered almost as we do when we say Amen, and they did this in a loud voice or noise, and then gave thanks to him, and must have

spoken words of flattery, to win his goodwill, and asked him to say what he had seen."
(*Apologética,* Chapter clxvi.)

Concerning the use of cohoba in the medicine of the *behiques* or *bohitis,* as Francisco López de Gómara calls them, who also gives them the name of "priests of the Devil," this chaplain and historian writes:

"They have great authority, as doctors and soothsayers, among all, although they never give an answer or treat anyone except people of consequence and the chieftains.

"When they prepare to find the answer to what has been asked them, they eat an herb called cohoba, either ground or ready to be ground, and they inhale the smoke of it through their nostrils, and this makes them go out of their mind and see visions. When the power and properties of the herb have worn off, they regain their senses. They tell what they have seen and heard in their council with the gods, and say that God's will will be done. Nevertheless, they answer in a way to please their interrogator, or in terms whose meaning is not clear, for this is the fashion of the father of lies.

"For medicinal purposes they also use that herb cohoba, which does not exist in Europe; they shut themselves in with the sick person, walk around him three or four times, foam at the mouth, make a thousand grimaces, and then blow on the patient and suck the back of his neck, saying they are drawing the sickness out of him there. After this they very carefully run their hands all over his body, even the toes of his feet, and then go out to throw the illness out of the house, and sometimes they show a stone or bone or piece of meat they have in their mouth and say that now he will get well, because they have taken out of him what was doing him harm; the women preserve those stones to help them in labor, as though they were holy relics. . . ."
(*Hispania Victrix, Historia General de las Indias.* Edition Biblioteca de Autores Españoles, Vol. XXII, p. 173.)

From this text of López de Gómara, which speaks of the habit of "eating" the cohoba "ground or ready to be ground," it might be inferred that the Indians of the Antilles, besides using the leaves of the Nicotiana plant for smoking or in powder, were in the habit of chewing them, a habit that also spread among the whites of Europe and is still preserved in certain countries among certain people. This reference to the "eating" of the *cohoba* gives verisimilitude to the opinion of Sven Loven

(*Origins of the Tainan Culture West Indies,* Göteborg, 1935)
who takes the position, following a brief linguistic disquisition,
that the Tainan word *cohoba* originally meant "to chew," and
not a plant.

According to Sven Loven, tobacco must have first been used
for chewing, and later the Indians learned to snuff it in powder
through nose tubes. The Carib Indians of the Antilles did not
use the tobacco in powdered form, but for chewing, with the
exception of their *piayes,* or witch-doctors, who smoked it in
cigars. The Caribs prepared the dried leaves of tobacco by soak-
ing them in sea water and then rolling them into a twist, as the
Europeans later did, making them into rolls, which were
known as *andullo,* taken from the sailors' slang, in which the
word means the roll of cloth with which harpings and blocks
are encased to avoid friction.

The passages from the cited texts of Christopher Columbus
and Father Las Casas are corroborated in part and amplified by
the chronicler Oviedo, who had a very precise manner of writing.

This is the passage which deals with tobacco, taken from the
great work of the first chronicler of the Indies, officially appointed such by the King of Spain, Captain Gonzalo Fernández
de Oviedo y Valdés, entitled *Historia General y Natural de las
Indias, Islas y Tierra-Firme del Mar Océano* (Part I, Book V,
p. 130). The first part of this work, containing the excerpt that
follows, was published by its author in 1535. The text I have
used is that of the edition of the complete work published under
the auspices of the Royal Academy of History, 1851, in Madrid:

"Among the vices practiced by the Indians of this island there was
one that was very bad, which was the use of certain dried leaves that
they call *tabaco* to make them lose their senses. They do this with the
smoke of a certain plant that, as far as I have been able to gather, is
of the nature of henbane, but not in appearance or form, to judge by
its looks, because this plant is a stalk or shoot four or five spans or a
little less in height and with broad and thick and soft and furry
leaves, and of a green resembling the color of the leaves of ox-tongue
or bugloss (as it is called by herbalists and doctors). This plant I am
speaking of in some sort or fashion resembles henbane, and they take
it in this way: the caciques and leading men have certain little hollow
sticks about a handbreadth in length or less and the thickness of the

little finger of the hand, and these tubes have two round pipes that come together, as in the drawing (Plate I, figure 7), and all in one piece. And they put the two pipes into the openings of their nostrils and the other into the smoke of the plant that is burning or smoldering; and these tubes are very smooth and well made, and they burn the leaves of that plant wrapped up and enveloped in the same way the pages of the court take their smokes: and they take in the breath and smoke once or twice or more times, as many as they can stand, until they lose their senses for a long time and lay stretched out on the ground drunk or in a deep and very heavy sleep. The Indians who do not have these tubes take this smoke through hollow stems or reeds, and that instrument through which they take the smoke, or the aforesaid reeds, are called by the Indians *tabaco,* and not the plant or the sleep that overtakes them (as some have thought). This plant is very highly prized by the Indians, and they grow it in their gardens and farms for the aforesaid use; they believe that the use of this plant and its smoke is not only a healthy thing for them, but a very holy thing. And when the cacique or leading man dropped to the ground, his wives (of whom he had many) would pick him up and put him in his bed or hammock, if he had so ordered before he lost his senses; but if he had not so said or ordered, he did not want them to do anything but leave him there on the ground until that drunkenness or sleep passed from him. I cannot understand what pleasure they get from this act, unless it is the desire to drink, which they do before they take the smoke or tabaco, and some drink so much of a kind of wine they make that they fall down drunk before they start smoking; but when they have had all the drink they want, they begin on this perfume. And many smoke the tobacco without drinking too much, and do as has been described until they fall to the ground on their back or side, but without swooning, rather like a man who has fallen asleep. I know that certain Spaniards now use it, especially some of those who have contracted buboes, because they say that while under its effects they do not feel the pain of their disease, and it would seem that the one who does this is like one dead in life, which I think is worse than the suffering they spare themselves, because it is not as though they were cured by it.

"Now many of the Negroes that are in this city and in all the island have acquired the same habit, and they raise this plant on the farms and properties of their masters for the purpose described and take the same smokes or *tabacos,* because they say that when they stop work and smoke the tobacco, it takes away their weariness.

"It seems to me this is like a bad, vicious habit the people of Thrace followed, among other of their revolting vices, according to Abulensis's account of Eusebio in *De los tiempos* (Abulensis, Book II, Chapter clxviii), in which he says they all had the habit, men and women, of eating around the fire, and that they enjoyed being drunk, or seeming so; and that as they did not have wine, they took the seeds of certain plants that grow there and scattered them on the fire, and these seeds gave off such an odor that everyone was intoxicated, without drinking anything. In my opinion, this is the same as the *tabacos* these Indians take."

FIGURE A. Exact reproduction of the sketch in the work of Oviedo (edition of the Royal Academy of History, Madrid, 1851, Vol. I) of the apparatus to inhale tobacco used by the Indians of Hispaniola

Oviedo supports his description with a drawing that appears in a plate, as an appendix to Volume I of the Academy's edition of his work. (See Figure A reproduced here.)

Oviedo's account has been questioned several times. Wiener denies the possibility of such an apparatus existing (*Africa and the Discovery of America,* Philadelphia, 1920, Vol. I, p. 107).

Recently, in Havana, so authoritative a person as Dr. Benigno Souza has manifested his incredulity with regard to the instrument in question (Dr. Benigno Souza: *"Historias y leyendas en torno al tabaco,"* Diario de la Marina, Havana, 1938). If the instrument as it appears in the drawing in Oviedo's work is enlarged in exact keeping with his sketch to meet the dimensions indicated by him, that is to say, a handbreadth in length, as it appears in the accompanying sketch (see Figure B), "the two arms of the Y would not rest in the openings of the nose, which are separated by a thin partition, but in the middle of the cheeks of the Indian who was using it, and a cursory examination of this device makes it apparent that the two arms of the Y, which in Oviedo's drawing form an angle of almost forty-five degrees, could not have been inserted at the same time in the nose of an Indian, or even in that of a huge gorilla." Moreover, it seems improbable to Dr. Souza that they should have inhaled the smoke through the nose and not through the mouth.

Nevertheless, the use by the Indians of this simple device, which is nothing but a tubular nasal inhaler, would seem to be borne out not only by Oviedo's detailed description, but by the observations of Christopher Columbus and Father Las Casas as well. The latter in one passage refers to the rolled *tabaco* he saw among the Indians, and in another he describes this very ap-

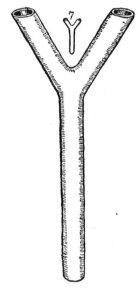

Figure B. Drawing to illustrate the article *Historias y leyendas en torno al tabaco* by Dr. Benigno Souza, with regard to the un-likely proportions of the bifurcated tube described by Oviedo

paratus as "the size of a small flute," all hollow, and "at a point about two thirds of its length it opened into two hollow tubes as when we spread apart the two middle fingers, leaving out the thumb, as when we shake hands." Then Las Casas goes on to explain how the two tubes are inserted in the nostrils and "the bottom of the flute in the powders in the plate" and then they breathe in and "snuff up with their breath the amount of powders they wanted to take."

It would seem certain, therefore, that the drawings of the ap-paratus in question in Oviedo's work are off scale; but the error in his measurements does not mean that the bifurcated device did not exist. With this reservation one must accept the statements of Christopher Columbus, Pané, Oviedo, Las Casas and Gómara. Today it is no longer possible to doubt the existence and shape of this little bifurcated device, for I have discovered one in the Ethnographical Museum of the city of Port-au-Prince, reproduced in Figure C.

As may be seen from the texts of the Admiral, Father Pané, Las Casas, Oviedo, and López de Gómara, there were two classes of tubes for taking cohoba, one a single tube or conduit and the other opening into two forked tubes or branches. With both these devices the Indians took in through their noses "with their breath" the smoke of the leaves that were being burned for that purpose, in the form of a nasal inhalation, and also snuffed up the powder. But it seems to me that the latter use was the more typical and frequent, and Columbus and Las Casas refer only to this custom of breathing in or snuffing up the powder with the instrument in question.

Certain texts, which we might describe as dealing with comparative ethnography, definitely confirm what Oviedo and other chroniclers have written about the inhaler tubes, including the forked type, that have given rise to so much discussion. This data is of special significance when it refers to Indians ethnically or geographically related to the Tainos of the West Indies and the countries from which they come. La Condamine, talking of certain Indians of South America ethnically related to the Antillean Indians, says:

"The Omaguas make great use of two plants, one called *floripondio* by the Spaniards, which has a flower shaped like a bell, and the other known by the name of *curupá* in the language of the Omaguas. Both are cathartic. When they take them these people induce in themselves a state of intoxication that lasts for twenty-four hours, during which time they see the strangest visions. They also take *curupá* in the form of powder, as we do snuff, but more elaborately. They use for this purpose a cane that branches in two, in the shape of the letter Y. They put one branch in each nostril, and snuff hard, making a grimace that seems very ridiculous to a European trying to relate everything to his own customs." (M. de la Condamine: *Relation abrégée d'un voyage fait dans l'intérieur de l'Amérique méridionale,* Paris, 1745, p. 73.)

This passage from an eighteenth-century text could not be clearer or more conclusive. There are other texts that completely bear out the foregoing. The Marquis de Wavris (*Mœurs*

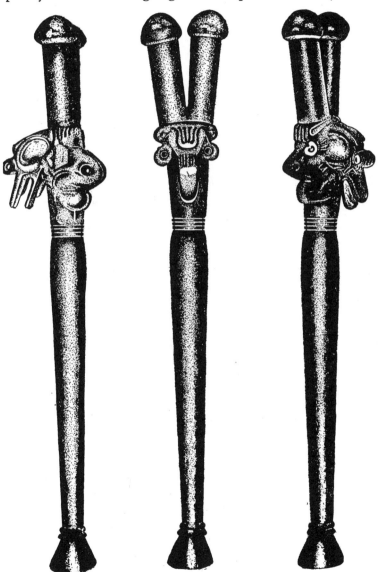

Figure C. Bifurcated pipe of the Antillean Indians.

et coutumes des Indiens sauvages de l'Amérique du Sud, Paris,
1937, p. 154) describes how on one occasion when white men
entered the territory of the Guaharibo Indians, they found near
their homes some long and slender reeds perforated throughout their entire length, which the Indians had hurriedly
dropped just as they were using them when they fled. He remarks later that these reeds had inside them traces of ash. One
of the men out of curiosity took a sniff of it and felt the same
intoxicating effects produced by *yopo.* The Indians had been
inhaling this ash to prepare themselves for battle. In Spanish
this substance was also known as *llopa,* and in the kingdom of
New Granada the word *enllopado* meant "drunk." These reeds
were simple single-tubed inhalers, such as the Eskimos still use
today.

Among the South American Indians the types of bitubular
and bifurcated inhalers are numerous and vary greatly. In Tiahuanaco (Bolivia) a curious old forked apparatus of this sort
was found. It was made of the bone of a llama, highly polished
and carved on the outside with symbols whose meaning could
not be deciphered. It is five inches long and an inch and a half
wide at the terminal point of the bifurcation. As can be seen,
this part of the apparatus could be inserted into the nostrils of
a human being. It is at present in the Museum of the University
of Berlin (Joseph D. McGuire: *Pipes and Smoking Customs of
the American Aborigines, based on Material in the U. S. National Museum,* Annual Report, Smithsonian Institution, for
1897, Washington, 1899, p. 365).

Max Uhle, who discovered this specimen (see Figure D, Nos.
1 and 2) considers it prehistoric. The designs on it, which
might be the stylization of different animals, a snake and a
small four-footed animal, at each end of the fork, "represented
a presumptive vitality in the powder to be inhaled." This seems
in agreement with the belief Humboldt discovered among the
Otomaco Indians that "they could not take their powders except through the tubes," which indicated that these were an integral part of the magic ritual. Max Uhle is of the opinion that
the same thing took place among the Indians on the east side of
the Andes.

In the symbolic designs on the inhaler tubes there was undoubtedly some interpretation of the mysterious power, the sac-
ripotence or *mana* that was taken in with the powder or smoke

and expelled with the smoke, the mucus, or the vomit. By analogy the same thing takes place with the carvings that adorn pipes in all countries, the straight as well as the curved variety. Particularly with the grotesque heads and figures as common on such smoking equipment as on the potsherds of the Taino

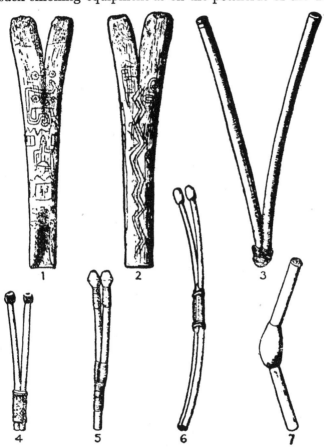

Figure D. Various types of inhaling tubes used by the Indians of South America.

Indians. These figures are often termed caricatures when in reality they are only fantastic stylizations by which the artist has tried to represent, at times with real genius, the ineffable, mysterious beings of a supernatural world. This tendency, which in a sense might be called "animistic" because it "animated" the ritual utensils with mythological personages, is originally to be found where the use of tobacco is only a religious and magic rite; later this "animism" is extended and perpetuated because of its initial traditional sense, and when this disappears it still persists for æsthetic reasons, catered to and fostered by the conspicuous luxury of a social class and by the business acumen of mercantile greed. In the last analysis, this original symbolic "animation" of the inhaler tubes helps explain the universal tendency of smokers to individualize the tobacco, cigars, pipes, snuffboxes, and other smoking accessories as deriving from a tradition inherited from the Indians.

This forked apparatus found in Tiahuanaco fits in, according to Max Uhle, with the information, set down by Garcilaso in his *Comentarios Reales* (Part I, Book II, Chapter xxv), that the ancient Peruvians used tobacco (*sairi*).

There is abundant proof of the existence of these bifurcated inhaler tubes in South America. Max Uhle ("A Snuffing-Tube from Tiahuanaco," *Bulletin of the Free Museum of Science and Art of the University of Pennsylvania,* Philadelphia, 1898, Vol. I, pp. 159–77) collected several different varieties to be found along the Ucayali River, from Chanchamayo (now in the Berlin Museum), among the Tecuma of the Amazon region (now in the Munich Museum), etc., which are reproduced and commented upon by George A. West ("Tobacco, Pipes and Smoking Customs of the American Indians, *Bulletin of Pub. Museum of the City of Milwaukee,* Milwaukee, 1934, Vol. XVII, Part II, Plate 12). See the different types in Figure D, numbers 3 to 7, reproduced here.

The Arawak tribes of the Purús rivers also used these forked tubes (J. B. Steere: "Narrative of a Visit to Indian Tribes of the River, Brasil," *Report of the U. S. National Museum* for 1901, Washington, 1903, p. 371). The Indians of the Cayari-Uaupés region kept their powdered coca in little receptacles made of sea shells and snuffed it through forked tubes (T. Koch-Grünberg: *Zwei Jahre unter den indianern Reisen in Nordwest Brasilien,* Berlin, 1910, Vol. I, p. 323). Baron Humboldt discovered that the Otomaco Indians took the powder of the seeds of *niopo* mixed with cassava flour and powdered lime, obtained from a certain shell, parched it on an earthenware griddle, and then made the mixture into balls. When they went to use it, says Humboldt, they pulverized the mixture and spread the powder on a plate five or six inches in diameter; then the Otomaco, holding the plate in his right hand, snuffed up the powders of *niopo* through a forked pipe made of bird bones, about seven inches long, putting the two ends of the pipe in his nostrils (A. von Humboldt: *Personal Narrative of Travels to the Equinoctial Regions of America,* London, 1852, Vol. II, p. 505). With this Humboldt confirms the information supplied a century earlier by Father Gumilla, who spoke of the intoxication produced among the Otomacos by powdered yopa and lime,

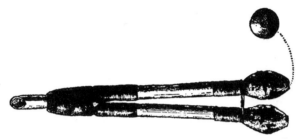

FIGURE E. Bitubular device for inhaling narcotic powders still used by the Guahibo Indians of Venezuela. Specimen in British Museum. (See *Handbook to the Ethnographical Collections,* 1910, Fig. 252, 2.)

snuffed through the nose, although he made no mention of the forked instrument (op. cit., Vol. I, p. 181). The Mauhé Indians also used powdered paricá, snuffing it through the nose through two pipes made of two vulture feathers tied together with a thread (H. W. Bates: *The Naturalist on the River Amazons,* London, 1892, p. 169). See Figure E, which represents the bifurcated powder-inhaler used by the Guahibo Indians of Venezuela (*British Museum Handbook to the Ethnographical Collections,* Oxford, 1910, p. 278). These double or bifurcated tubes, Sven Loven notes, are to be found in the tropical lands of South America for use in absorbing the powders of piptadenia in ceremonies of hallucination. Sven Loven is of the opinion that these forked tubes used to inhale powders for the purpose of inducing visions are a cultural feature that passed from South America, following the route of the Orinoco River, to Haiti. Probably this took place via the island of Trinidad, he says, where the use of snuff for this purpose is common.

Certain aboriginal Indians of what is now Costa Rica, the

so-called Huetares, whose typical culture is closely related to that of the Tainos of the West Indies, also took tobacco in powdered form through pipes with double tubes, as has been pointed out by Jorge A. Lines (*"Sukía: Tsúgur o Isogro,"* in *Anales de la Sociedad de Geografía e Historia de Guatemala,* 1937, p. 426).

An attempt has been made to classify these instruments into three types: the bifurcated, such as are described by Oviedo, Las Casas, and others in the newly settled Antilles and by Max Uhle in South America (see Figures D, 1 and 2); the bitubular, such as, for example, numbers 3, 4, and 6 of Figure D; the angular, such as number 7 of Figure D. According to Max Uhle, the bifurcated types have been found in regions completely remote from each other, on the outer circumference of the zone where powders were inhaled in this manner, such as Hispaniola, along the Orinoco, and in Tiahuanaco.

The Taino Indians of the West Indian islands differed somewhat from the other Indians of the basin of the Orinoco and the Cayari-Uaupes in that their forked inhaler tubes were not made of animal bones. Las Casas says that he saw tubes made of wood; Oviedo refers to "pipes or reed tubes" and certain "hollow sticks"; Columbus and Pané mention a reed. The technique of the Taino was therefore somewhat more advanced.

Somewhere I have seen a modern drawing of this forked apparatus to "perform the cohoba" in the hands of an Indian smoking in such a way that the ends of the fork were in his nose and at the other end was "twisted tobacco" or a cigar, as we would say today, giving forth smoke. In this drawing the two-pronged pipe to which the chroniclers refer was a sort of nasal "mouthpiece" for smoking the cigar. This conception is purely imaginative, without any basis to bear it out, but it is still to be seen in print.

On the other hand, there is no question that other varying manners of utilizing the device in question, or substituting for it something analogous, do exist. Among the Indians of the Tiquié River it was customary to put one of the branches of the fork in one nostril and the other in the mouth, and in this manner, simply by breathing in, the powder reached the innermost mucous membranes of the nose (Koch-Grünberg: *Zwei Jahre,* Vol. I, p. 324, Fig. 203).

Among the Guahibo Indians (Jules Crevaux: *Voyages dans l'Amérique du Sud,* Paris, 1883, p. 550) the inhalation of powdered yopa or paricá is customary, but these Indians have two procedures for these rites. In one the powder, which is placed in a holder consisting of a univalve shell covered with a bat skin, is snuffed through a tube made of two bird bones joined in fork shape. One is placed in the mouth, the other in the nose, and in this position the powder is drawn in. This method, says

Figure F. Double instrument for inhaling narcotic powders, used by couples simultaneously among certain Indians, according to Crevaux

Crevaux, is that used by the "selfish." The more "social-minded" Indians use two tubes joined in the shape of the letter X, and two friends, after filling the hollow bones with powder, blow them at the same time into each other's nose, thus being of mutual assistance (see Figure F). Among the Mura Indians of the lower Amazon, the powder of paricá is likewise snuffed up the nose through tubes, but the tubes they use are not forked, but straight, and the rite is practiced in couples, one Indian blowing the content of his into the nose of another, while he at

the same time receives a similar charge of powder, and thus
both become intoxicated (H. W. Bates: *The Naturalist on the
River Amazons,* London, 1892, p. 169). These variations of the
method of the Tainos are simpler and are evidence of types of
inhalers that antedate those of the Antilles described in detail
by the chroniclers, notwithstanding the error in drawing in
Oviedo's work, as pointed out by Dr. Benigno Souza.

Another Cuban author of great ability, Alvaro Reynoso, had
already refuted Oviedo's assertion. In Reynoso's opinion, the
chronicler confused and described as one two different customs
of the Indians: that of smoking tobacco and that of snuffing
up their noses certain powders that made them go out of their
head. According to Reynoso:

"It is evident that Oviedo never could have seen either Spaniards
or Negroes smoke *tabacos* through the nose, and, still less, perform
the cohoba rite. He could have seen them smoke through the mouth,
as they have always done, but never snuffing narcotic powders
through the nose to induce temporary madness as a means of preparing themselves to establish contact with the *cemís.*"

Speaking of the passage in Las Casas in which he refers to
the leaves rolled up like "muskets" which the Indians smoked,
he adds:

"Las Casas's narrative is accurate. The Ciboneys smoked through
the mouth, like all the other Indians, and it is amazing to see with
what care they carried a lighted faggot in their hand to relight their
cigar if it went out. The Spaniards who adopted this custom of the
Indians naturally smoked through the mouth from the start. Oviedo
(Book V, Chapter ii) has let his imagination run away with him,
and has given us a completely false description." (Alvaro Reynoso:
Agricultura de los indígenas de Cuba y Haití, Paris, 1881, pp. 72–3.)

Reynoso is severe with Oviedo:

"The deliberate falsification of the truth by Oviedo is all the more
bewildering inasmuch as he himself says that many Spaniards and
Negroes smoked in his day, and they did not do this through the
nose or to intoxicate themselves to the point of falling senseless to
the ground, either. As will be later seen in another passage, this
chronicler winds up by not knowing what he is talking about.

"It would have been impossible to smoke thus in the open air. Assuming that the operation was carried on in a closed room, the burn-

ing tobacco would fill the whole room with smoke, and the smoker would find himself in very unpleasant quarters. But the use of the tube to inhale the smoke through the nose, unless he used his hands to cover his nose, is an awkward arrangement, aside from the fact that the smoke would be mixed with air. The intoxication produced by tobacco is so disagreeable that nobody would smoke if the effects of the first cigar were repeated on each successive occasion. Even after having acquired the habit, each person selects the kind of tobacco he likes best, depending on his taste and temperament, to avoid ill effects. But why go on talking about a habit that never existed even in the mind of the author when he was in his senses?" (Op. cit., p. 74.)

Reynoso was not altogether just. There seems to be no basis for assuming that Oviedo was lying. Why would he lie about something that meant nothing to him, one way or the other? Moreover, López de Gómara had said the same thing in different words, although he might have been influenced by Oviedo, whom he followed on many occasions.

Reynoso did not question the existence of the forked device for inhalations through the nose. He was familiar with its use among the Omaguas, as described by La Condamine. "In our own days," wrote Reynoso, "there are Indians who still perform a kind of cohoba, making use of the same instrument employed by the natives of Cuba and Haiti." His criticism of Oviedo is based on other arguments. His contention rests on two premises, both plausible and deserving of consideration: (1) that what the Indians snuffed through the nose by means of the tubular device in question was not smoke but a powder; (2) that the plant that produced the leaves the Indians smoked through the mouth in the form of tight rolls was different from the plant whose powder they snuffed through the nose.

It seems that someone, without even having read Reynoso, must have taken the trouble to experiment and prove the impossibility of smoking with this Y-shaped device (A. Ernst: "On the Etymology of the Word Tobacco," *American Anthropologist,* 1889, Vol. II, p. 134).

"At first glance it would seem of slight importance," says Reynoso (op. cit., p. 72), "to stop to find out whether the Indians smoked through the mouth or through the nose. Nevertheless, the matter should be cleared up, for Oviedo's misleading account has led to the confusion of two different usages, making them appear as one and the same custom, and, besides, alleging the identity of two separate plants. According to Oviedo, the Indians of these regions smoked through the nose, when in reality they did it through the mouth. What they snuffed through the nose was a stimulant powder different from tobacco, whose effects had a religious significance for them."

"It must be clearly understood," adds Reynoso, "that this was real powder they snuffed through the nose, and not the smoke of the burning leaves."

The only chronicler who states that the Indians snuffed the smoke through the nose is Oviedo. López de Gómara merely repeats the information he has taken from him. The others who mention these narcotic nasal inhalations and state that they have seen them refer to the snuffing of powders, known as *cohoba*. This is the case with Christopher Columbus, Ramón Pané, Father Las Casas, and Peter Martyr d'Anghiera. Therefore Reynoso's assumption that Oviedo is mistaken cannot be lightly dismissed. Oviedo has no one to support him except Gómara, whose testimony is of scant worth since it has been said that he was never in the Indies. But Oviedo's statement should be considered an error rather than a lie.

It is also Reynoso's opinion that "although tobacco was smoked and also taken in powdered form through the nose, in neither case did it produce the effects the chroniclers describe" (op. cit., p. 75). It is indubitable that the powders in question were some narcotic drug that produced cathartic effects in addition to others such as inducing hallucinations, automatic talking, prophetic pronouncements, manifestations similar to those which formerly took place in persons possessed of the devil who were burned at the stake by the Inquisition, or such as occur today at spiritualist seances, with greater frequency and less scandal and without punishment. Similar scenes of hypnotic hysteria can be seen at religious gatherings of Afro-Cubans and certain Protestant sects, when "the spirit descends" upon the "faithful," black or white, to the rhythmic spell of the sacred drums or the organ hymns. It might not be out of place to point out that to perform the cohoba rite and induce its maddening effects, the Indians also beat a drum, according to Father Pané. All the chroniclers are in agreement on the abnormal excitation induced by the cohoba. Columbus says that while under its spell the Indians "go out of their mind"; Anghiera says they "instantly become delirious"; Las Casas that "they immediately go off their head"; Gómara that the herb "excites them greatly and makes them lose their reason"; and so on. But it is interesting to note that the chroniclers do not describe clearly the cathartic or visionary effects caused by the smoking of tobacco—that is to say, not the action of the powders, but of smoking the rolled tobacco—except for Oviedo and Gómara, who would seem to be mistaken on this point. Oviedo himself, as will be seen later, in describing a feast given by a Chorotega chieftain, speaks of the smoking of these cigars as a pleasure, as it is regarded the world over today.

This would seem to give weight to the opinion of Alvaro Reynoso, who believed that the cohoba, taken in powder form, and tobacco, used for smoking, were two different plants.

"What plant is the *cojoba?*" asks Reynoso (op. cit., p. 80), and he answers his own question:

"It would not be difficult to decide this, because we know of several that possess the stimulant properties mentioned by Las Casas; but to avoid the risk of falling into error, we shall wait to carry out certain experiments on the subject. La Condamine's noteworthy observations not only throw light upon the matter, but can serve as the basis of important deductions."

This traveler had referred to the plants called *floripondio* and *curupá*. Reynoso goes on to say:

"Martius (*Etnographie,* Vol. I, pp. 441, 631) has illustrated this subject and has classified the plants that are useful for this purpose. These are the *Mimosa acaciodes* (p. 441) and the *Acacia niopo* (p. 631). Without proposing to make a study of the plants used by the natives of the New World to induce temporary madness, we think it would be pertinent to mention three of the most outstanding: the *carapullo* of Peru (Frezier: *Relation du voyage de la mer du Sud aux côtes de Chili et du Pérou,* Paris, 1716, p. 213); the *coatlxoxouhqui* and the *peyotl* of Mexico (Fray Bernardo de Sahagún: *Historia general de las cosas de Nueva España,* published in Mexico, 1829, Vol. III, p. 241)."

In our own days J. Alden Mason (*Use of Tobacco in Mexico and South America,* Field Museum of Natural History, Chicago, 1924, p. 13), has written in his study of the use of tobacco that the habit of taking powdered tobacco, or snuff, is common among the Indians of central and southern South America. And just as coca is always taken in combination with an alkali, so snuff, or *niopo* or *iopo,* is prepared with powdered acacia or mimosa, yucca flour, and pulverized sea shells, and snuffed up the nose through forked or double tubes.

It may now be added that the *ñopo* or *yopo* used by the Indian tribes of the plains and valleys of the Orinoco—that is, the Guahibos, the Guayaveros, the Piapocos, and others—is composed of the powdered fruit of a certain tree mixed with a little ash. (Marquis de Wavris: *Mœurs et coutumes des Indiens sauvages de l'Amérique du Sud,* Paris, 1937, p. 154). M. E. de Ribero states that the Goahibos and Chiricoas carry these *yopa* powders with them as "their only provisions" (in *Colección de Memorias Científicas,* 1857, Vol. I, pp. 103, 104).

According to Filippo Salvadore Gily (*Saggio di Storia Americana,* Rome, 1780, Vol. II, p. 103), the *piayes* or priests of the Otomaco Indians of the Orinoco used powdered tobacco to communicate with the supernatural powers and to enable them to prophesy; but Sven Loven points out that this must be an error on Gily's part, for he did not know the Otomacos, and Father Gumilla (*Historia natural, civil y geográfica de las naciones situadas en las riberas del río Orinoco,* Barcelona, 1791, Vol. I, p. 204) states clearly that the powders used in their rites are of *yopa.* I might remark also that, according to Max Uhle, (loc. cit., p. 165), the odor of the powder of niopo is similar to that of tobacco, which might help to explain their substitution or interchange, depending on the circumstances.

Nor should we forget coca and marihuana, which were known to the Indians and whose effects are most deplorable owing to the spread of their use among certain inhabitants of the large cities. And the possible list of the stimulant plants of America does not end here. But all this information does not clear up the matter of which plant the cohoba was.

Therefore it was not the German Hartwich who first denied that cohoba was identical with tobacco (*Die menschlichen Genussmittel,* Leipzig, 1911), as Brooks supposed, but a Cuban

WOMAN OF THE
TOBACCO COUNTRY

ENTR'ACTE

ÉLÉGANTS

*OF THE PLEASURES
THAT ARE NO SIN,
SMOKING IS THE
BEST*

De los gustos sin pecar, el mejor es el fumar.

AFTER DINNER

Sobre mesa.

DE GUSTIBUS NON
DISPUTANDUM EST

Sobre gustos no hay nada escrito.

writer. Years later, in 1916, a North American ethnographer, Safford ("The Identity of Cahoba, the Narcotic Snuff of Ancient Haiti," *Journal of the Washington Academy of Sciences,* 1916, Vol. VI, p. 547; and *Proceedings of the Nineteenth International Congress of Americanists,* Washington, 1917, p. 27), upheld the same point of view as Reynoso, without referring to him or knowing it. Safford maintained that the Tainos did not use powdered tobacco for their nasal inhalations, but powdered *Piptadenia peregrina.* According to Safford, cohoba is the term applied by the Indians of Hispaniola to the same intoxicating powder of *Piptadenia peregrina* that in Brazil is known as *paricá,* along the Orinoco and the Marañón as *yupa* or *ñopa* and *curupa* or *curuba.* And he even adds that powders of closely related plants were used by the Indians of Argentina, to which the name *sebil* was given, and by the Quechuas, who called them *huilca.* Nevertheless, Sven Loven insists that these words do not refer to any clearly determined plant, but to a generic idea, just as the current terms *powder* or *snuff* do not signify exactly and exclusively "ground tobacco." Safford says that the *Piptadenia peregrina* is to be found among the flora of Haiti, Puerto Rico, and other islands of the West Indies, as well as in the Orinoco and Amazon regions; Sven Loven observes that it is impossible to ascertain whether the plant was pre-Columbian in those islands, but it is a fact that none of the chroniclers ever alludes to it.

Safford's opinion, in which he coincides with Reynoso, is that the physiological effects produced by the inhalation of the mysterious powders of the Taino medicine-men cannot be obtained from tobacco, and that the assumption that they were powders of piptadenia is inescapable. Nevertheless, this argument is not wholly convincing.

As Sven Loven points out, the stimulant powders of piptadenia are obtained by grinding the seeds, not the leaves, which the chroniclers of the Indies indicated was the case with the cohoba. Moreover, the piptadenia belongs to the family of the Mimosæ, not the Solanaceæ, and is a shrub, not a plant. And the color of its powders was gray (Koch-Grünberg, op. cit., p. 323) and not brown or "tawny" or cinnamon-colored or the color of ground henna that Las Casas speaks of (*Apologética,* p. 445). Sven Loven finds it extremely hard to believe in this hallucinatory effect attributed to powdered tobacco. But, he adds, one should bear in mind that the tobacco of the Antilles was much stronger than that in use today. Father Cobo, in his time, also mentions a wild tobacco "much stronger than the cultivated variety." Moreover, adds Sven Loven, the powdered tobacco was mixed with sea water and salt, thus reinforcing its narcotic properties. One of Sven Loven's observations seems to me of outstanding importance: he explains the intense narcotic effects induced in the Tainos on the basis of the fasts by which they prepared themselves for these rites, the fact that they were extremely superstitious and had an unquestioning faith in these visionary experiences. The inhalation of the powdered tobacco after debilitating fasts and accompanied by suggestive ceremonies probably produced, aside from the natural physiological effects, certain "conditioned reflexes," as they are known today. Something of the same sort can be observed today in Cuba and in Haiti in connection with such phenomena as "saints" who become possessed in the Africanoid religions, and in the trances of the medium in spiritualistic seances. There are persons who immediately become possessed while others are slower or immune, in spite of all the invocations, gestures, rites, and drum-beating. In psychic phenomena of this sort there are certain natural predispositions or individual idiosyncrasies whose influence quickly spreads, through repeated practice and mass suggestion, to a whole group.

Stoll definitely believes that *cohoba* is tobacco, supporting Sven Loven in this (Stoll: *Suggestion und Hypnotismus in der Volker Psychologie,* 1904, p. 134). At any rate, the possibility of hallucinatory hypnosis through the nasal inhalation of powdered tobacco should not be completely rejected, and it will be necessary to await the results of direct scientific experimentation before deciding the question.

In Father Las Casas's *Apologética Historia,* which has not been widely read, there is a passage referring to the Indians of Cuba that has been neglected and that seems important in clearing up this question of the cohoba. Bartolomé de las Casas implies that on the island of Cuba the Indians used powders of coca, from that same plant from which cocaine is obtained today, the alkaloid that is employed in medicine as an anesthetic of the mucous membranes, which persons addicted to its use inhale through the nose. There is this curious passage in Las Casas referring to the use of coca by the Cuban Indians as a habitual anesthetic in their visionary rites:

"They had another sacrifice, rite, or devotion, and this was a great fast, and it began among them in this way: Brother Ramón the hermit, whom we mentioned before when we were speaking of the gods of this island [Hispaniola], who came here five years before I did, says that it was generally believed and held in this island that a certain chieftain and king of theirs made a great fast in honor of the Great Lord who lives in the sky, from whom came the knowledge or opinion that there was a God of heaven; the fast consisted in staying shut up for six or seven days without eating anything except the juice of certain herbs to keep from collapsing altogether, and also washing the body with this juice. And these herbs must have been powerful, like the herb they call coca in Peru and others Pliny mentions and that we have spoken of above. During this fast their heads became so light that there came to them or appeared to them certain shapes or figurations of what they wanted to know, or, one cannot help thinking, that the devil conjured up and brought before them to deceive them, because even though the first chieftain or lord or lords who invented or began that fast and abstinence did so out of devotion to the Lord who is in the sky, and it was of him that they begged an answer to what they wanted, nevertheless those who continued with it did it in honor of the *cemís,* or idols or statues, or of the one who by this means worked to lead them away from the knowledge of the true God, and the evil one always made some progress among them because they lacked, as has been often observed, grace and knowledge.

"This can be inferred from what those of us who were in the neighboring island of Cuba saw of the ceremonies used there. In that island the fasts practiced by some of them, principally the *behiques,* priests or witch-doctors, were strange and horrible. They would fast without interruption for four months or more, without eating a thing except a certain juice of a plant or plants, which sufficed to keep them from dying, from which it may be assumed that this plant or plants have properties more powerful than those Pliny (Book XXV, Chapter 8) speaks of, which we mentioned above. And this is that same *coca* which is so highly esteemed in the provinces of Peru, as is borne out by the testimony of missionaries and Indians who have come from Peru, who saw it and recognized it in the aforesaid island of Cuba, and in great abundance. Mortified and tortured by that cruel, most severe, and extended fast which brought them to death's door, it was held that they were then prepared and worthy to have the *cemí* appear to them and see its face, who could have been none other than the devil; then he would answer them and advise them about what they asked, adding whatever he liked to deceive them, all of which the *behiques* repeated to the others and persuaded them thereof. This was the only indication and deception of idolatry that we discovered in this island of Cuba, for we found neither idol, nor statue, nor any other thing that smacked of idolatry." (*Apologética,* Chapter clxvii.)

"And in corroboration of this I can cite Pliny, who speaks of the plant *spartania,* which the Scythians use as being of great efficacy; when they have it in their mouth they feel neither hunger nor thirst. He asserts that the same effects are induced by that which the Scythians call *hippice,* which is equally effective with horses. He adds that with these two plants the Scythians can go for twelve days without eating or drinking. To be sure, more impressive than what Pliny relates is the fact, which some do not believe, that the Indians keep a ball of coca in their mouth because of the good effects they derive from it, not because of vice, but because of the great advantage they receive from the use of it." (Ibid., p. 181.)

Whatever the interpretation put upon this passage from Las Casas, it seems fair to infer from it that the Indians of Haiti and of Cuba were in the habit of using powders of highly stimulating properties distinct from those of tobacco, with which Las Casas does not confuse nor even compare them. Moreover, it has already been pointed out that diverse Indian tribes of South America employed the powders of certain plants that were not *Nicotiana tabacum* nor other allied species, and inhaled them through the ritual Y-shaped tube. And these cases are very significant because of the identity of the technique, the operation, and the religious purpose.

Today the opinion is widely held that "the custom of taking snuff was originally associated with plants other than tobacco; there were many kinds of snuff not made from tobacco leaf"; and that this custom was prevalent in the Amazon Basin, the Antilles, Peru, and Mexico (R. U. Sayce: *Primitive Arts and Crafts,* Cambridge, 1933, p. 199).

Brooks is of the belief that the use of powders for narcotic effect was common in the Antilles as well as in South America, but that the employment of ground tobacco for this purpose was more rare. It is curious to note, remarks Dr. Louis Lewin (*Les Paradis artificiels,* French translation, Paris, 1928, p. 352), that in the use of tobacco, as in that of coca and betel, people have discovered "instinctively" a means of strengthening their effects, which is really the best for this purpose—namely, the addition of an alkali to free the active agent, in this case nicotine.

It may be significant to observe, also, that many African tribes, which became acquainted with the use of American tobacco through the European slave-traders in the sixteenth century, were in the habit of mixing tobacco with ash or some alkali powder. Even the African Negro slaves living in Seville, according to Dr. Monardes in 1575, also used some kind of narcotic powder. Brooks infers from this that the Negroes must have learned from the Indians not only the use of tobacco, but to mix it with stimulants or inhale it so deeply that it produced in them those same symptoms of seizure the chroniclers describe in the Indians. Brooks recalls that the Spanish doctor Monardes (1575) also refers to the Indian habit of taking powdered tobacco mixed with calcined snail shells. He points out, too, that Amerigo Vespucci in 1499, during his second voyage to the new Indies, saw the Indians chewing tobacco mixed with other powders on the island of Margarita. Brooks (op. cit., Vol. I, p. 19) is of the opinion that the powders used in the *cohoba* were a mixture of tobacco with some other ingredient of greater toxic effect.

Therefore the possibility should not be excluded that the powders taken by the Antillean Indians in the cohoba contained something besides the powdered leaf of the plant known today as tobacco and that they were mixed with other substances that increased their hypnotic or stimulant properties.

I think it might not be out of place to call attention to the phonetic and possibly semantic similarity between the word *cohoba* and the Indian word *cobo,* which is still employed in zoology and in the popular language of Cuba to describe a kind of large marine snail from which trumpets or horns are made. The *cobo* was highly important in the mythology and religious practices of the Taino Indians and their predecessors, the Ciboneys; it was often used to make such utensils and tools as jars, awls, chisels, scrapers, and so on. In Cuba there was a "shell age" and the *cobo* was part of a cultural cycle. It was an archaic substance for the Tainos, of deep mythological significance, linked, in my modest opinion, to the religious interpretation of that great meteorological phenomenon of the tropics, the hurricane. Would it then be out of keeping for the Indians to use the alkaline powders of calcinated cobo in their inhalations for the purpose of assimilating its supernatural power? Might it not have been powdered cobo or *cooobo* ("the middle syllable long," as Las Casas would say) that, according to the authors cited, was mixed with tobacco, thus intensifying its effects on the organism, aside from fulfilling a magic purpose? Bear in mind that the cobo may have been the ancient receptacle used in the cohoba rites, into which the mysterious powders were put to be inhaled and the aromatic leaves to be burned; and that the cobo, in a small size, could also have served as a pipe—that is to say, a container holding smoldering herbs whose smoke was drawn in through an opening in the pointed end of the spiral. Some years ago I saw, but cannot recall where, an old drawing of a poor Negro slave who was smoking his tobacco in a pipe of this sort made of a small univalve shell, which is proof that the idea of a pipe of this sort is not merely imaginary or hypothetical.

Moreover, as will be seen in a study that has nothing to do with these annotations, it seems possible to connect in the same complex of religious culture the hurricane, god of the air, the sea shell or cobo, which, aside from other peculiarities, seems to evoke its noise, and tobacco, especially when mixed with powdered cobo, which rises in the air in smoke, the visible form of the mythical power, *mana* or *cemí,* subtle and intangible, which penetrates to the realm of the god Huracán, reaching him with its waves and clouds. The Aztec god of rain, Tlaloc, smokes tobacco, pouring out winds and clouds through his mouth (J. D. McGuire, op. cit., p. 365). Among the Indians of modern Mexico this "intimate connection between tobacco smoke and the rains" still exists. As part of their rites they send puffs of smoke to the four points of the compass in their ceremonies to bring rain (J. A. Mason, op. cit., p. 8).

Besides, the taking of powders and, better still, the smoke of tobacco and powdered cobo were a means of establishing communion with the god Huracán, transubstantiated in these two varieties of powder, one from the depths of the sea, the other smoke of the storm clouds' color.

The same root is present in other elements of the Taino mythology, such as cobo, jobo (turpentine tree), caoba (mahogany), coaibai ("the place of the dead"), cacibajagua ("the grotto
from which human beings emerge"), etc., but this is not the
place for a digression of this sort.

I am inclined to believe that the confusion that arose in the
minds of the chroniclers regarding the powders of the *cohoba*
and powdered tobacco was due to the fact that both of them
were supposed to be inhaled through the nose, either together
or separately; and that whereas in everyday and semi-profane
use they were snuffed without ceremony and in pinches, in the
liturgy of their tribal rites they were inhaled by the caciques and
behiques, who were the civil and ecclesiastical authorities. Naturally the tobacco powder produced different effects, depending
on whether it was taken pure, as in snuff, or mixed with other
pulverized ingredients. And it is more than likely that this is
what took place on solemn and religious occasions to join for
greater magic efficacy to the stimulating properties of tobacco,
symbolically expressed by the spirals of smoke rising upward to
the gods, the mysterious attributes of other alkaloid substances,
evocative of supernatural beings.

Another passage from Las Casas, sometimes regarded as mere
information on Indo-Cuban therapeutics, can also throw some
light on the subject. The medical bibliography of the sixteenth
and seventeenth centuries has a large section devoted to tobacco,
and it would be out of place to discuss it in these notes. But I can
speak of another use the Indians made of tobacco, which was as
an emetic, and not only for curative purposes, but in their religious customs and ceremonies. I believe that this reference will
be helpful in clarifying the magic-religious significance of the
rites involving the use of tobacco, including smoking. It is as
follows:

"The Indians had another practice, which might seem a vice, but
they did not do it out of vice, but for reasons of health, and this was
that when they had finished eating (and a slim meal it was), they
put certain herbs in their mouth, which, as we said before, resemble
our lettuce, and which they first wilt over the fire and then dip in a
little ashes and put in their mouth like a lump of food without swallowing them. This upsets their stomach and, going to the river, of
which there is always one near by, they throw up what they have

eaten, and after washing themselves they come back and eat again; and as all their eating, whether by day or night, was so frugal and consisted of such few things, it seems clear that they did not do this out of gluttony, but to relieve themselves and live more healthfully. Not all did it for this reason; at least one that I knew, one of the Spaniards, who was a very upright person, was said to take these same herbs, which had the same effect on him as on the Indians, so he could eat again. He was of those who to satisfy their gluttony divide their gorging into four parts: breakfast, lunch, dinner, and compensation or collation, as we say." (*Apologética,* Chapter cciv.)

This use of tobacco as an emetic recalls the habits of certain gluttonous Roman patricians who after surfeiting themselves at their banquets retired for a few minutes to the vomitorium and, putting their fingers down their throat, brought up the undigested food and returned to begin all over again.

Undoubtedly tobacco is a very effective emetic. The herders of Cuba used to take an infusion of tobacco leaves mixed with sugar as an emetic. The emetic effects of tobacco are so great that "if the bruised leaves are placed upon a wound, the toxic effects of nausea with vomiting and prostration are soon felt" (J. M. Dalziel, *The Useful Plants of West Tropical Africa,* London, 1937, p. 430). For this reason the Negroes of the Guinea coast use tobacco as a cathartic. Perhaps this custom was transmitted to them by the Indians, or by the slavers, or perhaps they learned it by chance experience. But among the Indians these emetic effects Las Casas refers to, aside from the immediate relief they afforded, had a religious and magic ritual significance. Compare this other passage from Las Casas, referring to certain Indians of the mainland:

"If the pain is slight, their doctors take certain herbs in their mouth and put their lips on the ailing spot and suck with all their strength, giving the impression that they are drawing out the evil humor; then they go out of the house with their cheeks puffed out as though their mouth were full of this evil humor, and spit it out and curse it many times, and say that the sick person will get well at once because by that sucking the illness was drawn out of his veins. But if the illness is more serious, like high fevers or some other grave sickness, they cure it in a different way: the *Piacha* goes to see the sick person, carrying in his hand a twig of a certain tree that he knows how to use to induce vomiting, and he puts it in a dish or glass of water to soak, and sits down beside the sick man, saying that the devil has entered into his body, and all believe him, and all the relatives beg him, since this is the case, to effect a cure; he licks and sucks the whole body of the sick man, muttering certain words between his teeth, by which he says that he is drawing the devil out of the marrow of the bones; then he takes the twig that is soaked in water and brushes it back and forth over the palate to the uvula, and then down the patient's throat until he makes him vomit and throw up all he has eaten. He gives great sighs, shakes all over, quivers and groans, then gives great bellows as though he were a bull that was being pricked with many goads; the sweat runs down his chest for two hours, like rain water running down a gutter, together with other sufferings he endures for this purpose. When our friars asked why so much suffering and anguish was required in their medicine, he answered that all that was necessary to get the devil out of the marrow of the bones of the sick man, by means of those words that rob the devils of their power, and the sucking and other efforts he makes.

"After the Piacha had mortified and tortured himself in this way, he brought up a certain amount of thick phlegm and in the middle of it was a round and very black thing, and as the Piacha was half dead, they extracted from the phlegm that black object and went out of the house shouting and threw it as far away as they could, repeating these words many times: '*Maytonoroquián, Maytonoroquián,*' which means: 'Away, devil, go away from us, away, devil, go away from us.' When all this was over, the sick person and all his relatives and friends felt sure that he would get completely well very soon." (*Apologética,* Chapter ccxlv.)

Father Pané in his relation mentions these ritual vomitings of the Indians on Hispaniola. He says:

"When someone becomes ill, they take him the *buhuitihu,* which is the doctor. The latter is obliged to follow the same diet as the sick man, to assume the expression of a sick person, which is done for the reason you shall now know. It is necessary for the doctor to purge himself the same as the sick man, and to do this he takes a certain powder, known as *cohoba,* which he snuffs through his nose, and this so intoxicates them that afterwards they do not know what they are doing, and so they say many senseless things, alleging that they are talking with the *cemíes,* and that these have told them the cause of the illness."

Farther on, Father Pané adds:

"When they are alone they take some leaves of *gueyo,* which are broad, and another plant, wrapped in an onion skin, half a hand's breadth long; the leaves of *gueyo* are what they ordinarily use; they squeeze them into a paste, and then they put this ball into their mouth at night to vomit up what they have eaten so it would not do them harm. Then they begin to sing the song referred to, and, taking a torch, they drink that juice. When they have done this, after a little while the *buhitihu* gets up, goes over to the sick man, turns him over twice as he thinks best, then stands in front of him, takes hold of his legs, and feels them from the thighs to the feet. Afterwards he gives him a sharp tug, as though he were trying to pull something out of him, goes to the door of the house, closes it, and speaks, saying: 'Go at once to the mountains or to the sea or wherever you like,' and blows as though he were blowing away a straw. Then he returns, folds his hands, and closes his mouth; his hands shake as though he were cold; he blows upon them, draws in his breath as when one sucks the marrow out of a bone, and begins to suck the patient in the neck, the stomach, the back, the cheeks, the breast, the belly, and other parts of the body. When he has done this, he begins to cough and make a face as though he had eaten something bitter, spits into his hands, and takes out the thing already referred to that he put into his mouth in his house or on the way, a stone or bone or bit of meat, as has been explained."

López de Gómara also mentions this vomiting (op. cit., p. 173):

"After they had entered the temple they vomited by putting a little stick down their throat, to show the idol that nothing bad was left in their stomach."

M. R. Harrington found in Cuba some little canes ornamented with mythological figures which the Indians used to induce vomiting in certain rites. Harrington applied to them the name "swallow sticks." (See M. R. Harrington: *Cuba antes de Colón,* Spanish translation, and Fernando Ortiz: *Historia de la Arqueología Indocubana,* Vol. II, Havana, 1936.) This brings us to a consideration of the religious significance of these self-induced vomitings as expurgatorial rites and their connections with tobacco.

Ethnographers today are well aware of the fact that vomiting among many primitive peoples is a way of confessing their sins, expelling evil, purging the conscience, a kind of internal "house-cleaning." There are the studies of Raffaele Pettazzoni on this subject (*La confessione dei peccati,* Bologna, 1929). The confession of sin takes several forms, concommitant or substitutive. As a rule, among peoples of a lower order of civilization, the confession of sin, aside from the oral and auditory form, is accompanied by complementary symbolic manifestations, such as ablutions, baths, sprinklings, blood-letting, flagellation, fumigations, vomiting, threats, charms, exorcisms, amulets, and witchcraft. The "evil thing" must be driven away. The Catholics use for this purpose the sign of the cross, holy water, scapularies, medals, votive offerings, vows, relics, and exorcisements, in addition to prayers and litanies, with set formulas reminiscent of magic rites. All these lustral usages are morphologically different, but functionally the same, all designed for eliminatory effect, as much or more a catharsis of the soul than bodily medication. It was only with the development of metaphysical philosophical thought that the great pre-Christian religions began adding to the cathartic effect of the confession of sins the penitential and ethical concept (Assyria and Egypt) and then that of contrition with its regenerating effect (Judaism and Christianity), aside from the permanent political interest on the part of the sacerdotal caste to learn people's secret thoughts as a powerful *instrumentum regnum* and an indispensable wheel in the machinery of government.

In the texts quoted from Las Casas and Pané some of these depurative methods have already been mentioned, such as vomiting, rubbing, tugging, blowing, sucking, coughing, spitting, and adjurations to the evil spirits. There are still other methods mentioned in other texts of the chroniclers. Las Casas speaks of the importance of ablutions among the Indians, not merely for the purpose of personal cleanliness, but for the cleansing of the conscience. He has this to say of the therapeutic procedure followed with a sick person:

". . . They wash them, for the treatment they mostly used was washing the sick, even if they were at death's door, with cold water; which they did, either because of their habit, when they were well, of washing themselves every hour, or out of superstition, because they believed that water had the power to wash away sins and bring bodily health. . . .

"They had another way of curing the sick on that island (Hispaniola); this was that the priests or wizards who, as has been said before, are called *behiques,* took their arms from the shoulders down between their hands, rubbing them and blowing upon them, and the same with their legs and all the body, almost as though with that rubbing and blowing they could drive out the sickness, and I think this is what they made the simple folk believe; and they also said certain words, calling upon the devil, with whom they probably had a pact. . . .

"They performed a ceremony that was like a penance when they thought they had been guilty of some sin, and this was that they went to the river and undressed and washed themselves all over. They believed . . . that the waters had power to remove or wash away their sins . . . so frequently and repeatedly did they all wash, not only when they were well, but when they were sick and as the first and last cure. And in this island and islands this ceremony and custom was very widespread and habitual. If the sinner felt that his sin was very great, as penance and expiation he would burn the clothes he had been wearing when he committed it." (*Apologética,* Chapters cciv and clxxxii.)

It was a religious method of healing, a sacramental lustration of anointing and blowing, similar to that employed by Catholics when they apply the holy oil which purifies and consecrates. It was a kind of exorcism, like the adjuration of the ministers of the Church against the devil; and that nothing might be lacking, aside from certain differences in the liturgy, even the devil talked, as happened on several occasions with the exorcising ministers in America, and Father Las Casas, who was a bishop and a Dominican and in close relations with the Holy Inquisition, knew this well. Oviedo also says:

"I have observed that the Indians, when they know that they have an excess of blood, bleed themselves in the ankles, the arms, the elbows, and near their hands, in the broad part above the wrists. . . ."

That is to say, the Tainos employed blood-letting as a means of ejection. And López de Gómara (op. cit., p. 173), besides mentioning the vomiting provoked by the aforementioned "swallow-sticks," gives a more detailed account of the methods of the witch-doctor *behique,* as follows:

"To cure a sickness they also take some of that plant cohoba, which does not exist in Europe; they shut themselves up with the patient . . . foam at the mouth, make a thousand grimaces, and then blow upon the patient and suck at the back of his neck, telling him that they are extracting all the sickness from him there. Then they run their hands carefully over his whole body, even to the toes of his feet, and then they go to throw the illness out of the house, and sometimes they show a stone or bone or bit of meat they have in their mouth and say that he will soon get well, for they have removed what was causing the sickness."

In this account we see how "the evil thing" expelled by the patient as the result of spells, massage, shaking, sucking and fumigation was symbolically located by the behiques in the material manifestation of a little stone they pretended to have removed from the patient or bewitched person.

"The sin that was the objective of these practices, which could be destroyed by fire, washed away with water, expelled in vomit, with blood, etc., was conceived of as something possessing actual consistency; in other words, it is evil felt as a painful experience and exteriorized in the idea of a force-substance that produces it." (R. Pettazzoni, op. cit., p. 53.)

All these lustral rites of active magic and therapeutics involved tobacco. The mysterious cohoba appears in them. And that other plant referred to by Pané appears in the different translations consulted as *gueyo, giogia,* and *digo,* which goes to show how dubious was its nomenclature and how justified one is to translate these different names by "tobacco plant, *Nicotiana tabacum,*" as did the Cuban Dr. Ernesto López in his valuable monograph: *"Medicina de los siboneyes" (Revista Cubana,* Havana, Vol. VII, 1888, pp. 193 et seq.). This plant, *gueyo* or whatever its name was, was sacred and had a place in the Indians' cosmogony as well as in their vomitory rites. For these reasons Dr. López reached the following conclusions:

"To induce vomiting they used a mixture of tobacco and a kind of onion mashed together, and Friar Román Pané adds that, for the same purpose, they used a sacred plant they called *gueyo.* It may be that this plant was none other than tobacco, which is an emetic but not purgative.

"The sacred plant whose use was taught by Bohito II was called *gueyo;* since we know that of all the plants known to the Indians at the time of the discovery of America tobacco was the most important because of its many properties, and as it is said that it was also used in religious practices, it might well be that *gueyo* was the sacred name of the plant, or of the growing plant, whereas *tobacco* was used to designate the dried leaves of this plant intended for smoking, as

well as the device through which the smoke was inhaled; and, finally, cojoba was the drink made from the juice of the green leaves
of tobacco, which was offered to the *cemíes* to win their favor and
which is so often mentioned in the Indians' religious and medical
practices. They must also have used tobacco as a diaphoretic, since it
is almost as valuable for this purpose as for an emetic." (Op. cit., p.
206.)

Be this as it may, throughout the ritual uses of tobacco there
is clear evidence of a complicated practice of purification,
through the assimilation of the sacred power, mana or *cemí,* of
the plant and through the elimination or expulsion of something from within the human being. The smoke, the powder,
the chewing tobacco, and its infusion are acts of internal purification through contact with the divine mana of tobacco. By
these means an increase of saliva is induced, which is spit out as
a bad thing, vomiting, which cleanses the stomach, and the se-
cretion of nasal mucuses, or "runnings of the head," as was
the expression of the times. The warm smoke itself, taken in
through the mouth and expelled through the nose or the
mouth, after having censed the inner passages, and the leaves,
which are spit out after their essence has been extracted by
chewing, are likewise rites of purification, for casting out "the
evil thing." All these are lustral, cathartic, or expurgatorial
rites, which are frequent among the Indians of all America.

Mayas, Aztecs, Incas, Chibchas, Arawaks, and many other
Indian peoples were familiar with the oral and symbolic confession and practiced it so faithfully that when the Christians
arrived and the Indians learned that they, too, confessed themselves at times, they began to ask for this sacrament with such
fervor that the missionaries could hardly attend to them, and
this gave rise to grave abuses, as various ecclesiastics of the period have recounted. For example, Friar Gerónimo de Mendieta
in the chapters on the subject in his *Historia Eclesiástica Indiana*
(Mexico, 1870) alludes to "certain manners of oral confession
that the Indians used though infidels, and how the sacramental
confession of the Church suited them." These confessions of the
Indians were carefully noted by the missionaries converting
them to Christianity, and if some of them regarded them as
snares of the devil, others, especially the Jesuits, interpreted
them as "in part the foresight of the Lord to have allowed this

past custom that confession may not be difficult for them"
(Acosta: *Historia Natural y Moral,* V, 25). So deep-rooted were
these pagan confessions that they were practiced centuries after
the preaching of Christianity, and even today one can observe
traces of them beneath the superficial Catholicism of the Indian
masses.

I do not intend to go into prolix digression, but I do wish to
point out that in the confessional rites formerly followed by the
Indians of the most advanced civilizations in America there
were practices involving the use of powdered sea shells, mixtures of powders of tobacco, infusion of tobacco, tobacco smoke,
together with that of other plants, and the burning of aromatic
herbs and odorous resins, the inhaling of smoke, spitting with a
definite purpose, and so on. Mendieta tells how the Indians of
Mexico confessed and traveled long distances to the missionaries
to seek their help in the magic rite of their confession.

"Even children barely seven years old, when they are ill, ask their
parents to take them to the church to confess. It is a wondrous thing
for which we should praise the Lord that as soon as an Indian gets a
fever or a headache he asks his relatives to take him to confession."
(Op. cit., p. 283.)

To Father Mendieta this expurgatory rite of the heathen confessions must have been all the more wondrous when the Indians concluded them with the chewing of tobacco and the inhalations of their typical powders, for he says:

"They also use some kind of communion or receiving of the sacrament, and what they do is to make tiny little idols of the seed of wild
amaranth or ashes or some other herbs, and they take them as the
body or in memory of their gods. Others say that some of them regard a plant they call *picietl* and the Spaniards tobacco, as the body
of a goddess whom they call Ciuacouatl. [In the margin of the page
there is this annotation: 'female snake.'] And this plant, which has
medicinal properties, should be considered suspect and dangerous."
(Op. cit., p. 108.)

In ancient Peru the confession of sins was very frequent too,
especially in the region of Cuzco, whence it probably spread
when Cuzco became the co-ordinating center of the Incan civilization. The *ichuri,* a kind of soothsayer or priest, heard the con-

fessions of all the members of his *ayllu* or community. The penitent carried with him powders of different colors, such as powdered sea shells (*mullu*), cinnabar (*paria*), and certain green powders (*llaxa*), as well as coca, chicha (a fermented liquor made of corn) suet, and corn bread. According to Molina, the sea-shell powders carried by the penitent were of different colors, as were many grains of corn. These multicolored grains of corn and sea shells were ground to powder by the confessor himself as, sitting on the ground before the communicant, he spread the powders on a stone placed between them. Whereupon the penitent, after an invocation to the mountains, the valleys, and the birds of the air, confessed all his sins aloud, holding in his right hand a ring or little ball of powdered shell. When the confession was ended, the *ichuri* assigned the sinner's penance and handed him the stone with the varicolored powders on top, so that the penitent could blow them away and disperse them, or else he blew them away before beginning his confession, according to Molina: "The chieftain or *Huillac uma* in the temple, holding in one hand a sheaf of hay, flowers, and aromatic herbs, confessed his sins to the supreme being (*Illa Tici Uiracocha*), afterwards praying the god that the smoke might carry with it his sins, and with the same object he cast the ashes into a brook." This manner of confession was also employed by the humblest of the community, although in their case this confession took place before the *ichuri,* and he took advantage of this oral confession to order obedience to the ruler and other precepts of a political and social nature. The name applied to the confessor was the word *ichuri,* which comes from *ichu* or *hichu,* which means "straw, grass, or reed." *Ichuri* means "the one who uses straw," as this was employed in certain of the ceremonies performed by the confessor (R. Pettazzoni, op. cit., p. 122). See also Pablo José de Arriaga: *Extirpación de la idolatría del Pirú* (Lima, 1621), p. 29; Antonio de la Calandra: *Crónica moralizada de la O. de San Agustín en el Pirú* (Barcelona, 1639), Vol. II, Chapter xii, p. 376. In connection with the rites of the *ichuri,* Pettazzoni reaches the following conclusion:

"This scattering of the powders in question, which is analogous to ritual practices in confession among certain Negroes of Oriental Africa and the act of throwing away certain pebbles and shells among the Arawak and Ijca Indians, is evidently one of those rites of an

LOS AMORES DE VENTANA

"WINDOW-LOVES"

ESTE ES EL TABACO QUE FUMA TU AMO? QUE POBRE CASA

*"IS THIS THE TOBACCO YOUR MASTER SMOKES?
WHAT A POOR HOUSE!"*

VISTA DE UN INGENIE.

VIEW OF A SUGAR MILL

VISTA DE UNA VEGA DE TABACO.

VIEW OF A TOBACCO PLANTATION

eliminatory nature with which confession is accompanied among primitive peoples." (Op. cit., p. 124.)

López de Gómara (op. cit., p. 173) refers to a method employed by old Indian women in their cures, which is very similar to that used with tobacco:

"Many old women were doctors, and they administered the medicine through the mouth with hollow tubes."

Between these tubes of the old wives' medicine and the tubes, canes, hollow reeds, or inhalers of the tobacco rites, was there any essential difference?

Tobacco was a complex source of magic-religious ritualism. Interpretations may vary as to whether it was a propitiatory offering to the gods, a simple procedure for benumbing the senses of the initiate, or a rain-making ritual, but the fundamental feature of the cultural complex of tobacco among the Indians would seem to have been its purifying magic, its mental stimulation, and its sedative and cathartic effects, either to induce mystic states or to allay nervous tension. The mere enunciation of this theory of psychic, physiological and religious catharsis through the medium of the Nicotiana plant goes to show how complicated the phenomenon of tobacco was in the culture of the Cuban Indians, which has become much simpler today in the process of its transculturation among white peoples.

Tabacos or cigars were smoked, not only in Cuba, but in the other Antilles, in Central America, and in the northern and central regions of South America. But, in the opinion of Sven Loven: "The habit of smoking cigars reached far greater proportions among the Taino Indians than among the Indian nations of the Atlantic coast and tropical regions of South America."

The geographical origin of the tobacco plant is unknown, and the same thing is true with regard to the various forms of using it. There have been different opinions on the subject. Some think it was first used in the form of powder, then in a pipe, then in cigars, and later as cigarettes. But this morphological scale is hardly tenable.

Montandon suggests that the pipe preceded the cigar (*Traité d'ethnologie culturelle,* Paris, 1934, p. 287). The dried leaves of tobacco were lighted and burned in it. The cigar, in his judgment, represents an advance; into the whole leaves others were rolled to form a cylindrical packet, and this was stuck into one end of a pipe. Later, with the improved technique of making cigars, the pipe or holder was no longer necessary, and the cigar could be smoked alone. This theory of Montandon's would seem to be contradicted by other observations. This can be deduced from the simple, elementary method of smoking tobacco among certain Indians of Darien, as related by Lionel Wafer in his *Travels in the Isthmus of Darien* (1699). He says that the tobacco leaves were duly dried and cured by the natives and that then:

". . . spreading out two or three leaves one on top of another, they roll them up, leaving a little opening. Around these they roll other leaves, pressing them tightly together, until the roll is as thick as the wrist, and two or three feet long.

"The way they smoke when a number of Indians are gathered together is as follows: A boy lights the end of the roll, which burns like an ember, dampening the part near it so it will not burn too fast. He puts the other end in his mouth and blows the smoke through the roll into the faces of all those present, even though they be two or three hundred. Then, squatting in the usual manner on their legs, with their hands close together they make a kind of funnel around their mouth and nose into which they receive the smoke that is blown toward them, inhaling it with all their might until they are breathless, and then they blow it out, and it gives them great satisfaction."

There are those who have thought this way of smoking was the origin of the cigar (F. W. Fairholt: *Tobacco, Its History and Associations,* London, 1859, p. 214). But I do not hold that this type of cigar is a necessary forerunner of the kind Columbus discovered in Cuba and Father Las Casas described, although it does seem sufficiently significant to invalidate Montandon's hypothesis.

According to J. Alden Mason, there is another probable evolution in the smoking of tobacco. In his opinion, the cigar, consisting of some tobacco leaves wrapped in another of the same kind, came first; then the cigarette, of tobacco wrapped in the leaf of another plant; and then the tubular pipe and later the elbow pipe. But as Mason himself points out, there is no historical evidence to prove that this was the evolutionary process of the typical methods of smoking.

The morphological evolution of the use of tobacco is complex and apparently there is no longer sufficient evidence to trace it with any assurance. It must not be forgotten that the mysterious power or substance of tobacco was assimilated in four ways: by chewing, in powdered form, in liquid form, and in smoke. And for each of these different ways techniques and variations were invented, which were not necessarily serial or successive. Probably these different ways of using tobacco, and even the variations from the specific types within each form, did not appear one after another, in a definite sequence. Most of them probably sprang up independently, among different peoples who often had no connection with one another, and chance migratory contacts either spread them over wide areas or limited them to small regions, even though such diffusion is not indispensable in every case. In this as in almost every aspect of human culture, arguments can go on indefinitely between those who hold that the same things appeared independently as the individual invention of human genius in different places and moments, and those who defend the theory of the diffusion through the world of every original invention, unique to begin with, but carried from one people to another in all lands and all times. Bearing this in mind, we can examine certain hypotheses concerning the different methods of using tobacco and their possible successive connections.

It might be thought that the use of tobacco among the Indians began with the use of brews or infusions of this plant as an emetic with medicinal effects, which were interpreted as magic phenomena of purification; the plant contained a mysterious power by which the evil within the bowels was expelled, leaving them free of all ill. From the infusion and emetics it was an easy step to the chewing of the leaf, and by swallowing the saliva saturated with nicotine the same sedative and stimulating effects of tobacco were induced, which, in a measure, serve to expel certain nervous discomforts. The discovery of the peculiar purifying properties of tobacco through the inhalation of the powders and the smoke, especially this latter form, represented new magic perfection of the rustic, antiquated methods of cleansing by means of emetics and chewing, more subtle, characterized by purer ritualistic and emblematic qualities, of a more transcendent effect.

I have spoken of Sven Loven's opinion regarding the priority of chewing over the other forms of imbibing the essence of tobacco. The expellant method of smoking must have been a later derivation, a combination of the hollow tube and the individual cigar. These cigars with a hole in the middle throughout their full length recall certain cigars or cheroots manufactured in some countries with a straw in the middle, but never in Cuba, where the Indians were unaware of these developments in tobacco morphology or their possible antecedents.

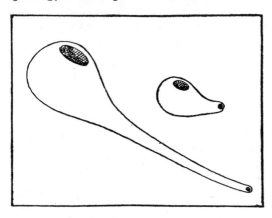

Figure G. Gourds used by the Makusi Indians to blow powdered pepper into the nose, according to Roth

Following another morphological line of development, one arrives more easily at the pipe from the absorption of powders and liquid. Roth (loc. cit., p. 247) describes how the Makusi Indians take a mixture of water and powdered pepper. A squash or gourd of elongated shape cut in the form of a pipe is used for the purpose (Figure G) and is known as a *kassakra*. The gourd is filled with the hot liquid and the stem end placed in the nose of the passive subject of the rite so he can snuff up the contents. It seems easy to understand how the morphological type of this apparatus designed for the absorption of a liquid could be transformed into a pipe through which to inhale the smoke of the burning leaves, once the technique of making it out of clay or fire-resistant wood—that is to say, of practically incombustible materials—was discovered.

The Taino Indians of the Antilles were acquainted with an-

other mode of using the leaves of tobacco: the cigarette. Las
Casas speaks of a way of preparing tobacco for smoking by
wrapping the leaves in the leaf of another plant. He refers to
them as *mosquetes* or cylindrical tubes made of the leaf of a tree
with a certain herb inside them. His description is as follows:

"In this island of Hispaniola and in the neighboring ones they
have another kind of plant that looks just like lettuce, and they dry
this in the sun and over the fire, and from the dry leaves of a tree
they make a roll and put inside it a little of that plant and light the
roll at one end and through the other they suck or draw the smoke
into their lungs, and in this way they feel neither hunger nor fatigue.
These rolls are known as *tabacos,* the middle syllable long. Some of
the Spaniards say that the Indians keep this herb in their mouth
more out of vice and bad habit, or because they imagine it does them
good, than because it really does, but they say this because they do
not know about certain herbs that produce the same effects and are
equally useful." (Las Casas: *Apologética Historia,* Madrid, 1909, p.
181.)

In connection with the Indians of Hispaniola, Girolano Benzoni, the Milanese, who was in these Indies from 1541 to 1556,
recalls these methods of preparing cigars by rolling tobacco in
a corn leaf (corn being indigenous to the Antilles) with the
same or even greater precision; for almost a year he was in
the town of San Cristóbal de la Habana in 1544, shortly after
it had been sacked and almost completely destroyed by a French
corsair. Benzoni speaks of the high esteem in which certain
leaves similar in shape to those of the walnut tree, but larger,
were held by the natives and the Negro slaves. He says:

"When these leaves are ready, they pick them from the plant, tie
them into bunches, and hang them up close to the fire in their homes
until they are well dried; and when they want to use them they take
one of the husks from an ear of corn, and putting one of the leaves
inside it, they roll them up together in a tube; then they light it at
one end and, putting the other in the mouth, draw through it, and in
this way the smoke drawn in goes into their mouth, their throat and
head, and they hold it there as long as they can, for in this they seem
to find pleasure, and they fill themselves so full of this cruel smoke
that they lose their senses. And there are some who take in so much
smoke that they drop to the ground as if they were dead, and remain
unconscious the better part of the day or night. There are some of

the men who inhale only enough of this smoke to make themselves a little dizzy, but no more. It is evident what a disgusting and abhorrent poison of the devil this is." (Girolano Benzoni: *La Historia del Mondo Nuovo,* Venecia, 1572, Book I, p. 547.)

Benzoni (of whose work, written in Italian, I might say, in passing, there is as yet no Spanish translation, although it is available in Latin, French, German, Flemish, and English versions, probably because it contains certain statements and judgments unfavorable to the Spaniards and their churchmen) adds at the end of the text I have translated that in the language of Mexico this plant is called *tabaco.* The Italian traveler picked up the term in current use among the Spaniards of all America, but the word is not Mexican, nor was tobacco so called in any of the languages spoken in New Spain, aside from that of the Spaniards, who often incorporated into their vocabulary the Indian names by which the natives of the Antilles called the things the conquistadors discovered, such as *cacique, juracán, yuca, jamaca, mamey, naguas,* and certain others.

The Mexican writer Alfredo Chavero gives this opinion: "The word *tobacco* comes from the islands, and was introduced here by the Spaniards" (in *México a través de los siglos,* Barcelona, p. 305).

Benzoni's observations could hardly be clearer. Two of these cigars can be seen in a drawing from his work (see Figure H), which shows one *behique* treating a sick person, and another with a cylindrical roll like those described by Las Casas sitting in front of a patient who is lying unconscious, with another object similar in shape to the above, but with lines on it to indicate that it was twisted. These have been believed to be cigars, although it would be necessary to accept them with certain reservations, in view of their imperfect form and because of the absence of smoke, which would be characteristic. It must be borne in mind, however, that these illustrations, like those in Oviedo's *Historia,* were drawn and engraved in Europe, without the supervision of the authors of the books, several years after they were written, and perhaps even without sketches, merely following the scanty information of the text and the imagination of the European artist, who had no direct knowledge of the objects he was trying to represent graphically. In Benzoni's illustrations this is frequent, as in those of Oviedo when dealing with trees or dwellings unknown in Europe.

This method of smoking tobacco wrapped in the leaf of another plant which is the typical form from which cigarettes derived, was to be found among the Indians of the coastal regions of the Caribbean Sea. In Mexico this type of cigarette wrapped in a cornhusk is still used, although the filler is generally peyotl rather than tobacco. For this reason there are those who have thought mistakenly that while the cigar was of Cuban origin, the cigarette was exclusively Mexican.

Figure H. Curative treatment among the Indians of Hispaniola, from an engraving in the work of Girolano Benzoni. In this composition there are, apparently, two cigars: one beside the sick man in the foreground, and the other in the hand and mouth of the *behique* treating him. In the background can be seen another *behique* caring for a sick man stretched out in a hammock.

In Brazil tobacco was used in this form, but rolled in a tube of palm leaf, according to Father Thevet, the ex-Carmelite friar, who spent the year 1555 there and who took credit to himself, probably justifiably, for having introduced tobacco into France, some ten years before Nicot (A. Thevet: *La Cosmographie universelle,* Paris, 1575, Book XXI, Vol. II, p. 926). Thevet states that the Indians in Brazil dry the leaves of petun and "wrap a number of them in a big palm leaf, making a roll the size of a candle; afterwards they set fire to it at one end, draw in the smoke through the mouth, and blow it out through the nose, for the purpose of attracting and distilling the superfluous hu-' mors of the brain; moreover it stops thirst and hunger for a while, and for this reason they use it all the time; even when they are talking with you, they draw in the smoke and then keep on talking and do it again in the same way more than two hundred times; it is very useful to them when they are going to war, to eliminate the vapors and other bodily disorders that may arise on the way; the women also use petun, though not so frequently."

Thevet's *Cosmographie* carried a drawing the proportions of which are inaccurate, but which shows a palm leaf rolled up, containing tobacco for smoking. An almost identical passage is to be found in another work by the same Protestant author (*Les Singularités de la France antarctique,* Antwerp, 1558; Paris, 1878, p. 157). In similar terms the Frenchman J. de Léry describes the method of smoking petun (*Histoire d'un voyage fait à la terre du Brésil, autrement dit Amérique,* Rochelle, 1578). In 1587 the custom of smoking among the Brazilian Indians was described in the same way by the Portuguese Gabriel Soares de Sousa (*Tratado descriptivo do Brasil em 1587,* Rio de Janeiro, 1879, Chapter clxiv). Even today certain Indians of South America, who are still in a very retarded state of culture, preserve this mode of smoking tobacco, rolling the dried leaves in those of a different plant. This is true of the Indians of Rio Negro and the upper Orinoco, who roll the tobacco in a very thin wrapper taken from the bark of a tree called *tabarí.* The Napo Indians smoke tobacco rolled in a cornhusk or also in some very thin bark. The Colorado Indians wrap it in cornhusks or banana leaves (M. de Wavris, op. cit., p. 152). The same usage is followed by the Indians of the Guianas, who use the leaves of kakarali or cakarillo or sapucaia (*Lecythis ollaria*) or of mani-cole (*Euterpe oleracea*) or *Couratoria guianensis,* according to E. F. Im Thurn (*Among the Indians of Guiana,* London, 1883, p. 318), W. G. Farabee (*The Central Arawaks,* Philadelphia, 1918, p. 46) and W. E. Roth (*An Introductory Study to the Arts, Crafts and Customs of the Guiana Indians,* 38th Annual

Report, Bureau of American Ethnology, 1916–17, Washington, 1924, p. 241).

Some of these wrappers required careful previous preparation to give them the necessary flexibility, and among the Uaupés Indians the cigar is held, while being smoked at their festivities, in a kind of ritual tongs that recall those little tongs of gold or silver employed by gentlemen of fashion, priests, and ladies to keep their fingers from becoming stained from constant smoking. (See Figure I.)

It may be taken for granted that these rolls of dry tobacco leaves smoked by the Indians did not have all the finish later

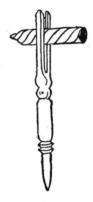

FIGURE I. Cigar-holder used for ceremonial smoking by the Uaupés Indians today. The holder is two feet long, and is used to hold the cigar and pass it from person to person so each one may take a puff. When they are not smoking, the holder is stuck into the ground until they begin again. According to A. R. Wallace.

given cigars in Cuba through the use of a wrapper carefully selected for texture, body, color, and quality, carefully trimmed with a sharp blade, carefully rolled about the filler, and shaped to conform to an elegantly fashioned design or special shape. In the beginning the rustic smoker did not always have on hand whole leaves, dried but flexible—that is, wrappers in which to hold and give body to the twisted leaves in such a way that the cigar would not come to pieces in the hands while it was being smoked.

Often the leaves used for filler had to be tied around with thread to hold them together so they would burn slowly and the smoke would pass through them so it could be taken in puffs. We have Indian antecedents for these twisted cigars "which are a handbreadth in length, and as thin as the finger, and are made of a certain leaf rolled up and tied with two or three threads of thin fiber." These are the words in which Gonzalo Fernández de Oviedo describes them in the account of a great feast given by the Chorotega chieftain Nambí, of Nicoya, in the land of Nicaragua, in honor of the Spaniards, August 19, 1529 (*Historia General y Natural de las Indias,* Book XLII, Chapter xl). Such, then, must have been the primitive method of wrapping the leaves of the Nicotiana prepared by the Indians to be smoked in their sacred rites. And this roll of leaves, the simple product of the rustic technique of the wild Indian, which came to be substituted for the ritual inhaler tube, and was known as *tabaco,* according to Oviedo in one passage of his *History,* might have received this same name through evident analogies, as Oviedo himself states in another place.

There is also the possibility that a transition was made from the original monotubular inhaler to another type more suitable for absorbing the mixed ingredients—that is, a tube such as a reed or a palm leaf, for instance, that would hold them because of its firmness, or because it was fireproof, or at least fire-resistant, such as a tube made of shell or bone. There is evidence of the existence of such tubes in the history of Yucatán and Mexico. Oviedo mentions that in 1518, when Juan de Grijalva was exploring and marauding along the coasts of Yucatán, he was on several occasions honored by caciques with censing and smoking, as in the following case:

"This Indian was an old man . . . and he poured out many perfumes before the idols that were in this tower, and loudly intoned a certain song in a monotonous voice, and he gave to the captain and each of the other Spaniards a kind of reed which when they set fire to it burned slowly, like *pivetes,* and gave off a sweet fragrance" (*Historia General y Natural de Indias,* Book XVII, Chapter ix).

Later, on another occasion, the Indian chief seated Grijalva, the captain general, and his men on *bihaos,* which are a kind of broad leaf spread out on the ground after the manner of rugs, as though to honor them:

"And when the general and those the Indian chief had chosen were seated, he gave the general and each of the Spaniards who were with him a little hollow tube lighted at one end, which are made in such a way that after they are lighted they burn to the end without making a flame, as do the *pivetes* of Valencia, and they smelled very good and the smoke that came out of them; and the Indians made signs to the Spaniards not to let that smoke float away or be lost, as though they were smoking tobacco" (op. cit., Book XVII, Chapter xiv).

And still farther on, Oviedo says that the Indian:

". . . gave sticks of incense to the captain and the leading Spaniards, *pivetes* like those mentioned before in the first visits. And the general ordered the chaplain of the fleet to say Mass . . . and when he began they brought in a clay vessel with certain sweet-smelling incense and set it beneath the altar, and another similar one in the space between the priest and the people." (Op. cit., Book XVII, Chapter xv.)

Pivete in the dialect of Valencia is equivalent to *pebete,* which means "paste made of aromatic powders, as a rule in the form of a little stick, which when lighted gives off a very fragrant smoke."

Father Las Casas also makes mention of this experience of Grijalva's:

"They gave them blankets of colored cotton with great display of pleasure and happiness, as though they were their own brothers, and among other gifts, which they are in the habit of making to their visitors as we know from experience, they gave each Spaniard a lighted tube, full of aromatic things, very fragrant, like muskets made of paper, and they drew in the smoke from them with their breath, and let it out through the nose" (*Historia de las Indias,* Chapter cxii).

In Mexico, too, the *pivetes* make their appearance, although the fact that they were "painted and gilded tubes" suggests that they must have been fireproof, or almost so.

In Bernal Díaz del Castillo's account of the splendor in which the Aztec Emperor Montezuma lived, speaking of the solemnity that attended his meals, he says:

"They also set upon the table three painted and gilded tubes containing liquidambar mixed with a certain plant they call *tabaco;* and when he had finished eating, after they had sung and danced for him, and the table was cleared, he took the smoke of one of those tubes, and little by little with it he fell asleep" (Bernal Díaz del Castillo: *Verdadera historia de los sucesos de la Conquista de la Nueva España,* Chapter xci).

The same author speaks in another passage of the trade that went on in the great *tatelulco* or market of the city of Mexico, and among the things sold there he saw "little pipes containing liquidambar full of tobacco"—that is, similar to those Montezuma smoked after his meals (Chapter xcii).

Father Bernardino de Sahagún in his *Historia General de las Cosas de Nueva España* (1576) speaks of the use by the Mexicans of certain little tubes containing aromatic herbs, among them *picietl* or tobacco. Father Toribio de Benavente, "Motolinía" (*Historia de los Indios de la Nueva España,* Chapter xv), makes mention of the *picietl,* saying that with this "medicinal herb" the Indians put to sleep or stupefied vicious and poisonous snakes. Sahagún makes a great point of the fact that these tubes were gilded or painted in colors, and adorned with figures of flowers, fish, birds, or other designs; and that sometimes they were covered with black or white powder, beneath which certain drawings were hidden that became visible only when they were heated. This detail is not too clear in Bernardino de Sahagún's text, but it would seem that as the heat from the combustion inside the tube reached the surface containing the drawings, it dried the moisture holding the powders to the outside of the tube, which was fireproof, and that then the designs, which were probably of religious or magic import, appeared.

Dr. Francisco Hernández de Toledo, who was sent out by Philip II to study the flora of Mexico, speaks of these Mexican tubes in his work printed in 1615, identifying them with the *tabacos* of the Antilles. He writes as follows:

"In the island of Hispaniola they give the name *tauacos* to certain hollow pieces of cane, a handbreadth and a half long, which are smeared with coal dust on the outside, and inside are filled with *tauaco,* liquidambar (or *xochiocotzol*)" (F. Hernández: *Cuatro libros de la naturaleza y virtudes medicinales de las plantas y animales de la Nueva España,* Morelia, 1888, p. 136).

Hernández refers to certain cane tubes used in Hispaniola as *tabacos*. As he had never pursued his investigations in the Antilles, only in Mexico, it is probable that this piece of information is erroneous as far as the description of the Antillean device is concerned, but not as regards the name *tabacos*. From Las Casas we know that in Haiti they smoked the prototype of the modern cigar, made completely of tobacco leaf (wrapper and filler), and also others similar to them but wrapped in the "dried leaf of a tree," probably a piece of palm leaf with herbs inside. In his *Apologética* Las Casas compares them both to *mosquetes;* but he does not say that, in addition, there existed in Hispaniola another type formed of "hollow pieces of cane," smeared and painted like the typical "smoking tubes" of Mexico. In any case this report of Hernández's confirms the other information from Mexico about these smoking tubes or "canes of smoke."

To which morphological phase in the evolution of the manners of smoking do these tubes filled with powdered or shredded tobacco and other similar substances belong? Do they proceed from some earlier type of pure tobacco that in order to be mixed with certain more powerful or aromatic ingredients had to sacrifice the wrapper of its own leaf and adopt one more solid and tubular? This is the opinion of Brooks (op. cit., p. 15), who believes that these tubes filled with tobacco represent the intermediate step from the cigar to the pipe. Dunhill, too, believes that the pipe was subsequent to the cigar (*The Pipe Book,* London, 1926, p. 29). But it does not seem to me that this transformation was indispensable.

Is there not the possibility of another hypothesis—namely, that this tube filled with a mixture of ingredients preceded the cigar? This similarity between the cigar and the tubular pipe is very suggestive. As Father Labat said, the cigar "is a kind of natural pipe comprising both tobacco and the apparatus in which to smoke it" (op. cit., Vol. VI, p. 321).

For this reason the cigar is usually placed, in a hypothetical morphological sequence, before or after the pipe; but this is not necessarily so. Rather it would seem that from the original single cane tube, for inhaling or exhaling the powder and smoke of tobacco, two types evolved: one, the "cane of smoke" —that is, the fireproof tube filled with tobacco, cohoba, or balsams, like that used by Montezuma, from which it is easy to arrive at the pipe—and another, the rustic, everyday simplification of the tube, reducing the whole thing to a roll of tobacco leaves, which is the cigar that has come down to us.

I do not feel that any completely dependable explication of the evolution of the different methods of smoking exists. Aside from the theories that have been examined, still another hypothesis might be put forth. I must say first of all that it is not necessary to assume that the snuffing of powdered tobacco inevitably preceded the smoking of tobacco. Nor that it followed it. These are two distinct processes, even though there are analogies between them. The fine powder is like a solid smoke that has not yet been volatilized; the smoke is like a powder so fine and delicate that it becomes intangible and loses its gravity. The powder may be so fine that its effects resemble those of smoke. Probably powder and smoke were used indifferently by the Indians. A case in point exists. During his fourth voyage, in 1502, when Columbus was sailing along the coast of the mainland off Cape Nombre de Dios he had dealings with certain Indians. When they saw the notary putting down mysterious signs on paper, they thought it was witchcraft and fled; but to render such spells powerless, they drew near the Spaniards again and "threw powders into the air in their direction, and they lighted these powders and saw to it that the smoke was wafted toward the Spaniards" (Las Casas: *Historia de las Indias,* Book II, Chapter xxi).

These powders and smoke of tobacco had the same magic function. Such powders could have been taken through the nose in pinches, as the dandies of Europe did later on, or by putting some small container or smooth object in the nose, like certain Negroes of the Congo, who put a portion of snuff on the blade of a knife and snuff it from this into the nose with a deep breath. But, without doubt, the use of an inhaler tube facilitates the absorption, particularly in collective ceremonies where each participant takes the powder, one after another, from a common receptacle in which it has been prepared for the rite. Moreover, the inhaler tube plays a part of a ritual nature. By means of it the consecrated powders filled with a sacred power pass into the nostrils of the officiant or the patient without being touched by anyone's hands, and any contact that might weaken the magic power of their supernatural attributes is avoided. Nevertheless, the expedient of an inhaler tube would seem to have been more necessary when it was a question of the smoke given off by burning leaves than in the case of the dry powder of inert matter.

Snuff can be taken without the use of a tube, but without a tube of some sort to convey the smoke it is impossible to smoke. In order to smoke—that is to say, to absorb the smoke of burning tobacco—one must find a solution to the basic problem, which is that of conveying the smoke from the leaves that give it off to the person who wants to enjoy it, while at the same time keeping the smoker at a sufficient distance from the fire that produces the smoke to keep him from getting burned. A separation and a union; keeping the fire from the smoker, but joining them by means of the smoke which goes from the burning leaves to the one who burns them. The essential feature of every smoking device is carrying the smoke from the fire to the smoker. There is an essential difference between tobacco in powder and in smoke, which is combustion. This gives greater mystery to the rite of tobacco, adds the new element of fire, and by means of the smoke permits of a greater range, almost unlimited and of the most subtle symbolism, in its magic-religious applications. Whereas the action of the powders is limited to their being absorbed through the nose and inducing cathartic effects of a physiological and psychological nature, the smoke of tobacco is inhaled, goes through the mouth to the vital organs, and then is expelled through the nose or mouth in whatever direction the smoker desires, toward a concrete object or toward the powers on high, in allegorical columns or spirals that can carry a message to the invisible numens. For this reason the burning of tobacco for the sake of its smoke reached a greater development than any other form of utilizing the powers of the plant. Tobacco has been used for ritual purposes chiefly because of its smoke. It is this more than anything else that gives it its subtly diabolical attributes.

The simplest form must have been censing—that is to say, the use of certain plants for religious, magic, and medicinal purposes, as is still done today by means of incense, fumigation, and burning perfumes. Aromatic or balsamic substances as an offering or to please the senses; foul-smelling or disagreeable substances to drive off evil spirits; narcotic substances to attract the spirits; special substances to aid in the treatment of diseases. . . . In the smoke, more than anywhere else, the visible manifestation of the sacred mana of tobacco is to be found. Through the smoke the sacred powers of the plant are transmitted to persons and things by their being enveloped in it. If the object of the fuming is malignant, the smoke renders its evil powers harmless; if it is powerless for a certain purpose, it isolates it or revitalizes it through such censing; if it can be propitiated, the smoke establishes a magic link of communication with it.

I have given an instance of the first case in the episode that took place between the Spaniards and the Indians of Nombre de Dios, when the latter used powders and smoke to destroy the supposed spells and witchcraft of the former. We have an example of the second case in the witch-doctors among the Indians of Brazil, who fumigate their *tammaraka* or *maracas* with tobacco smoke when about to use them for soothsaying purposes, as referred by H. Stade (*N. Federmanns und H. Stades Reisen in Südamerica, 1529 bis 1555,* edition of K. Klupfel, Stuttgart, 1859, p. 183). And there are numerous instances of the third case among the narrations of the chroniclers I have spoken of who tell how the Indians burned the leaves of aromatic plants to propitiate their idols and win their help in difficult undertakings and in the satisfaction of their economic needs, their crops, and their harvests.

It is evident, therefore, that *in illo tempore* the Indians burned tobacco leaves in their religious rites, in their magic-making, aggressive or defensive, and in their cathartic, purification rites. And once the rite of ritual censing existed, it was a natural step to think of spreading the smoke about in the collective rite and absorbing the fragrant smoke for the purpose of internal purification, of body and of soul, and to assimilate the supernatural power present in the smoke.

Smoking is the act of taking in and blowing out smoke or puffs of smoke. To smoke, the plants whose smoke was desired were laid upon a brazier or vessel and the smoke they gave off was breathed in. This is the simplest method of smoking. This is the fuming, similar to the inhaling of finely pulverized powders and of equivalent efficacy.

We should recall that in their inhalatory rites the Taino Indians put their powders in "a round, smooth, fine dish, which would not have been much more beautiful if it had been of gold or silver; it was almost black and shone like jet" (Las Casas: *Apologética,* p. 245). Certain idols found by students of archæology in the Antilles, which are now in museums, have the remains of such dishes for powders on their heads.

This simple form of utilizing the smoke was followed by various improvements devised by human ingenuity. Above all, by a rudimentary principle of economy: to make as much use of the smoke as possible. It was not always possible to have on hand the leaves, seeds, resins, or other materials necessary for producing the smoke, and they could not be prepared for every occasion with the solemn traditional rites. The plant for the fuming was a precious substance not to be wasted.

Two improvements had to be sought above all others. First, to be able to take in more smoke more easily, and, second, to make the ceremony last longer. For the first the devices for smoking by inhalation or exhalation were invented; for the second, methods of slowing down and controlling the combustion.

The substance to be consumed, let us say tobacco, was put into a receptacle of some sort (shell, gourd, wooden or clay dish), the leaves ignited, the smoke produced, and from this it was inhaled by the officiants or participants in the rite. At first it was enough to bend over the smoke and open the mouth to inhale it deeply, perhaps helping to take it in by placing the hands in front of the mouth to form a funnel, or by placing a dried pumpkin or squash shell or palm leaf over the burning leaves with the broad side down to act as a funnel for the smoke and inhaling it through an opening in the stem end; or, among fishing peoples, using the natural funnel of a large univalve (as for example a *Strombus gigas*) with the end broken or perforated. And, finally, utilizing any sort of tube, made from a cane, a feather, a bone, wood, or stone, specially perforated.

It is possible that the first inhaler tube for the powders or smoke of tobacco consisted of a length of the stem of the plant itself. (We owe this valuable suggestion to the Cuban engineer, Francisco García y Álvarez Mendizábal, who is versed in the cultivation of tobacco and the folklore connected with it.)

The stem of the tobacco plant is straight and cylindrical and has a soft white pith, which some country people call fuzz. When the pith dries, the stem is left practically hollow, and of a tubular form. Thus the very nature of the tobacco plant may have supplied the Indians with the first inhaling device they knew to draw in the smoke of the burning leaves or the pulverized powder of the leaves. This prototype of the inhaler tube must have been gradually replaced by another, imitative of the first, but made of more durable material, for tobacco is an annual, and its stem, though somewhat woody, is not too lasting unless it is specially prepared.

It is also very possible that the typical forked inhalers were likewise made of the tobacco stem, using a section of it having a natural bifurcation at one end where two branches grew out, and the tubes formed by these two shoots could be easily connected with the tube of the main stem if there were no knot or dividing wall to separate them. I have made the test myself, constructing one of these single tubes and one of the forked ones with sections of the stem of the *Nicotiana tabacum,* giving them the dimensions indicated by Oviedo, of a handbreadth in length, and with two short arms opening out to be inserted in the nostrils (Figure J).

These ramifications of the tobacco plant are rarely seen, for as a rule, the growers as they go over the plants are careful to nip them off to keep all the plant's strength in the leaves and not let it go into unnecessary growth. For the same reason the flowers and seed of the tobacco are unfamiliar to many, for the growers pinch them out when they top the plants to check their growth and send their strength into the valuable leaves.

The acceptance of the theory that the prototype of the bifurcated inhaler tube is to be found in the natural fork formed by the stem and two lateral branches of the tobacco plant itself does not affect the other considerations regarding the different uses of the plant, or the morphological variations of the inhaler tubes in question. But it does seem to add weight to the theory of the actual inhalation of tobacco smoke or powder into the nose through the ritual forked tube. The use of the tobacco plant for the manufacture of the forked device in question would seem to point to the fact that the powders inhaled through it were also of tobacco. It does not seem likely that they would pick the tobacco plant to make the inhaler of if they were not already interested in the plant itself for the sake of the smoke or powder of its leaves. The act of inhaling the tobacco in some form must have preceded the use of this device to perfect the operation. The prototype of the tubular cigarette may also have come from a piece of the tobacco stalk, by filling it from one end with a mixture of shredded leaves. This conjecture is supported by the fact that at times the country lads of the tobacco-growing regions of Cuba play at smoking cigarettes, which are made of little tubes of tobacco stem filled with the plant's shredded leaves. But there is no proof that the Cuban Indians did anything analogous. However, one must not lose sight of the "canes of smoke" used by the Aztecs, according to Bernardino de Sahagún, which were nothing but hollow canes filled with aromatic herbs. The analogy is apparent. The cane tubes might well have been tubes of tobacco stem filled with tobacco in places where tobacco was more abundant than canes. But this is merely a hypothesis.

It is not too far-fetched to suppose that the use of these simple inhaler tubes, made of natural forked tobacco stems, was adopted by the Indians not only because they were so easily made, but because, being of the same nature as the powders, the latter would lose none of their magic power by coming into contact with another matter, but, on the contrary, their original sacropotency would be enhanced. There is no doubt that these inhaler tubes shared in the sacred nature of the rite as instruments used for it, which can be easily understood in view of the magic emblems with which we know they were at times decorated, as well as because of the materials of which they were made (the feathers of flying birds, the bones of animals, the stems of certain plants, etc.) and the figures with which they were decorated (sacred animals, mysterious designs, etc.).

Besides, it is worth recalling that there was a time when a certain kind of snuff was made in Cuba by grinding the tobacco stems, which when dried became hard, and it was said that for this reason the snuff made from them was finer and better than that made from the leaves. Be this as it may, this gives us reason to believe that the Indians may have done the same.

The method of inhaling the smoke given off by the burning tobacco through a simple hollow reed or tube may have been the earliest, and was subsequently modified for ceremonial use

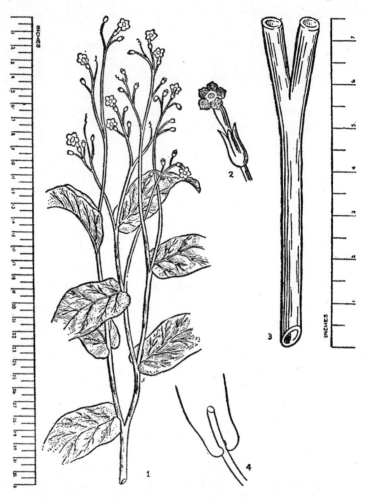

FIGURE J. (1) Upper section of a plant of *Nicotiana tabacum,* 33 inches tall, as indicated by the scale. Note the several natural bifurcations presented by the branches, contrary to the general conception of it. (2) Tobacco flower. (3) Forked tube, hollow and a handbreadth in length, as indicated by the accompanying scale, made by the author from a section of the tobacco plant portrayed here. (4) Manner in which the leaves are attached to the stem. (Drawn from life. F. Ortiz.)

through the introduction of double tubes or a forked tube, which required the special preparation of such a device and considerable work that can be explained only on the grounds of its ritual importance. The use of this inhaler tube was preserved in religious gathering-places and sacred ceremonies by force of tradition. This greater elaboration of the inhaler tube would seem to run counter to the principles of economy; but this was compensated for by the greater suitability of the device to its ends, since it could be placed in both the nostrils instead of only one, by the æsthetic element which is generally joined to all ritual acts and religious emotion, and by the addition to the tube of certain symbolic and magically animistic elements. The difference in elaboration between the single- and double-tubed inhaler is analogous in this sense to that which exists between a simple clay pipe fastened to a stem and a pipe carved of ivory, with beautiful mythical figures and ending in an amber mouthpiece. A similar process of evolution is evident in the tubular cigars of the Aztecs; the earliest were made of simple reeds, or perhaps of tobacco stems, then later of polished cane decorated with paintings and even specially prepared so that, as they were smoked, the exterior decoration became visible. Following this technological trend one comes gradually to the holder and the pipe, both the straight and the elbow or angled variety.

Another economic problem was that of making sure that the smoke-yielding materials would not burn too rapidly, and quickly become a heap of ash. It was discovered that, to avoid this, it was helpful to press the leaves tightly together. There are five methods of doing this: (a) to pack the leaves into a small receptacle or brazier with a single opening; (b) to press the leaves together, rolling them against each other to form a cylindrical mass of several leaves wrapped in another that covers them completely in the manner of a containing tube; (c) rolling the leaves as in the preceding case, but putting them in a tube of tobacco stem or some other combustible material; (d) preparing the leaves in the same manner, but inserting them in a fireproof tube; and (e) placing the packed leaves in a vessel or brazier and inserting the inhaler tube into the lower part.

By method (a) the smoke produced by the burning is absorbed without an inhaler, merely by using the hands as a funnel or employing some inhaling device. In connection with this type (a) the inhaling device came into being by itself in its different forms, monotubular or bitubular, reedlike or cylindrical, funnel-shaped or conical. By method (b) inhalation of the smoke was achieved by lighting the roll of leaves at one end and drawing in the smoke at the other in such a way that it traveled the length of the inside of the tubular covering. That is to say, the smoke was not inhaled directly from the air, at a moment when it was dispersed and evaporating, but by means of a suction current that penetrated the mass of burning leaves to the point of the flame or embers of its incandescent zone, causing them to burn more brightly for a moment and thus producing smoke that was inhaled at the very moment it was produced. In this way the smoke did not even pass through the outer air, but came directly from the flame to the mouth, without losing the aromatic breath of its supernatural power. This represented a saving of leaves, of work, of fire, of smoke, of aroma, and of magic power. By method (c) the inhalation was carried on in the same way, with the leaves burning at one end of the tube and the smoke being absorbed at the other. By method (d) the process was the same, but the tube did not burn and could be used over and over again. And method (e) represented an improved variation of the preceding in that a better recipient was made available, which could be more easily filled, held a larger amount, had a broader combustion surface, and the advantage of a narrower and longer tube, which made it easier to handle. And as the whole apparatus was fireproof, it could be used indefinitely.

So far as I know, we have no example of type (a). But it is a simple variant, a smaller version of the dish used by the Taino *behiques* in their rites, which they set before their idol or on its head.

Type (b) has been very successful. It is what we call today the cigar or Havana. It was used predominantly by the Indian peoples of the Antilles and of the Amazon, Orinoco, and Magdalena river basins.

Type (c) was the one the Spaniards found among many Indian peoples. It is considered the forerunner of the cigarette, which today is made of shredded tobacco wrapped in a little roll of paper. It was used by the peoples of the river regions of the

Caribbean, and more than any other form among the Indians of the Western United States, Mexico, and Central America.

Type (d) is the "cane of smoke" prevalent in Mexico, where the chroniclers saw it at Montezuma's banquets, at the religious rites, and in the city's great market-place. This type (d) in a way represents the combination of the single-tube inhaler with the idea that gave rise to types (b) and (c)—that is, the rolling of the leaves in the form of a twig or little cane and covering them with an outer wrapper or hard container. This type (d), when made of cane, as happened with the *acayetl* of Mexico, comes to occupy an intermediate position between the vegetable tube, made of cornhusk or tree bark, and the fireproof variety, made of some other material.

These types (b) and (c) took care not only of the economic problem of frugal smoking, but also of that of eliminating the need for receptacles, dishes, and inhaler tubes.

Out of type (d) a composite type developed that was later given the name *boquilla* (mouthpiece), which was a fireproof tube containing neither leaves nor shredded tobacco, but into which the tobacco prepared as for types (b) and (c) was placed; in this way one had a tube cigar or a tube cigarette—that is, a real tubular pipe in which the cigar stayed on the outside.

When the Indians did not have on hand the reeds for inhaling the smoke, or when in the daily round of their life they felt a need for the aromatic and stimulating smoke (perfume, Oviedo calls it) apart from the sacred gatherings and without the religious and magic ceremony of the liturgical tubes, these smokers settled their problem in simple fashion, rolling together by hand some leaves of the aromatic plant and wrapping them in another leaf, either of the same plant or of some other suitable one at hand, thus forming a vegetable and combustible tube, inside which the smoke traveled from one end to the other, from the fire to the mouth. It is in speaking of the ceremonies of the *behiques* that Christopher Columbus and the chroniclers of the Indies allude to the cohoba and the tubes for inhaling the powder or the smoke. It is in describing the travels of the Indians by canoe or overland that the Admiral discovers the leaves of the plant, and the rolls of these leaves known as *tabacos* that the natives carried with them. In this way the cigar came to be, after a fashion, an object of ordinary use to carry out a religious rite without the intervention of the priesthood with its sacramental solemnities.

However, Sven Loven goes too far when he asserts that the art of smoking among the Tainos, where it was very demotic, had completely lost its esoteric religious character. It would be more exact to say that, in addition to its liturgical and magic uses, tobacco had become an article of general consumption but without losing its religious connotation; in the same way that among the Indians, as among white men, amulets, little figures of idols, medals, and reliquaries containing bones of saints, warriors, or animals worn about the neck are common and popular, without on this account having lost their religious significance; on the contrary, in this way the mysterious efficacy of their supernatural powers is extended to all the acts and moments of everyday life.

Type (e) has been the most successful; this is the pipe, or more exactly the angle or elbow pipe, with its numerous variations and subtypes.

I must insist on the fact that these types are not serial; that is, they do not represent a process of successive morphological transitions. Except for the first, type (a), which because of its simplicity would seem the oldest and the natural prototype, the others might have come about in different ways. In any case, these genetic problems of the morphology of the methods of using tobacco have yet to be definitely cleared up.

The desire to enhance the effects of tobacco by mixing it with other more stimulating ingredients could have been satisfied through any of the five types of smoking I have indicated; the use of mixtures is possible in any one of these methods. It would seem more feasible, however, to obtain the medicinal and narcotic effects through the admixture of other substances, sometimes mineral, by utilizing tobacco in other forms, either in bulk, in liquid, or in powder.

It should at the same time be pointed out that just as in the Antilles different forms of using tobacco coexisted, the same was true, to a greater or lesser degree, in various other countries. In Mexico, for instance, we have already seen that:

"The Mexicans smoked tobacco in two ways, either rolling the leaves upon each other, to which they gave the name of *pocyetl,* or crumbling them up and mixing them with other aromatic plants such as liquidambar or *xochicocozotli,* and putting the mixture into a hollow cane, to which they gave the name of *acayetl.* They also took it in the form of powder through the nose. . . . The mass of the people, who could not afford such luxuries, sought this intoxication in the peyotl, in the seed *ololiuhqui,* which induced hallucinations, or in toadstools, *teonanacatl,* mixed with honey." (Alfredo Chavero, in *México a través de los siglos,* Barcelona, Vol. I, p. 805.)

The existence in the same country of different modes of using tobacco probably indicates the convergence of different exogenous influences on the same human group. That is to say, these different manners of employing tobacco may be due to different phenomena of cultural diffusion, and not to the development of a primary type into others in an inevitable sequence; but, I must repeat, it is impossible to arrive at any definite conclusion.

The only thing certain would seem to be that even though throughout America some of the different methods of taking tobacco for religious, ceremonial, narcotic, medicinal purposes, or simply for pleasure, were employed simultaneously, nevertheless in the complex culture of tobacco several territorial areas, characterized respectively by the instrument or typical fashion of using tobacco that prevailed in them, can be defined. An interesting map of this sort was drawn up by the ethnographer Wissler (Clark Wissler: *The American Indian,* 2nd ed., New York, 1922, Fig. 6). It would be necessary to indicate in this American geography of tobacco those zones where the powder of the ground leaves was used and those where the plant was taken in infusion. But it does not seem prudent flatly to deny the existence of the custom among the American Indians of inhaling powdered tobacco, by means of nasal tubes or without them, and pure or mixed with other ingredients, nor should the fact be lost sight of that Indian tribes still exist that take tobacco in brews ritually prepared, which they take with them to their ceremonies, as among the Arawaks of Colombia, according to Mason; and among the Caribs, according to Karsten, and other tribes of South America.

It might be possible to attempt to draw some conclusion with regard to nomenclature. From what has been said, it is evident that the word *tabaco* has nothing to do with Tabasco, a city of

Mexico, nor with Tobago, one of the islands in the Lesser Antilles. Both cases are instances of ingenuous and mistaken folklore etymologies, even though they have been accepted by authors of great repute.

According to Oviedo (Book V, Chapter ii), the Indians applied the term *tabaco* indistinctly to the "sticks or tubes with two openings" and to the "reeds or canes"—that is to say, to the devices for inhaling the powder or smoke, whether single tubes or the bifurcated or forked ones—"and not to the plant or sleep that comes over them (as some thought)." But in another place (Book V, Chapter v) the same chronicler refers to the "cohobas or smoke that the Indians take, which they also call *tabaco,* as was said in Chapter ii." But in the passage referred to, Oviedo also writes: "certain smokes that they call *tabaco,*" and farther on: "they take the smoke or *tabaco,*" giving both these terms as synonymous with *tobacco.* These texts seem to reflect the use of the word *tobacco* with different meanings. Which is the original one?

Perhaps the word *tobacco* among the Tainos of the Antilles discovered by the Spaniards already had a derived meaning and was of remote origin. The word *tabaco,* according to Max Uhle, seems to come from *taboca,* a Tupi Indian word meaning "a certain variety of reed or cane." In this case the word would not really be Taino, but would come from the Tupis of the South American continent, and was transmitted by them to the Arawaks and Tainos. According to Ernst (loc. cit.) *taboca* in Guarani means "a tube made of the bone of a tapir used to inhale certain powders."

From these antecedents it would seem possible to infer with a certain amount of assurance that *cohoba,* whatever may have been its primitive meaning, had become the Taino name of a plant, or at least of the powder made from its leaves, or of the powders with which these were mixed, and of the function, ceremony, or sacrifice of inhaling these powders; and that *tabaco* was the name of the tubular instrument for inhaling the powder or the smoke drawn in with the breath, and also, perhaps more precisely, the name of the roll of leaves that today the Cubans call *tabaco* by antonomasia and other people *cigars.*

We must agree here that if the cohoba was some plant other than the Nicotiana, and if *tabaco* was not the name of a plant but of an inhaling instrument, then we do not know what the Tainos called the solanaceous plant to which we give the name of tobacco today.

In any case, the word *tobacco* prevailed for the plant and for the cigar; and the fumings, the magic smoking, and the word *cohoba* of the Tainos were forgotten, as have been the many other words of different origins that were applied to the Nicotiana plant in those centuries in the Indies and in certain countries of Europe. When, in the eighteenth century, Juan de Solórzano published his treatise *Política Indiana* (Madrid, 1736) and expressed his opinion concerning the legal regimentation of the use of tobacco, after quoting various authors who were in disagreement, he wrote these paragraphs (op. cit., Book II, Chapter x, Vol. I, p. 102) summing up the matter:

"And this which has been said of coca, that Indians should not be employed in forced labor to plant it and cultivate it, should be observed and practiced with great zeal in the cultivation of another plant, which was first found in the Windward Islands, and later spread through the West Indies, and even through the other provinces of the world, whose name is tobacco, although there are those who call it *Peto,* others *Nicosio,* and others *Yerva Real."* (*De his & alijis nominibus Tabaci, & eorum causis vide omnino Antón de León.*)

But the name *tabaco,* applied to what the Spaniards later called *puros* and cigars, did not originate among the Spaniards with that meaning. There is no doubt that it was the Tainos of the Antilles themselves who, according to Oviedo, called the tube for inhaling the smoke *tabaco,* and in any case gave that name to the rolled cigar that has come down to us, for Oviedo himself states this in a passage that, because it does not refer to the Indians of the Antilles, except for an incidental allusion, seems to have escaped the notice of the historians of the Antilles. Oviedo writes as follows of the entertaining celebration given by the chieftain Nambí of Nicoya one summer night in 1529. I quote the text in its entirety because of its pertinent bearing on our theme:

"One Saturday, the 19th of August of 1529, in the village square of Nicoya, Don Alonso, the cacique of that province, also known as Nambí, which in the language of the Chorotegas means dog, two hours before nightfall on one side of the square as many as one hundred and eighty Indians began to sing and dance an *areito;* these must have been the common people or populace, for on another side of the square sat the cacique, with great signs of pleasure and enjoyment, on a *duho,* or little stool, and his leading men and as many as seventy or eighty other Indians, each on his stool. And a girl began to bring them in little gourds, like bowls or cups, *chicha,* or wine that they make out of corn, which is very strong and a little sour, which is of the color of chicken soup when a couple of egg yolks are beaten up in it. And as soon as they had begun to drink, the cacique himself brought a bunch of *tabacos,* which are a handbreadth in length and the thickness of a finger, and are made of a certain leaf rolled up and tied with two or three threads of thin fiber, which leaf and the plant thereof are raised by the Indians with great care for the sake of these *tabacos,* and they light them a little at one end, and they burn slowly (like a stick of incense) until they are all consumed, which takes a day; and every so often they put the end that was not lighted in their mouth, and drew the smoke in for a short time, and then took it out, and kept their mouth closed and held their breath for a while, and then breathed and that smoke came out through their mouth and nose. And each of the Indians I have mentioned had one of these rolls of leaves, which they call *yapoquete,* and in the language of this island of Haiti or Hispaniola is called *tabaco.* And they keep on drinking, men and women coming and going with that beverage, and between times they bring other gourds or big cups of boiled chocolate, which they are in the habit of drinking (but of this they take only three or four swallows, passing it around, and first they take a swallow of one, then of the other, and in between times they take those smokes, and some beat a drum with the palms of their hands while others sing), and they kept on that way until after midnight, by which time most of them had fallen to the ground unconscious, drunk, like topers. And as drunkenness acts in different ways, some seemed to be sleeping without moving, others wandered about crying, others yelling, and others falling over themselves. And when they were in this state their wives and friends or children came and took them away to their homes, where they slept until noon of the next day, and some until the following evening, depending on how much they had drunk." (Oviedo: *Historia General y Natural de las Indias,* edition of Madrid, 1855, Vol. II, pp. 96-7.)

Oviedo's description, though brief, is complete. He mentions the music, the dances, and the singing, the steady libations of alcohol, the frequent quaffings of chocolate (served, like the *chicha,* in individual gourds), the smoking of tobacco in between, the state in which the cacique, his leading men, and the other prominent people sat, each on his own chair or *duho,* apart from the lower classes. One would say the account was from the society column of a modern newspaper if the language did not reveal the period, and if it added to the names of the guests the customary flattering adjectives. Those Indians were Catholics—that is, they had been baptized—but they did not lead a very Christian life, as still happens among people and periods far removed from those newly converted Indians. And if the account does not make any allusion to the stylish attire of those present, it is because the chronicler had already alluded in earlier paragraphs to the "beautiful headdresses and breeches" and the "fancy doublets, embroidered and of many colors," which the Chorotegas wore on such occasions and which made them look "as well dressed as any German soldier." While in the "main square" and in the cacique's palace the guests were fittingly regaled and the crowds took their pleasure in the dances, the *areitos,* and the *chicha,* the splendid host, Don Alonso Nambí, treated his Spanish guests to cigars, rustic in shape, and with "bands" of fiber, but real cigars, made of carefully tended leaves, of rich aroma and fine quality, even though they were called *yapoquetes* and were not Havanas in either origin or quality.

And the words of the reporter of that festival in 1529 are explicit: "in the language of this island of Haïti or Hispaniola is called *tabaco.*" Oviedo says "of this island" because he was there when he wrote his history, which he concluded in 1548, a little before his restless life came to an end. Father Las Casas is also explicit: "these rolls they call *tabacos,* the middle syllable long" (*Apologética,* p. 181).

If the term *tabaco* applied to the cigar is of Indian origin, the term *cigarro* is of Spanish origin, used to denote the "cigar of rolled tobacco," discovered among the Cuban Indians, which because of its shape and coloring resembled the large locust— *cigarro* or *cigarrón*—that abounded in Andalusia. And the word *cigarro* of the Andalusians was adopted by the merchants of Seville and Cádiz, who acted as agents of the English and French, who put the word into circulation in the rest of Europe. When the French Dominican Father Labat mentions the *cigarros* smoked by the Spaniards, he translates the word *cigales,* which means locusts. Romey employs a variation of this Spanish etymological version, according to which the Spaniards who had been in the Indies brought back to their native land the seeds of the Nicotiana plant and sowed them in their gardens, either for ornamental purpose or to make use of their leaves, rolling them into cigars; and as the gardens were known as *cigarrales,* and the *tabacos* of one *cigarral* were compared with those of another, the term *cigarro* came into use (José Vilardebó: *El Tabaco y el café,* Havana, 1860). This etymology has found favor among writers who lack a knowledge of Spanish (Carl Werner: *A Textbook on Tobacco,* New York, 1914, and W. W. Young: *The Story of the Cigarette,* New York, 1916); but this interpretation is farfetched, improbable, and unnecessary. *Cigarral* was not the general term for garden in Spain, but was and is a localism of the province of Toledo, is not even Andalusian, and refers to an enclosed pleasure garden. The origin of the word *cigarro* need be sought no further than in the metaphorical comparison of the cigar, by reason of its size, shape, and color, to certain insects known as *cigarros* or *cigarrones,* which are abundant in the Andalusian countryside and also in the Indies, to judge by Father Juan de Torquemada, who says of the Indians of Honduras that "they ate *cigarros,* winged ants," and other vermin (*Monarquía Indiana,* Part I, Book III, Chapter xli). It may be that this book is responsible for the mistake of attributing the word *cigarro* to the language of the Mayas (G. Montandon: *Traité d'ethnologie culturelle,* Paris, 1934, p. 288).

I should point out, in conclusion, that Oviedo, the author of the *Historia General y Natural de las Indias,* which deals at length with tobacco and its peculiarities, says nothing of it in his *Sumario de la Natural Historia de las Indias,* which he himself published in Toledo around 1527, several years before two editions of the first part of his great work were brought out in Seville (1535) and later in Salamanca (1547). This led Wiener (loc. cit., Chapter i, p. 114) to believe that this detail was clear proof that tobacco was not known in America (*sic*) until some time after the discovery. But it only proves that tobacco, although its use spread quickly and it was noticed at once as something curiously exotic, did not seem to Oviedo to warrant the same attention as other discoveries of the New World, since it did not possess then, in the first third of the sixteenth century, the great economic importance it was later to acquire as this custom of the Indians became adopted by the Europeans on the other side of the Atlantic, thus giving rise to a profitable trade.

7

On the Transculturation of Tobacco

The history of tobacco affords an example of one of the most extraordinary processes of transculturation, by reason of the rapidity and extent to which the use of this plant spread as soon as it became known to the discoverers of America, the great opposition it met and overcame, and the very profound change operated in its social significance as it passed from the cultures of the New World to those of the Old.

This contributed to the oft repeated opinion that there was a strong diabolical element in the nature and past of tobacco. Grave theologians affirmed this, and many who were learned in demonology were inclined to believe in the intervention of the powers of darkness in the affairs of tobacco and in the episodes of its history. Even modern philosophers of the "as if" school might even accept the demoniacal hypothesis as admissible for lack of another experimentally demonstrable, for in the swift and turbulent spread of tobacco throughout the world everything has happened "as if" Satan himself had been in charge. Tobacco "ran through the world like wildfire," says one historian. As though tobacco had been fire out of hell that burned noiselessly inside the brain to ignite the spirit. To what factors were these strange phenomena of the spread of the Nicotiana plant due?

No one knows how the use of tobacco began or who was its real discoverer. Neither do we know the name of those who discovered fire, tools, clothing, agriculture, the domestication of animals, the wheel, writing . . . the most important elements of human culture. It is idle to conjecture about it. As Lord Raglan has said: "To hit upon an invention is in itself of slight importance; for it to be put into use a society with means and the stimuli for its adoption is necessary" (*How Came Civilization?* London, 1939, p. 43). What, then, were the stimuli and the means in the societies of the American Indians, among whom tobacco originated, that gave to their invention such enduring force?

As in every other invention, there had to be certain natural and certain social factors that were closely linked. The natural factors that gave rise to the use of tobacco by man were the unique physiochemical properties of the plant and its physiological effects on the human organism. If man began to use tobacco and became habituated to it, it was neither because of its food value nor for economic reasons, but because of his experience of certain effects produced by the external action of the plant. These effects were various and can be reduced to two: the sensual pleasure derived from it, and its medicinal utility.

Tobacco was pleasing to the senses and relieved nervous tension. Moreover, tobacco cured real or imaginary ailments. These simple natural factors suffice to explain the use of tobacco among any people, whether primitive or civilized. But these natural factors did not manifest themselves among the American Indians in so simple and clear-cut a fashion, nor isolated in each case for their objective consideration. These motives were always involved with a number of other factors of varying kinds. When the smoker today enjoys and takes pleasure in tobacco, his motivating and almost always only reason for this individual act of smoking is his enjoyment. Pure hedonism. When the Cuban countryman puts a tobacco leaf on his temples to cure a headache, he is applying an old folk remedy, without further implications. Pure utilitarianism. But when the Indian used tobacco to give him pleasure or as a medicine, these reasons did not appear to him simple or direct, but of great complexity.

Those very stimuli of sensual pleasure or medicinal value which tobacco afforded the Indians cannot be said to have been merely individual. For the Indians, no matter what selfish advantages they looked for from the uses of tobacco, these were only phenomena related to the tribe and to the mysterious forces the plant contained. Tobacco led the Indian to act upon these, and the social group, knowing the transcendent powers of the plant, both benevolent and baleful, governed its use for the benefit of the collectivity. Tobacco was more than a source of pleasure; certain aspects of its use had nothing agreeable about them and were, on the contrary, repulsive. Aside from the merely sensual incentive, and even in combination with it, the Indian experienced the magic-religious stimulus that led him to use tobacco as a magic producer of satisfaction, as a medicine, as a precaution, as a prayer, as a link with the supernatural. And at the same time the Indian felt that there was a social bond in the use of tobacco. Tobacco was not a whim; it was almost always and in almost every way a duty, for the individual could not use tobacco without taking account of the social tradition that determined the occasions and the ceremonies for its use. If the individual act of smoking tobacco was occasionally free, as is today the act of imploring the favor of the unknown by kissing a reliquary, the form of the rite was not a matter of indifference, for to render it efficacious it had to be carried out according to certain prescribed forms, with certain utensils, at certain times, with the body in the proper position, and, at times, after indispensable and painstaking preparations. It does not matter that the Indian enjoyed the use of tobacco; that has no bearing on the fact that this was a social duty. The Indian liked what he did when he smoked tobacco ritually, although he did not do it just because it gave him pleasure. And at times the rite was performed without any sensual satisfaction, as in its use as an emetic. But, as a rule, duty and pleasure were present together in tobacco, as in the dancing and singing of the *areitos*.

Religion among primitive peoples is "the cement of their social life," to use the phrase of Malinowski, and tobacco linked all the individual life of the Indian to that of his society. Tobacco was the inseparable companion of the Indian from birth to death; the Indian lived enveloped in the spirals of tobacco's smoke, as the ceiba is enveloped by lianas. Tobacco was socially an institution. It satisfied not only certain individual desires of the Indian, but also those of his group. Among the Indian people tobacco formed an integral part of their mythology, their religion, their magic, their medicine, their tribal ceremonies, their statecraft, their wars, their agriculture, their fishing, their collective stimuli, and their public and private habits. . . . The use of tobacco or the utilization of its power was not a superstition or a heresy, but an orthodox, unchanging, religious institution. The rites connected with tobacco were obligatory social acts, in the performance of which, individually or collectively, the whole human group was interested, and demanded their observance on the occasions designated by the tribal conscience and in keeping with the exact and sacred forms established by tradition.

As it was a social institution, tobacco among the Indians was connected with their economic life as well. Like their songs and dances, tobacco to them was a part of the religious-social rites bearing upon the fulfillment of their most important economic activities, such as food, hunting, farming, the weaving of a net, or the fashioning of a canoe; but tobacco in itself was not fundamentally an economic product.

Among the Indians of the Antilles tobacco was, without doubt, an element of their culture in which the sense of the supernatural predominated. Tobacco impregnated their whole system of magic and religious rites, as I have already indicated with numerous quotations, data, and commentaries (see Part II, Chapter v). For this reason the European conquistadors, who were Christians, upon observing the sacred attributes of tobacco, formulated the idea of its diabolic attributes. When the Catholic clergy witnessed the—to them—grotesque rites of the Indian *behiques,* in which they made great use of tobacco, they decided that as this was something religious but not orthodox, the sacred character attributed to it must be the work of Lucifer's wickedness. In the magic-religious rites of the Indians tobacco was the aromatic smoke raised to their deities as a sacrifice or as a symbolic message of the longing for the remission of their sins on the part of the faithful, like the incense used by the Christians in their temples; or it was the divine essence that the communicant consumed, taking it into the most profound recesses of his being like a purifying gift of grace, like the drinking of the consecrated wine to receive the benefits of the sacrament; or it was the mana or sacro-potent force that was blown out in puffs to operate its mgic, like the exorcism performed by the sprinklings of holy water. Father Gerónimo Mendieta, in his *Historia Eclesiástica Indiana,* devotes several chapters to a study of these "intimations" of the holy sacraments possessed by the American Indians. It is he who tells that among the Aztecs the tobacco plant was connected with "some sort of communion or receiving of the sacrament" and was held to be "the body of a goddess known as Ciuacoualt." The complete name of the goddess was Chalchilmitlicue, and she was the wife of Tlaloc, the rain god, who produced clouds by blowing smoke toward the sky. The Jesuit, José de Acosta (*Historia Natural y Moral de las Indias,* Seville, 1590), speaking of the *teopatli* or "divine balm" of the Aztecs, which was composed of tobacco pounded in a mortar with ashes of spiders, scorpions, salamanders, and snakes, says this was an invention of the devil's to imitate the holy oil used by the Catholic priests in the sacraments. It was this same "rain god" whom the Indians symbolized by a cross, as the Spaniards who accompanied Juan de Grijalva observed to their amazement, particularly their chaplain, Juan Díaz, when they reached the island of Cozumel on their first expedition from Cuba to Yucatán (Bartolomé Leonardo de Argensola: *Conquista de México,* Chapter iii).

Tobacco was also a sacramental method of communion that linked human beings to each other and to the divine powers in bonds of peace and solidarity, like the sealing of an oath with an exchange of blood and drinking wine from the same cup. Perhaps this was the purest religious expression of tobacco's social significance. And also the most widely extended, for it was to be found among the Indians in connection with all the uses of tobacco. It is believed that the natives of Guanahaní offered tobacco leaves in token of friendship to Christopher Columbus. This was the first relation between America and Europe, the very day of the discovery. Certain Indians of South America offer one another in friendly exchange the juice of tobacco, which they carry with them in containers. Others exchange chews of tobacco. The Chorotegas Oviedo spoke of, and probably the Tainos, offered their guests cigars in sign of friendship. The Indians of North America sealed their treaties and alliances by smoking from the same pipe, "the peace calumet."

All these ritual practices of the Indians seemed sacrilegious to the Christians, who believed that these were carefully thought-out schemes of the devil to deceive the natives of the New World. The Spanish priests were of the opinion that the devil, knowing that one day the religion of the true God would be preached in America, took the precaution to suggest to the Indians certain beliefs and rites, ridiculous and grotesque, but similar to the Catholic, to confuse them and delay their approach to the Church, which was the custodian of the truth. Thus the devil would have an increase of profits, which in the infernal economy, as calculating, insatiable, and cold as that of human capitalism, means a greater and uninterrupted supply of souls for the fires of hell. This was how the Catholic priests interpreted certain rites and practices of their Indian colleagues, such, for instance, as the sacramental communion of the Aztecs, eating a certain god represented by an idol made of bread; the confession of sin in auricular, eliminatory, and expiatory form; the mystic ecstasies of supernatural possession; the adoration of images and even the symbol of the cross, and other commonplace ritual acts such as the use of incense, aspersions, and so on. It was very natural, therefore, that when they observed the Indian rites in connection with tobacco they should have considered them the inspiration of the devil, who, moreover, communicated with the Indians in their states of trance or ecstasy, and even talked with the Spanish priests themselves when, with prayers, the sign of the cross, and holy water, they exorcised the devils from the body of those they had taken possession of; they rushed out leaving a foul odor and howling "like lost souls." But, notwithstanding, the white conquerors could not always resist the temptation of consulting the priests of the conquered and taking part in their solemn practices, either out of curiosity or because of sinful sensuality. And especially when all medical treatment and the virtues inherent in sacred relics, prayers, and incantations had failed to rid them of some stubborn malady, they resorted in desperation to the spells of the Indian witch-doctors, trading the problematical future salvation of their souls for the relief of their immediate ills.

Every believer, and especially the officiant, affirms that his religion, whichever it may be, is "the only true faith," and that all other beliefs are mere superstition inspired by the spirits of darkness. Every candid priest sees a cursed witch-doctor in his colleague of a rival religion. The *behiques* were foul wizards, devil-inspired soothsayers, to the Catholic clergy, just as the latter seemed to the Indians white wizards who could foretell eclipses, make paper talk, grant victory to the invaders from across the sea, and achieve their cruel abuses with impunity only through the force of their magic, which was "the more powerful." In the atmosphere of religious transculturation, elements of the vanquished religion survived for a long time in the form of witchcraft. The gods do not die all at once. Even today, among the Christians of Europe, countless superstitions dating back to the most remote paganism still exist, which are regarded as witchery or black magic. In the Indies the beliefs and rites of the natives could not be destroyed with the same ease as their idols; and the supernatural beings of their mythology, rechristened with the names of saints or devils, continued their apparitions, miracles, and impishness and often talked with the new converts, with the unregenerate heathen, and with the fathers engaged in evangelical labors, many instances of this sort being recorded in the annals of the religious orders and in the reports of the missionaries. Even in Cuba, in the year 1682, there is the astounding case of the priest of the parish of San Juan de los Remedios, an officer of the Inquisition, who by his powerful exorcism was able to "drive out one of the many devils of which a negress of this city, by name Leonarda," was said to be possessed, "who said he was called Lucifer," and witnesses to this expulsion were the two mayors of the city, as they testified under oath and before a notary (Valdés: *Historia de la Isla de Cuba,* p. 97).

If the devil was not an unfamiliar personage in the everyday life of the Spaniards in those times, neither was he in the Indies during this period of great mystic fervor, when every attempt was being made to convert the natives to the religion of their conquerors. And if Most Catholic Spain, and even Rome itself, were full of witches and those possessed of the devil, astrologers and soothsayers who, despite the penalty of excommunication and the terrors of the Inquisition, were consulted by the victims of outrageous fortune, sickness, or love, the whites in these regions of the Indies had no hesitation in recurring to the *behiques,* with their ceremonies, prophecies, and charms, when they thought they might secure some benefit, satisfaction, or consolation through their mysterious power. The Europeans began to use tobacco, to inhale its smoke, snuff its powder, "bewitch" themselves with its fumes, and even swallow its infu-

sions, knowing full well that this was a sinful practice, a departure from orthodoxy, a heresy against their traditional teach-
ings, an act of daring for which they would be held responsible
—in a word, falling into the snares of Satan. When Father
Mendieta speaks of tobacco as a "sort of communion or receiving of the sacrament" of the body of the goddess Ciuacouatl,
he says of this plant of tobacco: "For this reason, as it has certain medical properties, it should be regarded as suspect and
dangerous, especially in view of the fact that it makes the person who takes it lose his senses and judgment."

To the Spaniards, as afterwards to the other invaders of the
West Indies, tobacco was nothing but witchery, a device of the
devil, but this did not hinder them from using it. Perhaps the
sacred and heterodox nature of its mysterious workings was tobacco's principal attraction for the Europeans who discovered
it. Unquestionably, there has always been a temptation about
all magic, just as there is always a pleasure in sin. Moreover,
there was an element of truth about the witchcraft of tobacco.
The devil's best snare is a half-lie, or a half-truth, which is the
same thing. The white Christians noticed that in spite of the infernal origins of tobacco, and perhaps because of its very connections with the powers of darkness, the use of the plant produced certain really agreeable and beneficial effects; at times it
cured them of some ill and at others gave them, through its
strange enchantment, the benefit of a comforting illusion; but,
besides, they always noticed a certain pleasant satisfaction of
the senses and, above all, a gentle, delightful well-being of the
spirit, as though a kindly angel had entered into them for the
time being, bringing hope or resignation, or a mischievous little devil who infused them with new energies and blew into
them new life.

Likewise, as occurs whenever religious proselytizing is
combined with political aims, the missionaries often tolerated
certain practices of the natives. The most ostensible and idolatrous manifestations of the pagan cults were destroyed, the images, temples, and priests were burned, but certain rites that
could be interpreted as ingenuous or allegorical acts, such as
festivals, confessions, penitences, floral offerings, perfumes,
music, songs, dances, and similar acts, were winked at.

At the same time the original sacred character tobacco had

possessed for the natives began to disappear among its new users, the immigrants from overseas. Even for the Indians the ritual connotations of tobacco must not have been the same in the important tribal ceremonies and in the grave moments of therapeutic or soothsaying magic as in the frequent occasions of daily life when, consciously or unconsciously, the frivolous sensory or stimulant pleasures of tobacco predominated, in the reasons for its use, over the metaphysical implications of the liturgy. Among the Christian conquerors, despite their strong religious susceptibility, the early interdict under which tobacco was placed owing to the Satanic and supernatural qualities attributed to it because of its use by the *behiques,* the "priests of the devil," soon came to be disregarded. In keeping with the beliefs of the conquerors, that same wine which ordinarily made those who imbibed it drunk and filled them with mysterious and ineffable delights became converted into the blood of the Redeemer and the essence of holy grace by virtue of the magic words of the Catholic priest. And despite this belief, those who held it sinned greatly in their profane moments, seeking sensual pleasure and refreshment in that alcoholic nectar, at times to the point of drunkenness, without thinking that the simple sacramental rite of their priests sublimated this wine to the absolute. And they had no scruples about sinning in the same way with the tobacco of the Indian rites, giving themselves over to its use and even abuse in order to enjoy those sensory pleasures and psychic stimulation without worrying themselves over the fact that a few words and esoteric gestures of the *behiques* converted the tobacco into an infernal substance of Christian abomination.

When, shortly after the discovery, the shrines, images, and accessories of the Indians' cult were burned in the Antilles, and not a few priests and prominent men with them, and, above all, when the Indians of the Antilles disappeared, as happened in a few decades, the religious significance of tobacco began to die out, the rites of cohoba, the inhaler tubes, and the dishes for the powders were lost, and the solemn practices of healing magic ceased to exist, except in certain individual applications and in the reminiscences of its formulas in the folklore of the countryside. Nevertheless, tobacco persisted and its use spread more and more among the new settlers of the Antilles. Not even among the Christians did the temptation of tobacco, deriving from its diabolic nature, lose its power, but to explain the extraordinary spread of tobacco among the white men, other factors based on the real value of the plant must be taken into consideration: the fact that the Europeans considered it adaptable to their customs, certain curious social repercussions I shall soon point out, and, above all, the new economic, commercial, and tax-yielding significance tobacco acquired in the white man's civilization. Tobacco, which, aside from its physicochemical properties and its individual physiological effects, formed part of a social structure of a predominantly religious character among the American Indians, took on among the European Americans and later among other peoples a structure that was principally economic by reason of a very curious, swift and complete phenomenon of transculturation.

The originally religious character of tobacco determined some of the special circumstances of its transfer from the Amerindian culture to that of other peoples. Tobacco was first feared by the invaders of America, or regarded with suspicion. Those mysterious practices with herbs and fire, those powders that made men lose their senses, those smokes that gave new life, those vomitings that cleansed body and soul at the same time, were things toward which the white Christians outwardly displayed doubt, contempt, and repulsion, although secretly tobacco attracted them and made them sin. Tobacco was taboo for them; it was "a thing for savages" and "a thing of the devil." This explains why the use of tobacco spread first among the lower ranks of the new settlers of the Indies rather than among those of higher social standing.

The Negroes adopted the use of tobacco from the Indians before the whites did. They lived closer to the natives, so much so that at times they joined forces against the whites and fled together to the hills to live free. The chroniclers clearly reflect this spread of the habit of using tobacco among the Negro slaves. This occurred not only in the Indies, but among the many Negroes who lived in Seville in the sixteenth century. Dr. Monardes of Seville, in his famous treatise on Indian medicine, speaks of tobacco without mentioning its use by the whites; but he says that the Negro slaves had become so habituated to the plant that they made themselves drunk on it and unfit for work, and for this reason the masters burned the leaves and plants whenever they came upon them, and the Negroes planted them again in deserted and out-of-the-way spots. Monardes tells of having seen Negro slaves in Seville drunk on tobacco and senseless, just as in the case of the Indians.

This is in keeping with the rapidity with which all Africa received the tobacco of the Indians of America. It was introduced into the north of Africa from Portugal in the first half of the sixteenth century, and the Jewish traders operating in the Mediterranean Basin, and especially on the southern coast, carried it to the peoples of the Levant. R. Harcourt in his book *A Relation of a Voyage to Guiana* (London, 1613) speaks of the widespread use of tobacco "more than anywhere else among the Turks and in Barbary."

The spread of tobacco through the region of Ethiopia was even more extraordinary. Tobacco entered the Dark Continent and fanned out rapidly, not only along the coast, where the white traders had their posts, but all through the interior, the seeds of the plant, along with the habit, passing from tribe to tribe and from witch-doctor to witch-doctor, even before the trader appeared to introduce the agreeable habit and thus create among the Negroes a new necessity by which he profited. The adoption of tobacco by the Negroes of Africa proceeded so swiftly that the Europeans at the beginning of the seventeenth century considered it something native to the Africans, both to the Moors, those dark-complexioned people of the north, and to the blackamoors of the sub-Sahara region. In a poem by Brathwaite he makes the poet Chaucer on his return from Olympus say: "Ye, English Moors," thus upbraiding the smokers of London for falling into this "late Negro's introduced fashion." And in England the sign of the tobaccionist was the statue of a little Negro, with a big cigar in his mouth or a roll of tobacco under his arm, as can be seen in the illustrations of books dealing with customs of the early seventeenth century.

Lord Raglan has written that after the introduction of tobacco into Africa, probably by the Portuguese in the early part of the sixteenth century, a century and a half later the Dutch found all the tribes of South Africa planting and smoking tobacco. "I have seen," adds Raglan, "Negro villages on the frontier of Sudan and Uganda whose sole agricultural product from time immemorial had been and was tobacco for trade and exportation" (op. cit., p. 68). And this same English sociologist says something even more significant bearing upon the transculturation of tobacco from America to the African continent: "There is not a single cultural element common to all the territories and peoples of Black Africa with the single exception of tobacco" (op. cit., p. 152). Another English ethnographer has said: "In all Central Africa tobacco grows wild" (H. Ward: *Chez les Cannibales de l'Afrique Centrale,* Paris, 1910, p. 270).

The Indians' characteristic manners of using tobacco are to be found in Africa. There are certain examples of curious forms of transculturation. I have mentioned the fact that the Negroes continued to mix powdered tobacco with an alkali of some kind to intensify the effect of the nicotine, just as the Indians did. The Negroes of central Congo mixed it with the ash of a certain hard wood (H. Ward, op. cit., p. 271). In certain regions of Africa the tobacco leaves were soaked in urine before being rolled into cigars (H. Ward, ibid.). In this they were only following an old sailor's custom, which as early as 1616 was recommended to English smokers (Gervase Markham, quoted by Brooks, Vol. II, p. 15). In the fabrication of pipes African art found a broad field, not merely for the satisfaction of the æsthetic tastes of the individual smoker, but to do justice to the social importance of the pipe for collective use. Each tribe of central Africa gave its pipe a special and distinctive form (H. Ward, ibid.).

A certain phenomena of the transculturation of tobacco among the Negroes is particularly interesting. In central Africa a very primitive, original, and ingenious type of pipe was found that, so far as I know, had no counterpart among the American Indians. According to H. Ward (loc. cit.):

"Sometimes the native Negroes dug a little hole in the ground and buried a long plant stalk in it in such a way that the ends projected from both extremities of the hole. The Negroes filled in the hole, stamped down the earth well, and gently pulled out the stalk so that when it was withdrawn it left a little curved tunnel in the space the stalk had occupied. At one end they placed a tobacco leaf with a live coal that kept it burning, and the smokers, one by one, squatted down and took a puff from the other end of this subterranean tube. This is the method of smoking used by the smokers when they are making a trip, in this way avoiding the necessity of carrying along a pipe and tobacco leaves, which can be found within easy reach wherever they go."

This highly original underground pipe of central Congo would seem to be an invention or a "reinvention" of the Negroes, with certain Indian touches. This example proves that the basic element of the pipe is not the receptacle in which the tobacco is burned, but the tube that conveys the smoke from the fire that produces it to the mouth; and that the economy of primitive peoples knew how to simplify the ritual methods of smoking, adapting them to the circumstances.

It is not surprising that the Indo-African transculturation of tobacco should have taken place in this way if one bears in mind the proximity of the cultural level of the Negroes, both the natives of Africa and those enslaved in Spain and America, to that of the Taino Indians of the Antilles. The transculturation of tobacco from the society of the Indians to that of the African Negroes was much easier than between Indians and whites. As I have said, in the culture of the Indies tobacco was a religious institution that took many forms; the cultural and functional complex of tobacco was deeply rooted not only in the life of the Indian as an individual but in his social life as well, his religion, his philosophy, his science, his medicine, his art, his statecraft, his war, his agriculture, his family, and so on. This was not the case, to be sure, with the Negroes and still less with the whites, for whom tobacco and its uses were novelties. But the Negro, even though he lacked in the Antilles those complicated social mores he possessed in his own country, and was completely severed from his native society and without the bonds and protection of its institutions, nevertheless, because of his traditional culture, was closer to the Taino Indians than to the white Europeans. And even though the Negroes in Africa did not know tobacco, or the custom of smoking it, they quickly grasped, by easy analogies, its religious sense, its magic significance, its use as a narcotic or stimulant to induce states of supernatural possession or in the magic rite of physiological and spiritual catharsis. And while they did not incorporate tobacco into their rites, which were archaic and did not admit of exogenous innovations, they had no hostility toward it. In the Negroes' conception, these American leaves which had such power and could afford such pleasure were, without doubt, from a supernatural and wonderful plant, and worthy of a strange deity whom they had not known in Africa, but who was not incompatible with the gods of their own pantheon. In the eyes of the Negroes, tobacco was not the plant "of the devil," as it was to the whites. For this reason in certain African tribes today ethnographers have found evidence of certain religious syncretisms regarding tobacco, though they are not of great importance and have not become an indispensable element in the tribal institutions.

The Negroes must have understood more easily than the whites the expurgatory sense of the Indian tobacco rites, because analogous rites of this nature were not unknown among them. There is no lack of folkloric vestiges of these escatalogical beliefs in the habits of the Antillean Negroes. For example, in the Negroid island of Guadeloupe during the Easter rites of resurrection the women and children immerse themselves in the sea, the rivers and pools at the pealing of the first Hallelujah. "Glory, glory!" they all shout. And they all run about throwing water on one another's head. Why? Because, according to them, this is a simple way of getting out of going to church to confess their sins. The ablutions of Holy Saturday, performed at the very moment the bells start ringing, cleanse the soul and drive out or drown the devils that torment them throughout the year. (Thérèse Herpin: *"Pacques Tropicales,"* *Journal des Voyages,* April 4, 1929.)

In Cuba similar beliefs of Negroid origin have been preserved. Among these are the "cleansings" or expurgatory rites of the Yoruba Negroes, the household ablutions performed by our lower classes on the last day of the year, when the floors, the walls, and the furnishings are scoured and the dirty water is thrown into the street to carry away the *salación* or bad luck, and other practices of the same nature. The Negro *ñañigos* (the members of secret Negro societies) of Cuba have a rite that is directly related to smoking, or, more exactly, to one aspect of it. When the "little devil" approaches them, they greet him by blowing gently upwards, as though they were sending some message or outpouring of their soul to heaven with their breath. Which is just what the smoker, in his moments of pleasure, does with the smoke of his cigar. This rite is like smoking without smoke. It is a ritualistic transmission of a spiritual exhalation with a religious meaning, just as the priest blows his breath upon the baby when he baptizes it, to transmit to it the essence of the faith he bears in his spirit.

The Negroes, even those living in Africa, did not wholly adopt the religious character of the tobacco they had recently discovered, nor did it come to form part of their mythologies. Thus, for example, the cultivation of tobacco in Africa was carried on exclusively by men, as were the other agricultural tasks learned from the whites, whereas the women performed the tribe's archaic farming duties as in the remote epochs when they had invented their agriculture. Neither does tobacco play any fundamental part in the contemporary, present-day Afro-Cuban cults. The Negro gods of Africa do not smoke, but the force of the syncretizing tendency, even though guarded against by defensive mimicry, has managed to incorporate tobacco as one of the secondary instrumental elements of their rites and magic. The Negroes did not receive their religious ideas from the Indians, and among the rites of the whites that they adopted there was none in any way connected with tobacco, as can be easily understood. Tobacco, therefore, was transmitted to the Negroes from the Indians without any religious connotations, and therefore without any radical connection with their social institutions. Nevertheless, the tribal character of the African nuclei at times facilitated the adoption of certain tribal usages of the Indians, such as the collective pipe, which was smoked successively by all the members of a gathering, passing from hand to hand, in sign of communal solidarity.

The transmission of tobacco from the Indians to the Negroes was facilitated also by the historical and economic conditions that brought them together, not only in their cultures, but in their social position vis-à-vis the whites, to whom both groups were forced to submit as their common masters. It cannot be said that there was no discrimination among Negroes and Indians due to their different racial origin. The social history of America affords repeated examples to the contrary; but if Indians and Negroes were often separated because of the different and contradictory social use the white man made of them on certain occasions, it was not always possible to achieve this functional separation, which was a source of bad feeling between

Indians and Negroes, and neither was it unusual for them to join forces against their common domination. For this reason, if the tobacco of the Indians was not "a thing of the devil" to the Negroes, neither was it "a thing of savages." But among the whites, made up in large measure of conquistadors, land-grant holders, and clergymen determined to reduce the Indians to a subhuman status, it was only logical that their most typical traits and habits should have been disdainfully dubbed "a thing of savages," a brutish thing "without reason, polity, or civility," malevolently devised by the devil.

It was not long, however, before the white conquerors of the Indians were in turn conquered by tobacco. Fittingly enough it has been called "the conquering plant" (A. Nezi: *L'erba conquistratrice*). In the middle of the sixteenth century tobacco was still being publicly denounced in high places. Father Las Casas called it a "foul vice." Benzoni described it as a "malodorous and vicious poison of the devil." Father Mendieta said tobacco was a "suspicious and dangerous" plant, recalling that the Aztecs performed "a kind of communion" with it, as with the "body of a goddess." But by this time many of the Christian settlers had made their peace with tobacco and had developed a sufficient taste for it to be able to distinguish, choose, and buy different varieties, select seeds for their plantings, and make of this miraculous plant of the Indians a daily habit, a profitable crop, and an article of foreign trade.

It is probable that in America the use of tobacco was transmitted from the Indians to the whites principally as a result of the magic and medicinal help the latter received from the natives in their problems and ailments and that they thus acquired the habit of taking the powder and smoking in evocation of the pleasure they had experienced. The Europeans probably went to "consult" the *behiques* about their sickness and worries just as today they go to the gypsy fortune-teller, or to the African witch-doctor to have him give them a magic *embó* or the charm of the snails of Ifá. They must have visited the *behiques* on the sly in search of their cohoba or tobacco; perhaps they said the Lord's Prayer several times before so the "true God" would not punish them for the sin they were about to commit in having intercourse with the false gods—but they went. And whether the outcome of their initiation in the uses of tobacco was favorable or adverse, the habit remained, like the sick person who begins to use a narcotic to alleviate his suffering when ill and then keeps on with it as a vice even after the pain has disappeared. The whites made their first acquaintance with tobacco through a superstitious urge, but after this initiation they were "charmed" by it for sensual reasons of a gustatory and physiological nature, finding it at once stimulating and soothing. Particularly was this true of smoking.

The Spaniards in the Indies took their tobacco stealthily at first, as was reported of one of the discoverers, Rodrigo de Jérez, who smoked in the privacy of his own room in his house at Ayamonte, and little by little they began to smoke more openly, as though it were a venial sin, a boyish prank, until finally they made no apologies for it. The white settlers were soon sowing tobacco in their back yards and in the gardens of their ranches, just as the Negroes did around their huts, to have a supply of the prized leaves.

In Europe the magic-religious motive of the Indians could not exist openly among the whites, and those who used tobacco there did it solely because of the sensual pleasure it gave them and following the advice of those who had returned from America. But this sensual motive could not be made to justify the introduction of tobacco into the customs of Europe. Its sensual appeal and its mysterious power over the spirit laid it open to the attacks of those who saw in tobacco only a diabolic temptation, a new sin, a danger for the soul, and a form of black magic as a result of the strange excitement of the mind produced by that mysterious smoke which emenated from blackened leaves brought from a New World and burned without flame in a cryptic kind of rite. It was not faith in the supernatural that led people to use tobacco in the Old World; on the contrary, tobacco's original religious connotations were now adduced to combat it. Other reasons had to be sought in Europe far more than in the Indies to disguise the basic pleasure-producing motive that caused the tobacco of the Indians' religious rites to spread rapidly among Christian peoples. For this reason tobacco's introduction in Europe was backed by two motives, both more imaginary than real, but plausible and loudly proclaimed: the æsthetic and the medicinal. The medical basis was the most strongly put forth because, in addition to being based on therapeutic realities or myths, it took care of the pleasure motive, calling the physiological symptoms of sensorial satisfaction provided by tobacco healthful. If, as has been said, the partisans of tobacco in Europe were divided into two groups, "hedonists and panaceists," the latter provided the dialectical arms, but the former were the real victors.

The extension of tobacco from the Indies to Europe was a phenomenon of transculturation of the most radical sort. Tobacco was as nothing to the whites; it had to be transplanted to their consciences before it could be adapted to their soil and their habits. If tobacco was accepted by the whites in a somewhat clandestine fashion, they soon attempted to rationalize its use, not for the true reasons, which smacked of diabolism and witchcraft in that feverish epoch of religious conflicts and intolerance which was the sixteenth century, but for reasons that could find justification in the morality and trends of the Renaissance. Tobacco was presented as a plant of decorative beauty and possessed of astounding medicinal virtues.

Tobacco began to be cultivated in Europe as an ornamental plant. Its broad, handsome leaves, "like those of lettuce," were pleasing to the sight. But we must agree that its appreciable æsthetic attributes did not extend themselves beyond the brief confines of the flowerbeds and gardens. The leaves of tobacco are very fragile, wilt easily, droop and fade, the plant is an annual that is also delicate and frail, and its small colorless flowers cannot compete with the roses, carnations, and other traditional beauties of the Andalusian gardens. The decorative lines of the tobacco leaves did not go beyond the garden. They were not perpetuated in the architecture of capitals like the leaves of the classic flora, not even in Cuba, where although it is true that we go on copying the thistle and acanthus of Greece without paying patriotic tribute to our own tobacco and corn, there was a time when we did utilize the ornamental motifs of our own flora, as in the Church of San Ignacio of Havana (today the cathedral). When the Jesuits rebuilt it in 1725, they copied the palmate leaves of the papaya in the shafts of the columns and the fronds of the Cuban pineapple in the capitals (Countess of Merlin: *Viaje a la Habana,* edition of 1905, p. 73). Nor in Spain were the leaves of the Nicotiana reproduced in the elaborate ornamentation of courtiers' attire, as they exist today in the em-

broidered dress coats of the Cuban diplomatic corps. If the tobacco plant was painstakingly cultivated in Spain, it was more
because of its exotic quality and the prodigious curative properties attributed to its aromatic leaves, the same as sweet basil,
mint, rue, lavender, marjoram, and other plants of similar fragrance and virtues that are to be found in Spanish gardens, than
because of its æsthetic qualities.

By the medical profession tobacco was received in Europe as
a panacea, something like the cure-all that was sought by the
alchemists. For this reason the excessive apologia of its medicinal properties, which made this wondrous plant of America resemble that dream of medieval alchemy, so open to suspicion of
heresy, must have strengthened the scruples of certain reactionary, moralistic, and ascetic spirits against tobacco. But, nevertheless, the "propaganda," as we would say today, was carried
on, attributing to the plant innumerable therapeutic qualities;
and even though there can be no doubt that tobacco was put to
certain medical uses, given the pharmacopœia of the epoch, neither is it difficult to understand that in this unusual medical
propaganda in favor of tobacco there was no little "rationaliza-
tion"—that is to say, the justification of a fact for reasons having nothing to do with the true ones. Pleasure sought tobacco,
the dislike of new things and austerity opposed it; but medicine
justified it for reasons of its own, and sensuality was able to hide
behind the cloak of curative science. In this way tobacco began
to penetrate and extend itself all over Europe.

If the spread of tobacco throughout Europe and the rest of
the world is surprising because of its spontaneity, its rapidity
and scope, it is not less so because of the obstacles it had to overcome. Its enemies combated it with savage virulence, even with
the death penalty; its apologists lauded it to the skies, attributing to it the most fantastic merits. The literature for and against
tobacco is voluminous. It has not yet come to an end. The bib-
liographical index cards in the polemic section of Nicotiana run
into the thousands, without counting the sections cataloguing
the unceasing publicity in all languages dealing with the nature, the cultivation, the manufacture, the trade, the taxes, the
duties, the monopolies, etc. I shall not attempt to give here even
an abridged version of the history of tobacco throughout the
world. I shall only sketch, in rapid and synthetic outline, its

propagation, indicating the route of its social directions, taking as our point of departure the religious background that tobacco possessed when it was found in these islands of America, which accompanied it on its journeys to the other countries of the globe until other threads of human interest became woven into it and gave tobacco a completely different social framework.

Tobacco's best propagandist was the devil, its inventor, according to some of the most learned students of supernatural matters in all worlds, the New, the Old, and the Other. There is no dearth of texts that can be adduced to support the theory of the diabolical nature of tobacco, in addition to the already quoted opinions of the historians of the Indies, clergymen, and those versed in the perfidious wiles of the indefatigable enemy of mankind. There were even doctors, like Monardes, physician to the Archbishop of Seville and exponent of the miraculous curative properties of tobacco, who considered it of infernal origin.

As has been observed, tobacco quickly became a source of temptation to man. It may be that our first father fell from grace because of it. The Abyssinian Church upheld the theory that Adam fell into temptation because of tobacco, which he could not resist, and this tradition was popularized from the pulpits of England (Brooks, op. cit., p. 8). A tradition of the Greek Church maintains that the most important case of intoxication on record, that of the patriarch Noah, was caused by tobacco and not by wine (G. Johnston: *Travels in Southern Abyssinia,* London, 1844, Vol. II, p. 92). If the tree of knowledge of good and evil was tobacco, Admiral Christopher Columbus was quite justified in looking for the Garden of Eden in these lands of the New Indies. But the most generally accepted belief does not support these theories, and mankind is coming to the conclusion that good and evil are to be found in all plants, and that if there was a Paradise it included the whole world, which must be completely made over if we are to find and enjoy it once more.

Nor is the mythological version of the origin of tobacco put forth by an English writer more convincing. Richard Brathwaite in his book *The Smoking Age* (London, 1617) says that Tabaco was the name of an illegitimate son of the goddess Proserpine by the god Bacchus. Tobacco "clearly" means "son of Bacchus," according to the poet's fanciful etymology. That natural son of the god of wine was cast out of Olympus after he was grown, and sent to the earth under the protection of the god of the lower world, the powerful Pluto. The latter gave him a piece of advice: to seek in the world the men of science to help him with their reasoning, the lawyers to help him with their sophistries, and the poets to help him with their fables. According to Brathwaite, the god Tobacco followed this shrewd counsel in his earthly career, and the world has turned into an immense "tobacchanal."

When tobacco was first introduced into Europe its odor was considered pleasant or pestilential depending on people's tastes, but always something emanating from hell. In the first attack printed in England against tobacco, *Work for Chimney-sweepers* (London, 1601), signed with the pseudonym Phila-retes, and from the pen, it was believed, of Bishop Joseph Hall, the invention of tobacco was attributed to the devil as the fruit and instrument of his heartless perversity. Before this the Span-ish chroniclers and clergymen of the Indies had discovered the pact between the devil and tobacco and were aware of its evil origin. Another English writer, John Taylor (*The Nipping or Snipping of Abuses,* London, 1616), published the text of the stirring address made by Satan in hell to all his devils to spread the use of tobacco among mankind.

To this infernal origin of tobacco may be attributed a certain persistent atmosphere or whiff of sin and evil that surrounds it to this very day and bars it from certain sanctimonious circles. The Christians, who during the first centuries of the faith had incorporated no small number of pagan deities and customs into the new religion, never included in their calendar of saints, even though there were evident analogies and symbolisms, those deities the Indians connected with tobacco, corn, potatoes, coca, yucca, and the other agrarian, meteorological, and nar-cotic elements of their agriculture, their sex life, and their medi-cine. The religion of the conquerors, although it was still evolv-ing dogmas in the sixteenth century, was no longer so fluid and plastic as during the first milennium. Perhaps there was no longer time for other and new celestial beings connected with Indian mythology to become incorporated in the popular tradi-tion; nor have the centuries of the post-Columbian epoch, which belong to the age of the printing-press, with their abundance of books, favored the ingenuous syncretic assimilations of the Middle Ages at their peak.

Nor does Cuba have those parish brotherhoods of workmen such as still exist in the old Spanish cities, like those of St. Isidore, patron saint of farmers, St. Crispin, patron saint of shoemakers, and others. If they existed, perhaps we should have had in Havana Our Lady of the Tobacco-strippers or Christ of the Cigar-makers, although the former and the latter could invoke as their own Our Lady of Sorrows and the Mother of the Desolate, and certainly the Creole Virgin of Charity, since they cannot lean too heavily on the prayer of the Righteous Judge,[1] which is considered slightly heretical. At any rate, there has not yet appeared among the popular appellations of the Virgin Our Lady of Tobacco, as they say there is Our Lady of the Palm, nor a St. Picker, although she might be as imaginary as various other saints that have engaged the attention of folklorists and the erudite Bollandist fathers of Antwerp in their hagiographic *Analecta,* their *Acta Sanctorum,* which at times were put on the Index despite the fact that they were the work of Jesuits. Neither have religious names ever been applied to commercial brands of cigars or tobaccos. This complete absence of Catholic appellations in matters referring to tobacco can only be explained by the latent tradition of its diabolical origin.

It will be said that these celestial denominations have not been used for sugar either, and that there is no Child Jesus of the Canefields nor a Holy Mary of the Centrals; sugar cane, like tobacco, originated among the heathen, and was introduced among Christian peoples by Mohammedans, Moors, and Jews. Here, too, the absence of religious appellations might be attributed to this, but the argument does not hold true. Catholic denominations are frequent in Cuba in connection with sugar and the names of the plantations where the mills are located. Many of them have saints' names, such as "San Antonio," "Santa Lutgarda," "Santa Rosa," "Santa Marta," and so on, some because that happened to be the original name of the estate, but most because of the faith and choice of the owner. Inasmuch as the law demands that the plantation bear a name for territorial, commercial, and juridical identification, and the

[1] See page 310.

selection is left to the owner, he sometimes tries in this way to ensure himself heavenly co-operation, a kind of supernatural partnership. It is true that, aside from this detail of the names, all religious appellations have disappeared from the vocabulary of sugar, but this is due to the fact that all sugars, being naturally and commercially identical, have no need of a name, sacred or profane, to distinguish them on the market, by reason of their origin or their manufacture. If there had been need of a name, no one would have had any scruples in baptizing them with some such name as "sugars of Santa Rosa" and "muscovado of the Good Shepherd," just as there are many pious names for other commercial products of different sorts, especially foods and medicines. In the commercial world of the Catholics, with or without connections with famous shrines, there are many holy or religious trade-names, some born of sincere devotion, others of the astute business sense of merchants. In Italy the Lachryma Christi wines are popular, in France, the monastic liqueurs, Chartreuse and Benedictine, and in Spain the candied egg yolks of San Leandro, just as in Havana, despite a touch of irreverence, the "rolls of St. Francis of Paola," the "tortillas of St. Raphael," the "candied fruit of St. Teresa," the "crackers of the Arm of Almighty God," the "bread-sticks of the Grace of Eternal God," and the "waters of St. Rita" are familiar in the markets, not to mention the "saints' bones," eaten on All Souls Day, the "bacon of heaven," and the "Pius IX's," which are commonplace sweetmeats. And there is no end to the drugs bearing the names of saints, even though they have no curative value. In Havana even the "Sacred Heart of Jesus" is the name of a liquor store and a café for night-owls. If brands of tobacco have not borne names of an ecclesiastical nature, it has been because of a recollection, however dimmed, of its infernal origins and not because of its sensual connotations. There are plenty of candies, sweets, and liqueurs that bear saints' names. In Cuba we have a Santiago de las Vegas, which is linked to our tobacco history; but this is a mere toponomical accident, as in the case of the Cristo de la Vega of Toledo, which inspired the poet Zorrilla.

Despite its traditional diabolic connections, tobacco continued to spread throughout Europe, even among those of "the dark cape." When tobacco appeared on the horizons of Christianity, the days of St. Francis of Assisis had passed, and the pupils of St. Thomas Aquinas were in the pulpit, but the reforms of Luther, of the Council of Trent, and of Loyola were close at hand. The followers of St. Francis would not have smoked tobacco, those of St. Thomas would have tolerated it philosophically, those of Loyola had no scruples about allowing it, defending it, and taking advantage of it. Every age has its own customs and ideas.

Tobacco reached the Christian world along with the revolutions of the Renaissance and the Reformation, when the Middle Ages were crumbling and the modern epoch, with its rationalism, was beginning. One might say that reason, starved and benumbed by theology, to revive and free itself, needed the help of some harmless stimulant that should not intoxicate it with enthusiasm and then stupefy it with illusions and bestiality, as happens with the old alcoholic drinks that lead to drunkenness. For this, to help sick reason, tobacco came from America. And with it chocolate. And from Abyssinia and Arabia, about the same time, came coffee. And tea made its appearance from the Far East.

The coincidental appearance of these four exotic products in the Old World, all of them stimulants of the senses as well as of the spirit, is not without interest. It is as though they had been sent to Europe from the four corners of the earth by the devil to revive Europe when "the time came," when that continent was ready to save the spirituality of reason from burning itself out and give the senses their due once more. Europe was no longer able to satisfy its senses with spices or sugar, which, aside from being rare and, because of their costliness, the privilege of the few, excited without inspiring, strengthened without lifting the spirits. Nor were wines and liquors sufficient, either, for although they nourished daring and dreams, they were often the cause of degradation and derangement and never of thoughtfulness or good judgment. Other spices and nectars were needed that should act as spurs of the senses and the mind. And the devil provided them, sending in for the mental jousts that initiated the modern age in Europe the tobacco of the Antilles, the chocolate of Mexico, the coffee of Africa, and the tea of China. Nicotine, theobromine, caffeine, and theine—these four alkaloids were put at the service of humanity to make reason more alert.

Coffee (*Coffea arabica*), Abyssinian Negro by birth, became popular in Mecca between 1470 and 1500 and spread from there throughout Arabia to the world of Islam, as far as Constantinople (Heinrich E. Jacob: *Sage und Siegeszug des Kaffees, die Biographie eines Weltwirtschaftlichen Stoffes,* Berlin, 1934; W. H. Ukers, *All about Coffee,* New York, 1935, p. 22). About 1510 tea (*Thea sinensis*) reached Cairo from China, and from there it passed to eastern Europe (Ukers, ibid.), and the traders of the East Indies brought it to England, Holland, and other lands. It is believed that by 1528 chocolate was known in England and France, having come in from Spain; by 1610 they had tea, thanks to the Dutch trade; and by 1615 coffee, which was spread by the Turks through Venice and the cities of the Danube. In London during the seventeenth century there were successively "tobacco clubs," "coffee houses," and "tea houses" (1657).

These four alkaloids, solace for the senses and subtle nervous stimulants, all arrived at the same time to prolong the Renaissance. They were supernatural reinforcements for those of revolutionary ideas.

Jacob, in his saga of coffee, exaggerates the historical importance of this drink brought into modern civilization by the Mohammedan Semites. "Coffee has transformed the face of history through the cerebal stimulation it has given mankind," says Jacob. Judea, Greece, and Rome, according to him, were civilizations based on wine. The juice of the grape inspired their mythologies. Noah, Dionysus, Bacchhus, Silenus. Christianity, which inherited these civilizations and carried them to their peak, raised wine to the sublimity of a divine essence. Islam came into being religiously as an anti-Bacchic civilization. The Mohammedans closed the taverns and destroyed the vineyards. When they entered Constantinople and spread up the Danube, they substituted for them coffee houses in order to combat the natural tendency of the Christians toward drunkenness. For Islam, too, coffee is a nectar of the Renaissance. In the preceding epochs there were no other products that acted upon the human nerves except narcotics and depressives, for to the physiologist alcohol is a slow poison. There were no pure stimulants that could keep the spirit alert against fatigue and sleep and could intensify mental activity by making it clearer and more lasting. In Jacob's opinion, coffee was culturally as important as the invention of the telescope and the microscope, which increased the dimensions of the brain's capacity. Coffee has given great analytical powers to many minds, which in the old civilizations were led astray by wine into rhetorical syntheses and drunken stupor.

This can be accepted in part, but Jacob, even though he mentions tobacco to say that it and coffee "are the two anodynes of fatigue in modern civilization," forgets all tobacco's marvelous history, its extensive literature, its effective mental action, and its unanimous and universal acceptance by all nations, of all continents and all cultures. Without question, coffee has been and is a stimulant for thought, as is the heaven-sent tea of China. When around 1554 the public coffee houses were opened in Stambul, they were called by the people *mektel-i-irfan,* or "schools of the learned," and coffee was given the name of "milk that nourishes thinkers." But tobacco, too, is the great friend of thought. "From the moment a man starts smoking a pipeful of tobacco, he becomes a philosopher," said the Englishman Sam Slick. According to Thackeray, tobacco "makes wisdom flow from the lips of the philosopher and shuts the mouth of the fool." When one stops to think of the influence these four alkaloids have had on the intellectual life of modern times, they should all be regarded as co-operating forces, although in different degrees, depending on the period and the country.

Perhaps the tempting qualities of all of them are the emanation of the same infernal still. It was known that the same alkaloid was present in both coffee and tea, trimethyloxypurin. But not long ago Professor Nottbohm discovered that these plants contain besides another alkaloid, trigonelline; and Hantzsch has just proved (Jacob, op. cit., Chapter iii) that this alkaloid is one of the principal constituents of nicotine, the tobacco alkaloid. It is noteworthy, too, that the four alkaloids in question, or devils, although varying greatly in appearance, resemble each other not a little in their social trajectory. In origin they were all exotic, from beyond the sea, and were introduced among the whites by "colored people"—copper-colored, black, yellow. All had the common property of stimulating sensual appetites. All had religious connotations, and attracted the suspicions of the clergy. Medicinal claims were put forward for all of them. As their use extended, they were all persecuted by governments, moralists, and men of religion and defended by doctors, poets, and merchants. And all finally won a swift and sweeping victory, not only because of their sensual solace and their medicinal promises, but because of their early alliance with capitalism, which made of them the stamp of fashion, of rank and wealth, and a rich source of profit for the individual and the state. It is amazing to observe that today the economic life of different regions, large provinces, and whole nations depends basically on tobacco, coffee, tea, or chocolate.

In modern times these four devils have fought together and been worshipped together on the altars of sensuality with the old and medieval alcohols, spices, and syrups; but tobacco has always led them all since the beginning of the sixteenth century.

The idea of the supernatural paternity of tobacco was present at the time in all literature. There was metaphor in it, but belief, too. Some said that tobacco was a gift of the devil, others of God. Tobacco was a divine or an infernal plant, but at bottom there was always the basic concept of its wonders, its powers, its supernatural properties—of its consecrated nature. It possessed the attraction of a mystery. In Cuba we would say of its *cocoricamo,* using an Afro-Creole word. Saintliness or devilishness, in the last analysis it is the same thing, and the difference lies in the point of view from which they are regarded.

The original Indian conception of the mysterious supernatural quality of tobacco was slow to disappear among the nations of Europe, given as they were to theological, mystic, and magic lucubrations. As late as the year 1648 the distinguished doctor and professor of the University of Pavia, Johann Chrysostomo Magnen, in his book *Exercitationes de Tabaco,* upheld the theory that tobacco possesses an inherent magic and that it gave the Indians prophetic power. He dedicates to this use of tobacco for soothsaying purposes one of his studies, referring to its esoteric, soporific, and intoxicating properties, comparing them to those of the ancient oracles.

European letters of the sixteenth century are full of references to tobacco. Not only the many writings devoted especially to the plant, which contained as much praise as reviling, but works of literature as well. Among Spanish writers tobacco was not the object of such violent attacks as in the other countries of Europe, nor did it have such ardent defenders. Poets and playwrights, moralists and satirists, in Spain displayed a certain benevolence toward tobacco and tended to criticize its abuse rather than accuse it of fantastic congenital powers of evil. When the muses of Spain "discovered" tobacco it had already acquired a hold on people and was enjoyed even by the clergy, which was the social group most obligated to exorcise its powers of temptation and free the Old World from this "infernal pestilence." Nevertheless, the theme of tobacco as a creation of the Prince of Darkness is also to be found in Spanish letters.

So keen a satirist as Francisco de Quevedo regarded tobacco as a thing of the devil, even though we wonder whether this sharp-tongued, sly moralist did not himself at times succumb to its temptation.

In his account of his travels in hell he found, along with many other devils who inhabited the halls of Pluto, the "devil of tobacco," together with the devil of chocolate, both Indian. The two little fallen angels told the mordant author that the *tabacanos*—that is, the smokers, snuffers, and chewers of tobacco— were like "Lutherans," which in the realms of His Catholic Majesty at that time was as shocking and abominable as to call them Communists today and led straight to the purifying fires of the Holy Inquisition, like being possessed of a devil, practicing sorcery, sodomy, backsliding into Judaism, and even being guilty of smuggling, when this involved dealings with heretics. Quevedo wrote as follows:

"There came the devil of tobacco and the devil of chocolate, who, although I had had my doubts about them, I never regarded as complete devils. They told me they had avenged the Indies against Spain, for they had done more harm by introducing among us those powders and smoke and chocolate cups and chocolate-beaters than the Catholic King had ever done through Columbus and Cortés and Almagro and Pizarro. For it was much better and cleaner and more honorable to be killed by a musket ball or a lance than by snuffling and sneezing and belching and dizziness and fever; the chocolate-bibbers idolize the cup that they raise on high and adore and go into a trance over; the tobacco addicts are like Lutherans: if they take it in smoke, they are serving their apprenticeship for hell; if in snuff, for catarrh." (F. de Quevedo: *El Entrometido y la Dueña y el Soplón*).

Quevedo was aware of tobacco's anesthetic powers, for he alludes to the use of the powder by the Negroes and ironically advises them to console themselves with tobacco in their controversies with the exponents of white racial superiority (*La Hora de Todos y la Fortuna con Seso,* Chapter xxxvii).

Quevedo was bitter on the subject of America, for he was aware of the corruption that existed in colonial life and of the no less harmful influences that were infecting the atmosphere of Spain. For this reason he makes one of his characters say:

"Observe that America is a rich and beautiful whore. The Christians say that Heaven punished the Indies because they adored idols; and we Indians say that Heaven will punish the Christians because they adore the Indies." (*La Hora de Todos y la Fortuna con Seso,* Chapter xxxvi).

So it is not to be wondered at that the moralist should display such wrath against the tobacco that arrived from the adored and prostituted Indies.

Other Spanish writers of the golden age refer to the Spaniards' habit of using tobacco, not only in snuff but for smoking, in both pipe and cigar. The Inca Garcilaso, in *Los Comentarios Reales* (Lisbon, 1609), speaks of "the herb or plant the Spaniards call tobacco and the Peruvian Indians *sayri,*" saying that "they take the powder through the nose to clear the head."

Quiñones de Benavente speaks contemptuously of chocolate and tobacco in these verses he puts into the mouth of Melisandra, one of the characters in the interlude of *Don Gaiferos*:

> *Please God a man back from the Indies may ill-treat you*
> *And make you drink that chocolate brew,*
> *And some foul knave*
> *Make you sneeze tobacco. . . .*

Lope de Vega, in Act III of *La Mayor desgracia de Carlos V,* writes this dialogue:

JORGE: By God, if I get hold of a stick—
MARÍA: What is monkey-face saying?

MARTÍN: Here, take a pinch of tobacco,
     It will soothe your ire.

And again, in these verses of *La Gatomaquia,* he writes:

    . . . give me of your tobacco
    enough, at least, to cover a cue.

Rodríguez Marín, commenting on these lines, is of the opinion that Lope himself took snuff for his attacks of "runnings from the head." In still another work, *Amar, servir y esperar,* the great poet refers to "tobacco of the wits"—that is, tobacco ground into powder, and smoking tobacco.

In the seventeenth century in Spain pipe-smoking was also known. Agustín Moreto y Cabaña mentions the tobacco pipe in his comedy *La Ocasión hace al ladrón.* Tirso de Molina had made mention of tobacco as early as 1620, in his play *La Villana de Vallecas* (Act I, Scene iv), which served as the model for the play by Moreto mentioned above. Tirso, describing the desserts of a dinner, refers to "a tube of tobacco"—that is, a *tabaco* or cigar. He says:

    A tube of tobacco
    To give the final blessing.

A more Christian evocation of tobacco could hardly be imagined, putting it at the end of the meal to coincide with the custom, general at that time, of pronouncing the blessing and giving thanks to God for having provided "our daily bread." There was nothing sacrilegious in this use of tobacco, like a censer scattering its perfume at the conclusion of a rite. And Tirso de Molina was a priest and knew what he was talking about.

Vélez de Guevara likewise alludes to tobacco in *El Diablo Cojuelo* and *Estebanillo González,* but there is no point to drawing these notes out indefinitely.

These quotations prove that the custom of using tobacco had become a habit among the Spaniards, and that the Spanish theater was not really hostile to it. As a matter of fact, tobacco did not find much opposition in Spain in spite of its diabolical repute. There was a Spanish king who, like other monarchs of distant lands, imposed the death penalty for the failure to observe certain legal restrictions on tobacco, but this was not because of any animus against the plant, nor because of pressure from the Church or fanatical prejudices, but for reasons connected with the exchequer, as I shall later explain. The Holy Inquisition, which condemned those having dealings with His Satanic Majesty to the stake, as well as heretics and even smugglers and homosexuals, did not burn tobacco addicts in its autos-da-fe.

The great number of Spaniards who went out to the Indies, the social prestige of those who came back to Spain with a fortune, the sensuality prevalent in the habits of all classes, not excluding the clergy, and the barefaced roguery dominant in the cities were contributing factors to this Spanish attitude of tolerance toward tobacco and its early and general approbation. But, above all, in this period two new social factors, both of a fundamentally economic character, had a bearing on the fate of tobacco in Spain, one evident in the plays of the day, and the other tacit, but the more important and decisive. This was that tobacco took on an implication of class rank and became con verted into a great economic factor.

To smoke a cigar or take snuff became the symbol of position and wealth. It may have been that the use of tobacco was indicative of rank even among the Indians, at least in certain ceremonial forms. Some of the chroniclers indicate the social category of certain tobacco rites, suggesting that they were limited to the chieftains and priests. Among Europeans the use of tobacco was the enjoyment of an exotic luxury that was totally consumed at a single using and reduced to ashes. Its high cost made such a wasteful and fleeting pleasure the privilege only of the wealthy. Its exotic nature, coupled with its cost, gave this luxury the quality of rare distinction. People smoked ostentatiously, the same way they displayed a little Negro slave, a cage of talking parrots, a mahogany coach, or a tortoise-shell cane. These were not merely signs of wealth; they were intended to be symbols of courtly pomp, won in far-off, somewhat fabulous exploits of war, authority, and power. And the desire for social distinction stimulated the taste for tobacco in order to attract attention by its use, just as the parvenu likes to drink the finest and most expensive champagne in public to satisfy his vanity and love of show. Thus what had once been frowned upon by society now came to be a sign of "elegance among fashionable people." Even today a person who smokes a pipe is something of a character of folklore. The social category conferred in those days by the mere smoking of tobacco is evident in the allusions made to it in the Spanish theater. It was brought to the table along with the desserts, with the exotic and delicious fruits of the Indies and Castile, "to give the final blessing."

In addition, during this same period tobacco acquired great economic importance for traders, statesmen, and even the clergy. It was no longer merely a source of pleasure, but of wealth as well. Toward the end of the sixteenth century the use of tobacco was so generally accepted that it had become an article of trade that was always negotiable, and its cultivation was a lucrative agricultural enterprise. That of the Indies was so highly esteemed that it became an outstanding article of transatlantic trade, as profitable as spices had once been. Its high price, the unceasing demand for it, and the sumptuary nature of its use made it an article that could bear very remunerative taxes, fiscal inroads of the severest yet most tolerated kind.

Because of the threefold economic interest in tobacco—the mercantile profit, the benefits to the public treasury and the revenue from the land—the Spanish clergy was not inclined to attack it. The friars in the convents and the priests in their homes probably felt the impulse to sow and harvest along with the other products of their garden this most desirable crop that tobacco was becoming. It is fair to assume that the clergy, too, derived good returns from the tobacco trade as the demand for the article grew, for in spite of their professional apostolic calling, it was not unusual for them to pursue, as a side line, business ventures and even smuggling. There were many cases of friars who carried on undercover dealings in frequent transatlantic voyages, so much so that papal bulls were issued against such abuses, forbidding friars to carry with them on their ocean travels gold, silver, and other articles aside from their necessary equipment and ordering that they be carefully watched in the ports because of their smuggling. On November 23, 1542, the King of Spain wrote his Ambassador to the Holy See to arrange for a renewal of these bulls and stronger punitive measures. Philip II complained on the grounds that it was harmful to "the propagation of the faith," even of the "trade in wine carried on by the heads of convents and others, both laity and clergy"

(*Cartas Antiguas,* in the Archives of the Holy Congregation of the Propaganda, according to Ayarragaray). They engage "in continual trade, the clergy as well as the laity," without spurning the horrible traffic in slaves, "selling Indians to Christians, who buy the Indians even though they have become Christians, and take them from one province to another." And there were priests who were traders in Negro slaves.

In 1602 the Governor of Cuba, Don Pedro de Valdés, notified Madrid that it was not only the laity that was engaged in smuggling, but the clergy as well; that the priest of Baracoa ". . . was one of the greatest ransomers of heretics and enemies to be found in all the Indies, and that all the other friars and priests of the island imitated him openly, without any attempt at concealment" (J. de la Pezuela: *Historia de la Isla de Cuba,* Madrid, 1868, Vol. I, p. 543).

The canonical prescriptions against the agricultural, commercial, and smuggling activities carried on by the clergy in the Indies with heretics and enemies of their country and their religion were not strong enough, and on September 27, 1609, another King of Spain had to write once more to his Ambassador in Rome instructing him that:

". . . although by the terms of a bull issued by Pius IV clergymen returning from the Indies are forbidden to bring money, silver, and gold with them . . . this is not observed and carried out as it should be, and many of the aforesaid clergymen commit great abuses in bringing money from the Indies for business purposes and other uses, which leads to great negligence in the service of Our Lord and to the corruption of the habits of the clergy in question."

For this reason the King ordered him to ask the Pope for another apostolic bull to prevent the continuance of such practices, which led the friars to break their vow of poverty, "make the monks the owners of property, and ruin the monasteries." As a result of this, said the King, in certain parishes there had come about ". . . a licentious life, because the transgression of the vow of poverty is evident, as all have become property-owners and by means of gifts and presents induce the Visitator to overlook this ownership" (*Cartas Reales,* in the Archives of the Spanish Embassy, from Lucas Ayarragaray: *La Iglesia en América y la dominación española,* Buenos Aires, 1920, pp. 41-3).

Tobacco has always been an article suitable for smuggling because of the high return for the risks taken, as is all merchandise of high price and small volume that can be easily transported and hidden. And it is safe to assume that those who cheated the royal treasury, breaking their vows and flouting the respect due their sacred calling by dealing with heretics and smuggling in "gold, silver, and other things," probably did the same with tobacco, which at that time was worth its weight in gold.

The taxable possibilities of tobacco were probably discovered by the Catholic Church in the Indies before any other social organization became aware of them, as soon as the settlers began to raise tobacco for commercial dealings. The economic basis of the Spanish Church, as of the Catholic Church in general, aside from its landed properties and other rich benefices, was the tithes—that is, the tenth part of all the mining and agricultural production of the country—that it received. The legislative system providing for this church revenue existed in Peninsular Spain before colonial Spain came into being, and when this event took place, all that had to be done was extend to these new possessions this old tax system of Castile, which was done by the Catholic kings in a royal edict promulgated October 5, 1501. As soon as tobacco became an article of agricultural value among the Spaniards who had settled these lands, by the mere fact of becoming a cultivated product it fell *ipso facto* under the tithe laws; that is to say, the tenth part of the value of the crop went into the ecclesiastical coffers. Thus the Spanish clergy of the Indies, where the settlers began to cultivate tobacco first for their own use and then for export, soon received direct profits and economic returns from the spread of the diabolical plant, just as the kings profited from it through the duties, taxes, monopolies, and sales taxes of every sort that were imposed on tobacco under threat of the most Draconic penalties.

Tobacco brought in more than tithes to the Church. As the cultivation of tobacco spread throughout the Indies, a new basis of economic support for the colonial clergy came into being. The tobacco vegas are small gardenlike properties, located in the river-bottom lands, where the centers of population had sprung up, and around these little farms, richly productive and the abode of family groups of peasants closely attached to the land, the religious institutions built up a great variety of sources of income, such as livings, annuities, foundations, Masses, and other impositions of a pious nature, whose profits went principally to the different monastic orders. The mystical beliefs and the pressure of the Church, so powerful in those days, made bequest and legacies almost obligatory for the establishment of livings, annuities, and other burdens by means of which men attempted to secure, through the payment or investment of money on earth, the assurance of eternal life in the hereafter. At the same time, during the several centuries of relative insecurity of immovable property in Cuba, especially rural holdings, the assessing of a tax against a farm in favor of the Church was a way of strengthening the owner's title to it, making the Church an interested party in the maintenance of the legitimacy and juridical permanence of ownership.

For this reason, when, in 1682, the Governor of Cuba wanted to levy a tax upon the vegas, the mills, and the manufacture and sale of Havana tobacco, the convents protested angrily (letter of September 6, 1683 from the Governor to His Majesty). The Governor, Don José Fernández de Córdoba y Ponce de León, could not have made his report to the King more explicit. He complained of the rebellious inhabitants of Havana, saying that ". . . the people who live in the city are by nature so opposed to doing what they are ordered and so enamored of their liberty that everything involves a great deal of trouble."

And he goes on to say: "And not the least of the obstacles to this is the meddling of those who use their holy orders as a pretext to interfere in such matters in different ways: the priests and friars who, to serve their own ends, display great concern over the poor, giving rise to no little mortification. . . ."

Clerical opposition to the taxes on tobacco did not yield ground. When in 1717 a royal edict ordered all tobacco in Cuba to come under government monopoly, representatives of the Dominican, Franciscan, Augustinian, and other orders, and the abbess of the nuns of Havana sent a protest to the King, stating that "in all the lands where tobacco is grown" valuable taxes are paid to the convents for their upkeep, and these would cease or diminish greatly under the proposed monopoly. According to Pezuela, the subversive movement, which culminated in a furious riot against the captain-general, was brought about by the traders or "speculators" in tobacco of Havana, who aroused the tobacco-growers against the monopoly the state planned to set up. I am not questioning the part of the "speculators" in the riot, but I have no doubt that there were monks and nuns among the "speculators." The ecclesiastical opposition was so violent and stubborn that a Dominican friar, Father Salvador Suárez, was accused of being the chief instigator, and the King, on November 21, 1719, was obliged to issue a royal edict to the "Venerable and Devout Prior of the Convent of Santo Domingo of the Order of Preachers of Santo Cristóbal de la Havana," warning him that:

". . . being so much to the service of God and myself that the clergy, both secular and regular, keep within the bounds of reason and their calling, without disregarding the obligations incumbent upon their state, nor spreading pernicious rumors that only help to increase unrest and rebellion from which such serious and harmful results ensue. . . ."

His Majesty charged the prior in question:

". . . to take special care that none of your subordinates disturb the peace of that Republic, nor upset or interfere with its proper administration and the fulfillment of my royal orders, and if any should attempt to do so or cause trouble, that you restrain, check him, and remove him from that city to wherever my Governor and captain-general of that island should think best."

Even years later, in a royal edict issued at Aranjuez, on May 26, 1721, the King ordered Father Salvador Suárez sent into exile "to some remote place where he cannot stir up the spirits of the inhabitants."

For three hundred years property in Cuba, rural as well as urban, was burdened with ecclesiastical obligations, aside from the tithes. As Pezuela said:

"The piety of many of the inhabitants on the one hand, and the influence of the clergy on the other, made the livings, the pious bequests, legacies, and ecclesiastical benefits amount to much more than the tithes."

According to the calculations of Jacobo de la Pezuela (*Diccionario geográfico, estadístico, histórico de la isla de Cuba,* Havana, 1863, Vol. II, p. 242), by the middle of the seventeenth

century the capital of the ecclesiastical foundations in Cuba
amounted to four million dollars.

"They had drawn off one third of the public wealth. Don Gabriel
de Villalobos, Marquis of Barinas, in a long report presented to
Charles II with the title *Grandezas de Indias,* made this statement,
which is perfectly exact: 'I hardly need to point out to Your Majesty
the grave disadvantage and burden from which the residents [of
Havana] suffer, because of which, instead of prospering, the city will
decline if this state of affairs is not remedied. Within fifty years all
property will be in the hands of the clergy, and the laity will have
only an insufferable burden to bear.'"

As a matter of fact, the tithes were growing lighter because
of the passive resistance of the taxpayers, the tremendous difficulties connected with collecting them in kind, and the corruption of the tithe-gatherers; but the pious burden upon the
landed wealth of Cuba continued to be crushing until the nineteenth century. During this long period the members of the
clergy who profited by this state of affairs saw to it that the economic productivity of the rural farms was not diminished by
new taxes, which would have militated against the profit-yield-
ing levies of private law, and the tithes, recognized by public
law, which the lands had to pay the Church. So it is not un-
likely that this levy upon tobacco, initiated in the Indies by the
Spanish clergy, should have been largely responsible for their
early attitude of tolerance toward tobacco, in spite of its supposed evil origin. In the last analysis, the devil, by extending
the scope of his temptations, was indirectly contributing to the
funds of the Church. And the interests of the devil and those of
the clergy, in spite of their deep-seated ethical contradictions,
came to coincide on the point of extending the use of tobacco
among the Christian nations.

In Europe tobacco was making its social way from the West
Indies thanks to the initiative of personal sensuality and to public medical propaganda, both that of home healers and that of
professional doctors. But the Indian plant was not completely
grafted on the trunk of the culture of white people, nor naturalized among them, until a wholly new use was found for it, that
of serving as a great source of tax revenue to satisfy the mone-
tary needs of government. With this, tobacco became a complete and social institution among the white nations. What

among the Indians had been a social institution of a magic-religious character became among the whites an institution of economic character, a characteristic phenomenon of complete transculturation.

The cultural transition of tobacco aroused great discussion. Progressive and reactionary tendencies expressed themselves with burning vehemence, the most absurd generalizations were put forward, positions were sustained to the death, and divergences were stubbornly upheld. Theology and science crossed swords, ignorance and technique, until finally the economic and hedonistic tendencies won out, but even today the struggle goes on, with other ideas and objectives, and nearly always influenced by economics. The devil, who is wise in the ways of human weakness, in order to conquer the nations across the sea from America more quickly, combined the original, physiological appeal of tobacco to the senses with the social stimulus of vanity. But these two temptations alone were not enough, so he set greed into motion. He discovered the way to translate tobacco into money. The original appeal of tobacco was transmuted into an economic interest with capitalistic and revenue-producing possibilities. And with this conjoined stimulus of three sins, all of them mortal (gluttony, pride, and avarice), the devil's triumph was quickly ensured, one might say, without meaning to be irreverent, in the time it takes to cross oneself, for, as has been seen, even the prelates contributed their share to the world triumph of tobacco, that arch-diabolical and subtle instrument of sensuality and thought.

In the history of tobacco in Europe these phases of its transculturation stand out in sharper relief. It was about the middle of the sixteenth century that tobacco became an international article of trade and its cultivation was undertaken all over Europe. It is very possible that the seeds of tobacco, being so numerous, so tiny, and so easy to transport and conceal, were brought into Spain by many of the sailors, merchants, clergymen, and officials of the crown who followed Columbus, when the Spaniards in the Indies had developed a taste for tobacco, and in this way the plant was easily disseminated through Europe by the ships returning from American waters. The Espasa Dictionary, among many other erroneous statements about Cuba, says that "the missionary Romano [sic] Pane, in 1518, sent Charles V seeds of this plant, which the Emperor ordered planted and cultivated with great care, and the introduction of the cultivation of tobacco into Europe may be said to date from this period." I shall leave to the dictionary in question full responsibility for this statement, for which no known basis exists. It has been believed that the buying and selling of tobacco began in Europe a few lustrums after the discovery, but there is no data to support this theory. Some years ago J. B. Thacher wrote (*Christopher Columbus, His Life, His Work, His Remains,* etc., New York, 1903, Vol. I, p. 561) that in a clause in Diego Columbus's will, dated May 2, 1523, he left a bequest to a "tobacco merchant in Lisbon"; but this is questionable, owing to an error in the translation of certain words of the text, which say: "Antonioto Baco, *mercader,"* which were incorrectly translated as: "Antonio—*Tobaco—mercader"* (*Raccolta,* Part I, Vol. III, p. 207).

Jacobo de la Pezuela is mistaken when he says that "there is no reference to the use of tobacco in Europe outside the Peninsula until 1605, when the Turks began to smoke it" (*Diccionario,* Vol. IV, p. 563). The first documentary evidence of the tobacco plant being carried to Europe is to be found in the work already mentioned of the French friar A. Thevet (1556) and in that of the Spanish doctor Francisco Hernández (1558–9), also cited; but there can be no doubt that before this tobacco was known and used for smoking in some of the port cities of Europe. The cultivation of tobacco was begun in Belgium around 1554 (according to Dodoens), about 1559 in Germany (according to Lewin), about 1561 in Holland (according to Schwarb), and in England around 1570 (according to Lobel), long before the year 1584 or 1586, generally, but erroneously, given as the date of its introduction into that island of Europe by the celebrated Sir Walter Ralegh (Quotation from Vice-Admiral W. H. Smyth, in his translation of the work of Benzoni in the Hakluyt Collection, p. 82, note). The statement that has been put forward to the effect that the tobacco plant was introduced into Europe via Spain, but that the custom of smoking it came from England (W. G. Freeman, in the *Encyclopædia Britannica*), is without basis, and even lacks verisimilitude. The Spaniards came to know smoking tobacco in Cuba less than a month after the discovery of America, and they and the Portuguese

soon familiarized themselves with the properties of the Nicotiana plant in powder and leaf form in different Indian lands
long before the English took possession of Virginia and saw
the natives smoking pipes there. It had been believed that
Walter Ralegh became acquainted with tobacco in Virginia,
but today it is thought that he learned it in France from the
Huguenots, and it is pointed out that, in spite of the connections of this favorite of the Virgin Queen with the Virginia
colony, he was never there, nor could he have learned to smoke
in America.

It is an established fact that tobacco "was cultivated in England in 1573, a year before the discovery of Virginia" (Berthold
Laufer: *The Introduction of Tobacco into Europe,* Chicago,
1924), for W. Harrison in his *Description of England,* published in 1876, but written in 1573, the same year that the famous seaman Francis Drake, the abominated Draque or Dragón
of the Spanish writers, returned to London from the West Indies, describes the plant *Nicotiana tabacum,* which was used to
cure head colds. But it has not been definitely determined when
tobacco was first known in England, nor from where it was
brought into the country.

John Spacke said that tobacco came to England in 1565,
brought there from Florida. The English writer Richard Hakluyt, whom we could call a "historian of the Indies," wrote in
1582 that "the seeds of tobacco were brought to England from
the West Indies." There are those who believe that the tobacco
plant was brought into England by John Hawkins, on his return from one of his voyages to the Indies. This English captain, who was engaged in smuggling, came in 1563 with a
shipload of Negro slaves and linen and silk cloth and other
articles from England, and was in port for only a few hours. In
the following years he returned to Hispaniola and had a series
of adventures in the ports of the Spanish Main. Brooks, in his
invaluable study (Vol. I, p. 44), maintains that in 1560 tobacco
was already known in England, but only as a panacea in high
repute with doctors and herb-sellers. By this time tobacco was
used for smoking by the sailors of Spain, Portugal, France, and
Flanders, and before 1565 it was a general practice among all
crews plying the route to the Indies. There is evidence to the
effect that before 1570 some of these sailors smoked publicly in

London (Pierre Pena and Matthias Lobel: *Nova Stirpium Adversaria,* London, 1570). By 1590 smoking was nothing unusual in England, but tobacco was scarce, so much so that its rural devotees resorted to a plant known as coltsfoot as a substitute for tobacco. This "coltsfoot" tobacco, *tabac de pota,* is still smoked by the peasants of Majorca, where it was introduced by the English during their domination of the island, and there we came to know its foul, at times nauseating, reek. In the last voyage of Drake and Hawkins to the Indies, when they raided Santo Domingo, they took aboard a great cargo of tobacco for England (Hakluyt: *The Principal Voyages,* London, 1599, 2nd edition, Vol. III). And the Nicotiana plant was one of the articles of trade most highly prized by the smugglers flouting the Spanish law in the Indies.

According to Laufer, English soldiers brought tobacco into Bohemia with them during the Thirty Years' War, and later their sailors introduced it into Turkey. But these statements cannot be accepted at their face value, for Constantinople carried on a flourishing trade with the ports of Algiers, Ceuta, and Morocco in the Mediterranean and with the cities of central Europe along the Danube. It did not need to wait for the English or other sailors to bring in tobacco; its own did. It seems more likely that it was the great Turkish or Moorish smugglers who, in contact with the Andalusians and Portuguese, and organized by the rich Jewish merchants expelled from Spain, spread products of American origin around the Mediterranean, the Black Sea, and the rivers that empty into them. There is the fact that some of these articles (corn, prickly pears, and turkey) received and still preserve the name of Turkey or Moor in Italy, France, Catalonia, and even England. The turkey or *guanajo,* which the Spaniards discovered on the Guanaja Islands, where they went to steal Indian slaves, still bears that name in Great Britain and in all Anglo-Saxon America. Corn is called Turkish grain (*grano turco*) it Italy, *blat de moro* in Catalonia, and so on.

All this evidence would seem to prove that the cultivation of tobacco was introduced into the European countries above the Pyrenees about the same time. There were several indirect routes that might have been used: that of Spain, that of Portugal, and that of Morocco and the Barbary Coast; but the first and most rapid route must have been the direct one, that which was opened by the French and English pirates in their early maraudings in the Indies, which began about the third decade of the sixteenth century.

In the non-Spanish history of tobacco the transitional phases were more manifest than in Spain, in so far as the introduction of the Indian plant and its cultural complex represented a more important economic innovation. The prejudice and aversion that surrounded tobacco in the beginning, like a metaphysical aura more subtle than its own smoke, aroused the ire of persons of highest rank outside of Spain. There was even a King of England, James I, who in 1603, although tobacco was already considered a fashion of great social distinction, made a personal attack on the American Solanaceæ, himself writing a book that he entitled, without any mincing of words, *A Counterblaste to Tobacco*. The attack of the royal Misocapnus ("enemy of smoke") is outspokenly aggressive and insolent:

"It is evident that smoke is better suited to the kitchen than to the dining room and yet it makes a kitchen also oftentimes in the inward parts of men, soiling and infecting them with an unctuous and oily kind of soote as hath bene found in some great Tobacco takers that after their death were opened. . . . A custome lothsome to the eye, hateful to the nose, harmfull to the braine, daungerous to the Lungs, and in the black stinking fume thereof, neerest resembling the horrible Stigian smoke of the pit that is bottomelesse."

There is the inevitable evocation of its diabolical antecedents. The son of Mary Stuart, unmindful of the indulgent attitude of Elizabeth, the Virgin Queen, toward tobacco, even denounced the great navigator Walter Ralegh, whom he held responsible for the introduction of tobacco into Great Britain, with this broadside:

"Now to the corrupted baseness of the first use of this Tobacco, doeth very well agree the foolish and groundlesse first entry thereof into this Kingdom. It is not so long since the first entry of this abuse amongst us here, as this present age cannot yet very well remember, both the first Author, and the form of the first introduction of it amongst us. It was neither brought in by King, great Conqueror, nor learned Doctor of Physicke. . . .

"With the report of a great discovery for a Conquest, some two or three Savage men were brought in, together with this savage custome. But the pitie is, the poore wilde barbarous men died, but that vile barbarous custome is still alive, yea in fresh vigor: so it seemes a miracle to me, how a custome springing from so vile a ground and brought in by a father so generally hated, should be welcomed upon so slender a warrant."

The wrath of the English King against tobacco was repeated and took more violent forms in distant lands. It has been said that tobacco was brought to Turkey in 1605, but Laufer (op. cit., p. 61) is of the opinion that it must have been earlier, at the end of the sixteenth century, from England and for pipe-smoking. No mention of either tobacco or pipe is made in the *Arabian Nights,* which was given final form in the sixteenth century, it is believed, in spite of the detailed descriptions it contains of Oriental Mohammedan life. In any case, when tobacco began to be known there, having been brought in by the Mediterranean traders who bought it in Spain and Morocco, the sultans prescribed the penalty of death for its users. Later the Czar of Russia and the Shah of Persia were barbarous enough to imitate the Sultan of Turkey. Before the end of the sixteenth century tobacco was to be found in India, Indo-China, and Java, where it had been taken by Portuguese traders and missionaries (Brooks, Vol. I, p. 41). Tobacco reached China in 1573, at the time of the Ming dynasty, coming in through the region of Fukien, brought by the Spaniards from the Philippines, along with peanuts, sweet potatoes, and other agricultural products of America. By 1638 anyone planting tobacco was ordered decapitated, in spite of the fact that it was known as "the kindly plant," "smoke that revives the spirit," and "pill of the five elements." But one day it was said that tobacco cured the colds of the soldiers in the army, and from then on, its cultivation and sale were permitted —in return for paying a tax. Portuguese traders brought tobacco into Japan in 1605, but before ten years had elapsed, the Emperor ordered all the fields burned that had been planted without his divine permission (Berthold Laufer: *Tobacco and Its Use in Asia,* Chicago, 1924).

It has been said that the persecutions suffered by tobacco were due not only to religious but to political reasons as well. With the custom of smoking, houses or "tobacco taverns" came into being where the habitués not only could buy the weed and pipes and other smoking accessories, but could sit and indulge their

vice and assemble and talk over and discuss matters of government, whose authorities did not welcome public discussions of
their policies or possible subversive activities. So it is not to be
wondered at that in Arabia, Turkey, and Russia these "tobacco
houses" were persecuted, as were the coffee houses, too, and that
such establishments were regarded with suspicion in other European countries. In England itself Charles II tried to prohibit
cafés in the middle of the seventeenth century. To be sure,
throughout Europe cafés were political centers until the end of
the nineteenth century, as were the "tobacco houses," which disappeared as the use of the Nicotiana spread, for the enjoyment
of this plant did not require the many utensils and complicated
operations needed for coffee and tea, such as vessels to brew the
drink, glasses, stoves, mills, cups, saucers, spoons, sugar, and, as
a consequence, tables to put all this on, chairs to sit on, and a
place to prepare and drink the beverage. These circumstances
aroused suspicion against these centers of opinion, regarded as
haunts of the devil, but they did not justify the terrible penalties imposed on tobacco, not only against the assembling of
smokers, but even against every act of planting, cultivating, selling, and using that plant.

The fundamental reasons for these cruel persecutions were
basically economic. Tobacco at that time was a much sought-
after article of trade, which was produced principally in Spain,
Portugal, and their American possessions and whose cultivation
had not been known before and had not yet been undertaken
in the old countries of the Eurasian continent. Tobacco was a
strange and unexpected merchandise, without known antecedents, which suddenly burst upon the world, penetrating boldly
with the attraction of its novelty into peoples' habits and arousing an inordinate appetite, like some superlative spice never
tasted before. This unexpected commercial phenomenon gave
rise to large and very expensive foreign imports, which were
not compensated for and upset the balance of domestic economy. But it also upset the groups of established merchants, giving rise to a new and lucrative trade, outside their control, by
which other dealers profited handsomely, generally foreign in-
truders, sea captains whose boats visited the distant ports of the
West Indies, the monopoly-holders of Cádiz and Lisbon or the
pirates and smugglers of Morocco and the Barbary Coast. Be-

cause of the high price brought by tobacco and its relatively small volume, the profits from its trade went to those captains of boats traveling to distant ports, and to a lesser, though still considerable, degree, to such of the sailors as had a share in the venture. This new branch of trade, so profitable and upstart, infuriated the rich local merchants, who were accustomed to having the profits of all usual commercial transactions go into their own pockets, who knew all the procedures, the risks, the tricks, and the gains. Naturally, the great merchants of the seaports could not view with favor this new lucrative merchandise, which changed peoples' habits and was as costly as it was sought after, and which others now sold outside their establishments. And they worked against tobacco until they finally brought it under their control. Once this was accomplished, and the exotic plant was finally grown within the realm, through the sowing of its seeds, which some undertook privately, or by its cultivation in the colonies, the agitation of the opposition died down, even though the tobacco of the Indies, and especially that of Cuba, always brought a higher price and there was always a margin for smuggled tobacco, which was brought in clandestinely and sold to rich smokers.

When tobacco was grown domestically, the profits that had formerly gone to the dealer in foreign tobacco remained within the country. But this was not enough to smooth tobacco's path. The cultivation of tobacco introduced into the country a dangerous ferment of social transformation. Because of its special conditions (which have been analyzed in the preliminary essay of this work), the cultivation of tobacco does not demand a capitalistic economic basis, nor large landholdings. For tobacco the vega suffices, the small vega, not the great plantation. Tobacco is cultivated like a garden product, intensively rather than on a large scale. Because of this agricultural circumstance, it could be grown by humble peasants, who heretofore had been brutally downtrodden, and it was possible for these poor devils to emancipate themselves and even become rich, or at least hope to. But this possible reversal of the respective and traditional roles of serfs and landowners which suddenly loomed up on the horizon like a thunder squall out of the clear sky was not to be calmly accepted by the powerful class of feudal-minded landholders. The appearance of tobacco in the old strongholds of trade set off an explosion of ambitions and fears with its possible threat to privilege; but the appearance of the vega in an old country might touch off a veritable agrarian revolution.

At the same time the governments, alert to their own interests, were annoyed at the sudden appearance of tobacco, without permission from them and without advantage to them. They were surprised by the intrusion of that exotic plant, with its aura of mystery and strange powers, almost a luxury, running counter to old and established customs, and were alarmed at the way the taste for it spread among all classes, as though at the behest of supernatural powers, at the amount of money spent on it for immediate consumption, and the fortunes risked in its trade, the haste with which its cultivation was undertaken, and the manner in which it quickly became the object of the most complicated economic interests, with great social influence and repercussions. And all this behind the backs of the authorities, without their intervention and without benefit to the public treasury.

This seemingly insignificant novelty of tobacco gave rise to various problems and a danger. It would have to be brought under control. If the appearance of tobacco in the old Eurasian world aroused surprise, curiosity, and desire, it also aroused deep suspicion and envy. Some feared that the serfs, by becoming the small planters of this rich crop, might manage to emancipate themselves. The free planting of tobacco on small plots aroused fear, because the vega was a subversion of the traditional order on the land. For this reason, once the tremendous mercantile and agricultural possibilities of tobacco were grasped, it was vigorously combated; and in those nations where an absolute form of government prevailed, accompanied by tyranny, feudalism, and serfdom, such as Turkey, Russia, Persia, China, Japan, and others of the same social texture, the persecution was merciless, with the death penalty and the most dire execrations.

In those countries where such sharp economic contrasts did not exist, tobacco, though hampered in the beginning until it became an object of taxation, did not suffer such drastic measures. In the sixteenth century, in England, Belgium, Holland, France, Germany, Italy, the cultivation of tobacco was introduced; but the climate of these countries was not suited to the cultivation of tobacco, and it was an insignificant crop that could not upset the social balance of the rural population. But in spite of this, even there tobacco was the object of restrictions imposed upon it by the mercantile interests, urban and more powerful than the rural classes, by the shrewdness of the ministers of the exchequer, or for the sake of favoring the development of the colonies, out of fear of opening the door to a new form of agricultural wealth.

Various episodes in the history of tobacco in America confirm this interpretation of the fortunes of the Nicotiana plant in the world beyond the sea. When the colonization of Virginia by the English began, tobacco became its chief agro-economic basis. It was almost impossible to employ small-scale cultivation there; there were not yet enough free settlers, and the growing of tobacco did not threaten any established interest. Moreover, there was a great desire on the part of mercantile interests in London for an abundant supply of expensive tobacco to satisfy the growing demand, and they had ample capital for investment in the project. By the time of the colonization of Virginia there was already a great taste for tobacco in Europe; it had become a precious article of trade, like cinnamon; it was now a business, and a lure for capital. Tobacco became the principal crop of Virginia, and came to be used as money. For this reason, the cultivation of tobacco, which in Cuba from the beginning had been a garden or small-plot crop, was undertaken in Virginia on plantations, giving rise to social consequences completely different from those that occurred in the Antillean countries under the system of *vegas.*

In Cuba itself, on a smaller scale, but for the same reasons and with similar manifestations, those aforementioned economic-social incidents produced by the introduction of tobacco into the old countries of Europe and Asia took place. We have seen that in the old cities of the Continent tobacco had to fight against the hostility of established merchants who were fearful that this new article of trade, so easy to produce, so greatly in demand, and so profitable, might be sold by foreigners or by ambitious peasants and members of the lower classes. A similar phenomenon took place in Havana, the maritime capital of Cuba and of all the American empire of Spain for the same social reasons. When the sale of tobacco began to be a good business for the Negress tripe-sellers, who sold it at a handsome profit to the crews of the ships that put into Havana for a long lay-over, the city council, in an order dated May 14, 1557, forbade them to continue the selling of tobacco and wine in their taverns, thus depriving them of the means of buying their freedom. And the lucrative tobacco trade from then on became the exclusive business of the Spanish traders, who enjoyed a monopoly.

I have also pointed out that the cultivation of tobacco had been forbidden in those countries dominated by feudal landowners to prevent the serfs and peasants from improving their social position by raising a profitable crop on small acreage. The same hostility between the owner of the great plantation and the peasant holder of the small vega took place in Cuba. The conflict did not assume the same transcendence in this island as in the Asiatic kingdoms and in Russia. Here the great landowner was not protected by forbidding the raising of tobacco, or punishing those who planted it with the death sentence, but the conflict existed. Even in the early part of the nineteenth century "the war" between the plantation-owners and the vega-owners was still going on.

I have mentioned, also, that in the Euro-Asiatic nations the opposition to tobacco, which went to the lengths of making its cultivation punishable by death, disappeared when tobacco was brought under a system of production and taxation favorable to the established social and political interests. Both Cuba and Spain saw this same drastic persecution of tobacco, even to the extent of capital punishment. By a royal edict of August 26, 1606, the King forbade all planting of tobacco in America for a period of ten years, simply because this was prejudicial to the interests of dealers in Spain and favored foreigners engaged in smuggling in American waters, and eight years later (royal edict of October 20, 1614) the same monarch permitted planting once more, but ordered that once the local demand for tobacco had been satisfied, all surplus stocks were to be shipped directly to Seville, all this accompanied by threats of the severest punishment, worthy of a Turkish sultan, "under penalty of death." And when, later on, the Spanish government set up a monopoly on tobacco, its wrath was so provoked by those who attempted to evade these regulations that in the years 1719 and

1726 it ordered capital punishment for them, as though they were counterfeiters, and this inhuman penalty against those dealing in contraband tobacco remained in force until 1830. So in Spain, as in Turkey, Russia, and Asia, the executioner played a role in the historical drama of tobacco. But an explanation is needed for a curious discrepancy in the number and social rank of the actors, depending on the theater where the drama was played.

In all these terrible persecutions launched in the countries of the Old World by the rulers and their governments against the infernal tobacco plant, the ecclesiastics of the different sects fulminated their anathemas against it in their role of interpreters of the wrath of God. But nothing of this sort happened in Cuba and in the rest of the Indies, or in Spain, or even in the other Catholic countries. What was the reason for this difference in attitude on the part of the Church toward the same plant in different countries? The clergy of the Old World was opposed, as might have been expected, to the introduction of customs brought in from America. There is a story to the effect that a certain Rodrigo de Jerez, on his return from Cuba to his home in Ayamonte, was accused of being possessed of the devil for smoking a cigar, and that for this reason he spent a number of years in the dungeons of the Inquisition, and when he was finally absolved and released, he found that his neighbors had acquired the Satanic habit of smoking (Count Corti: *A History of Smoking,* London, 1936, p. 50). It is possible that this incident is fictitious, but in any case it is more indicative of the ignorance of tobacco and its effects than a direct attack upon it. In Spain, despite the fanatical intolerance that characterized its clergy, there was an attitude of indulgence toward this devilish habit of the Indies, and its persecution was never carried to the lengths it went in other countries.

This was not the case with the Mohammedans. When tobacco appeared in Turkey the Moslem clergy was shocked and outraged. A legend sprang up to the effect that Mohammed, having been bit by a viper, sucked the wound and spat the snake's poison on the ground, and from this poisoned spittle tobacco grew. The Ulemas carried on the same bitter campaign against it as against coffee, which had recently emerged from black Abyssinia to tempt anchorites and peoples, starting with

Arabia, the land of Mohammed. According to the Mohammedans, tobacco and coffee were both contrary to the religion of the
Koran, which forbade the use of coal. Coffee when roasted became coal; tobacco when smoked became coal. Horrible penalties were promulgated against the users of tobacco and coffee,
and the sultans lopped off many heads, following the pious ad-
monitions of the Grand Mufti.

When tobacco reached Russia from Constantinople, the patri-
archs of the Eastern Church spoke of it as "the devil's plant"
(A. de Gubernatis: *La Mythologia des plantes,* 1882, Vol. II, p.
327). And the legend circulated that this plant had sprung from
the grave of an adulteress and that the devil revealed his actual
presence through its odor and smoke. The Grand Duke of Muscovy and later the czars issued hair-raising sanctions against the
users of the plant, and many a poor wretch lost his life in
those fiercely despotic lands because of a puff of tobacco or a sip
of coffee.

Nor was there a dearth of anti-tobacco diatribes among the
Protestants. The Puritan and Calvinist divines were the most
hostile. Certain Lutherans of the canton of Bern were so severe
that they added one more to the Ten Commandments: "Thou
shalt not use tobacco, either for smoking or in snuff"; and the
use of tobacco was put upon an equal footing with committing
adultery. Even today in the United States there are Protestant
organizations that work, without much success, to eliminate the
foul use of tobacco, comparing it to the use of alcohol.

The Church of Rome was fairly tolerant toward tobacco.
First of all, the sensual habits that characterized its clergy at
the time of the discovery of America and for a considerable part
of the following century would hardly incline them to pay
much attention to this mischievous plant of the New World.
Moreover, the missionaries that came out to convert the Indians
during the sixteenth century in general were not distinguished
by their asceticism. In the Antilles a few years after the discovery the Admiral Viceroy was already sending complaints to
the Spanish King to the effect that:

"In the island there is great dissoluteness among the clergy, because many of those who have come out lead wicked lives, and some
renounce their calling to live the life of the laity, jousting, going off
to the hills with the women who take their fancy. Only those having

HIS FIRST CIGAR

STRENGTH AND WEAKNESS

Efectos de su primer cigarro.

EFFECTS OF HIS FIRST CIGAR

Una familia de fumadores.

A FAMILY OF SMOKERS

a true vocation and have given proof thereof should be allowed to come out here." (Duquesa de Alba: *Autógrafos de Cristóbal Colón,* Madrid, 1892, p. 81).

It is on record that Hernán Cortés in Mexico requested Charles V not to send out priests, but only friars of austere habits, for the priests of the Indians were so strict in their conduct, simplicity of life, and chastity that if they realized the self-indulgent and licentious life the Spanish clergy led, if "they saw the lack of piety that characterizes them in these kingdoms today," they would come to consider Christianity a farce, and it would be impossible to convert them. The Coroneles brothers also advised Charles V that in order to convert the Indians "he should sent out ministers who received from them only their food and clothing, otherwise they will reap no spiritual harvest among them." In spite of this wise counsel on the part of those who were familiar with the corruption and greed of the Spanish clergy of those days, secular priests kept coming out to New Spain, winning for themselves a very doubtful reputation (See Fray Juan de Torquemada: *De la Monarquía Indiana,* Book XV, Chapter i, p. 10). And unless the description of the life of the friars by Fray Pedro Durán in a letter to King Philip II in 1583 is exaggerated, it would seem that the members of the religious orders were not much better than the regular clergy (J. T. Medina: *Historia de la Inquisición en México,* Santiago de Chile, 1905, pp. 11, 12).

In that first half of the sixteenth century neither in the parishes or convents of Spain nor in the see of Rome was there the austerity necessary to withstand the temptation of this "evil plant," as tobacco was termed by certain moralists. There were no papal decrees against tobacco ordering fasts or abstinence, as in the case of appetizing meats.

For this and other reasons the use of tobacco spread rapidly among the Spanish clergy, which gave itself over to the inhaling of its powder and smoke with an enjoyment more pagan than devout and, into the bargain, derived from its cultivation great profits, especially through the ecclesiastic tithes and the pious gifts from the small farmers. Notwithstanding, some measures were taken against tobacco by the Church, but rather to check certain abuses than to prevent its normal use. Smoking is not a sin, said the scholastics, but it may be, if carried to excess, just as gorging on some simple food to the point of sickening and dying of indigestion becomes the mortal sin of gluttony. The definition of sin is perhaps excess.

During the sixteenth century, and later in both America and the Peninsula, clergy and laity indulged equally in the use of tobacco. In Spain by the beginning of the seventeenth century tobacco was used inside the churches by men and women, parishoners and priests. So great had the abuse become that in 1624 the Pope, Urban VIII, in response to a written petition from the dean of the cathedral council of Seville, issued a bull threatening the users of tobacco on holy premises with excommunication. The papal threat must have been idle for a few years later a Spanish priest was fulminating against the excessive use of tobacco. Father Tomás Ramón in his *Nueva Pragmática de Reformación* (Zaragoza, 1635), was very outspoken in his moral campaign. The fact that in the service for the dead it says: *"Pulvis eris et in pulveris reverteris"* was no excuse, he wrote, for the priest to put tobacco dust in his nose. It is offensive to God, said Father Tomás Ramón, for priests using this filthy, profane weed to touch the bread and wine of the Eucharist. And he added that it was ungodly for clergymen unconcernedly to take snuff or smoke "papers" before celebrating the Mass. About the same time another Spanish priest was upholding the theory that the use of tobacco reduced men "to the state of beasts" (Bartolomé Ximénez Patón, in his edition of Hernando de Talavera: *Reforma de Trajes,* Baeza, 1638).

Evidently the canonical diatribes were not too effective, even though several historians testify that around 1692, as a consequence of these bulls, five friars were condemned to death and entombed in a wall in Santiago de Compostela for the crime of having smoked in the choir during the divine services (Brooks, loc. cit., p. 81). But this unbelievably cruel procedure hardly seems compatible with the atmosphere of tolerance encountered by tobacco in Spain, as gentle and kindly as the plain-song chanting of prayers.

From the Indies via Spain tobacco made its way into Italy and from there to the Levantine lands. Just like syphilis, and for this reason Father Labat remarks that tobacco was eyed somewhat askance. Tobacco bore this stigma; some said it was a cure for the disease, others distrusted their historical association. The customs and habits of the Peninsula spread through Italy, which at that time was strongly under the influence of Spanish imperialism, and among these were the use and abuse of tobacco. According to Antonio Nezi (*Vicende storiche di una pianta conquistatrice,* p. 477), cigarettes (known as *spagnolette*) and snuff (called *pulviglio de Siviglia*) became very fashionable. The vice of tobacco (was it a vice?) penetrated to the pontifical throne itself. Pope Innocent X, in the year 1650, ordered excommunication *ipso facto* for all who profaned St. Peter's by using tobacco, "Spanish tobacco." Neither did this anathema seem to carry much weight, for it had to be repeated by Innocent XI in 1681 and, it is said, even by Innocent XII. The memory of St. Joseph of Cupertina had to be stoutly defended by his sponsors against the charge that he was a great user of tobacco, preferred by the ecclesiastical court's "devil's advocate," opposing his canonization. And to the latter's denial of their candidate to sainthood's "heroic virtue" for not having resisted the vice of smoking, his skilled defenders argued that tobacco was of great help to men of the cloth in resisting the temptations of the flesh and preserving a pious peace of spirit (Brooks, op. cit., p. 81). An attempt was made to deny "heroic virtue" because he smoked and took snuff to one who, despite these charges, is now reverenced as a saint, St. John Bosco. And it is said that an equally unsuccessful attempt was made, not long ago, to prevent the beatification of that irascible Father Antonio M. Claret y Clará, who during the past century was Archbishop of Santiago de Cuba and confessor of Queen Isabel II, famed for her sensuality. There are many anecdotes about his excessive fondness for tobacco and chocolate.

This abuse of smoking and snuff-taking, to the point of profaning the sacrament of the transubstantiated body and blood of Christ by indulging beforehand in tobacco, in smoke or powder, was very widespread in those days. Antonio de León Pinelo (*Questión moral, Si el chocolate quebranta el ayuno eclesiastico,* Madrid, 1636), studies one hundred and eighteen current drinks and holds forth somewhat pedantically on the question of whether tobacco is a food or a drink or neither, and expresses himself against the excessive use of tobacco and chocolate by the clergy. The famous jurist and priest Don Juan de Solórzano y Pereyra, in his *Política Indiana,* says:

"Eduardo Vestono (Vestonus in *Theat. Vitæ civil,* Book w, Chapter xxxix, p. 314, V) like a prudent man points out, and like a learned man proves, along with others, the reprehensible nature of this vice in all, in general, but especially among priests and friars, who have no scruples about indulging in it before celebrating Mass. By so doing, in the opinion of Antonio de León, they break the natural fast, and in mine and in that of all who understand the Eucharist well: as was pointed out to them with penalties and censures by the councils of Lima, Mexico, and the Canary Islands (Concil. Limens, III, Act. 3, Chapter xiv, Mexican, Lib. III, tit. 15, S. 13, and Canariense, *anno* 1629); adding that they should not take it even two hours after having said Mass, both because of the indecency of so doing and because taking it often provokes vomiting or too much spitting or clearing of the mucus; all of which are in my opinion sufficient reasons for not considering it a good thing for those who give themselves over to this disgusting pleasure."

Nevertheless, there are the casuistic opinions of St. Alphonse of Ligorio and of Benedict XIII and the theologians quoted by the former in *Práctica e Instrucción de Confesores* (Tratado XV, punto iii), which uphold "that tobacco taken through the nose does not break the fast, even though a portion of it should descend to the stomach," nor "does the smoke of a cigarette break it," nor even tobacco chewed or "ground by the teeth provided the juice is spit out." In the opinion of the saint of Ligorio, however, it is "an indecent thing to do this before receiving communion."

In Cuba the improper use of tobacco and chocolate in the churches had to be forbidden. The diocesan synod convoked by the Bishop Juan García de Palacios in 1684, in the articles it drew up, forbade the "puffing of cigars in the aforesaid churches and their vestries" (*Tit. nonus,* C. II) and the priest's "taking the smoke of tobacco before celebrating Mass" (*Tit. decimus,* C. III). The synodal bylaws of the diocese of Havana in 1888 still ordered priests to refrain from smoking before Mass, as well as in public, and to take care to prevent the fingers *"quibus Sacratissiman Hostia tangere debant"* from becoming stained (Lib. Tit. I., C. V). Nevertheless, and especially in Cuban dioceses, the sin of smoking at dawn and at all other hours are always *peccata minuta,* which do not imperil the soul and can easily be wiped out by the pious use of certain indulgences.

It is evident, then, that the canonical measures against the use of tobacco boil down to mere recommendable rules of liturgical etiquette or simple ecclesiastical hygiene. It was not the Catholic clergy that fostered the animadversion against tobacco in Europe and America, as did the Protestants. The Holy Office, so unremitting in its pursuit of sorcery and the snares of Satan, took no action against tobacco. We have read that the Inquisition in the Canary Islands confiscated some snuffboxes carried aboard an English ship, but this was not a move aimed at tobacco or even against trading with heretics. It was because one of the boxes had painted on the cover a head surmounted by a tiara and an inscription reading: *Æclesia perversa tenet faciem diaboli* (H. C. Lea: *The Inquisition in the Spanish Dependencies,* New York, 1908, p. 178). This was antipapal propaganda against which the Inquisition had to take measures.

In Spain, even though they were kindly toward the other devil-sent companions of tobacco, too, Quevedo speaks of the harm that was done by bringing from the Indies "chocolate cups and mixers," and other writers also satirized chocolate; but the clergy did not combat it. On the contrary, *cacao* was known as *teobroma,* "food of the gods," and the beverage of the Aztecs, good and thick, as the Spaniards prepared it, came to be typical of the sacristies, as the refection after Mass, and of the convents, famous for good food and easy living.

Without doubt, coffee, too, is a diabolical nectar. Even in the nineteenth century the statesman Talleyrand drew on the infernal regions for metaphors to describe it, saying: *"Noir comme le diable, chaud comme l'enfer, pur comme un ange et doux comme l'amour."* But this simile is inaccurate; coffee is not sweet if one does not add sugar to it. The typical and most pleasing feature about coffee is its bitterness. If the "sweetness of love" in this definition were changed to the "bitter stimulant of doubt," it could be applied with equal fitness to both tobacco and coffee. For this reason when these boon companions, similar in tastes, ideas, and purpose, tobacco and coffee, met in Constantinople, they joined forces in their mission of tempting mankind, keeping his thoughts and his doubts aglow. But the clergy did not ban "black water," either, in spite of its Moslem and Ethiopian origin. Their experience with tobacco and the pleasurable taste of "Moorish wine," which was the name given coffee, counseled against excommunicating this inspiring black nectar, a kind of "liquid tobacco," to ensnare human beings and arouse them from the chill and lethargy of spirit. The Church did not need to concern itself about Asiatic tea, for it never was too popular in the lands over which Rome held sway.

It is not too far-fetched to assume that the monarchs who proceeded so severely against tobacco did so because of enmity toward Spain, because the plant in question was not one of the products of their kingdoms, and because of economic disadvantages and understandable human jealousy aroused among governments and merchants by the growing demand for this American plant.

It would take, perhaps, too much metaphysical audacity to put forward as a certainty the god Pluto's advice to his godson, the semigod Tabaco, to seek support on earth among the poets; but it must have happened as it is told, for the friends of the Muses were, as a rule, defenders of the son of Bacchus. "Divine Tobacco," says the English poet Spenser, as early as 1589 (*The Faërie Queene,* Book III, Canto 5). And many poets of all nations and races have found in tobacco a lode of inspiration, a helpful go-between in wooing the coy muses, which is an essentially diabolic service that the poets have often esteemed a favor of the god. For this reason the poets were spirited defenders of the semigod Tabaco and of the "tobacchic" rites. At the beginning of the seventeenth century someone hit upon the metaphor that the poet Chaucer was probably cultivating tobacco on the slopes of Parnassus. The poets' love for tobacco would seem to be a fatal consequence of their own genius. "Poetry comes from the devil," writes Maximilien Rudwin (*Les Ecrivains diaboliques de France,* Paris, 1938, p. 18), recalling that, according to Raymond de la Taihede, "poetry is nothing but revolution" and that the devil is the father of revolutions. The reason Dante and Milton painted the pits of hell better than the regions of paradise was because they were true poets and for that reason, without knowing it, on the side of the devil.

Nevertheless, despite Pluto's counsel, there were poets who

were fawning courtiers and faithless to Apollo and who, under the inspiration of sterile muses of withered austerity, raised their voices against the evil-smelling devil of the Indies. The following is a lyrical imprecation from the pen of Barchayo (*Sub nomine "Euphormionis," in satirico*), "written with no less elegance than truth," according to Father Juan de Solórzano, who quotes it in his work on jurisprudence:

> *Harmful and wretched plant,*
> *Whose foul vapor reeks of death,*
> *Not for nothing kindly Nature*
> *Kept you remote from us*
> *In distant lands. Who was the fool*
> *That in a sorry ship and hour*
> *Brought you here?*
> *Have we not ills enough,*
> *Wars, hunger, poisons to kill us?*
> *But who can count those you occasion?*
> *Your nauseous vapors infect*
> *The pure air like those of Avernus,*
> *And suffice to slay all they touch.*
> *The infernal Furies could not torment*
> *With suffering worse the unhappy shades,*
> *And had Cacus in his struggle with Alcides*
> *Breathed forth your reek, he would have won,*
> *And had Antiquity, instead of hemlock,*
> *Made use of you, born as you were,*
> *Of Cerberean foam, and ordered the cursed*
> *Son who with blood-stained hand*
> *Violated his father's hallowed age,*
> *Instead of fire and drowning (punishment light)*
> *To drink, the more to suffer for his crime,*
> *Your clouds of smoke, infamous weed. . . .*

Greater contempt could hardly be imagined: "Born . . . of Cerberean foam"—that is to say, of the slavers from the mouth of that mythological three-headed dog that stands guard at the gates of the lower world where His Cloven-footed Majesty reigns, whose foul slobberings were sucked up by the witches at their Saturday night gatherings, according to the lengthy and detailed accounts of the proceedings of the Holy Inquisition, which at that time acted as the health department against all manner of diabolical plagues.

The theological wrath that has always been opposed to the introduction of novelties did not prevent other poets, "no less elegant than truthful," from extolling the medicinal virtues, proved or imagined, of the mysterious plant. Ben Jonson in 1598 (*Every Man in His Humor*) called tobacco "the most sovereign and precious seed the earth has offered man for his use." If there were those who cursed it, calling it "plant of the Devil," others hyperbolically termed it "sacred plant," "divine plant." In 1603 Marbecke wrote that tobacco was "the wine of God." The poet Raphael Thorius, or Torio, asking for tobacco in his apologetic poem, written before 1610, begs that he "may absorb God in my head" (*Hymnus Tabaci,* Leiden, 1625).

In 1683 Molière would write that *"Le tabac est divin, il n'est rien qui l'égale."* In his *Don Juan* he says that "tobacco inspires sentiments of honor and virtue," that it is "the passion of honorable men," and, finally, that "who lives without tobacco does not deserve to live." And Cohausen agrees that "tobacco is the king of the vegetable kingdom, the lord that reigns over all parts of the planet," and says that in all nations "the nose is its slave." Tobacco, adds Cohausen, "is for work and for repose," is the companion of princes in their courts and of the peasant in his hovel, accompanies the armies on their campaigns and the muses in the home of the writer.

Shakspere says nothing about tobacco, which was already known in his country in his day, but centuries later Lord Byron was to be the first poet to sing the excellences of the cigar. That is to say, in England, for in Spanish literature the friar Tirso de Molina had already praised the "tubes" of tobacco, "to say the final blessing."

The god Pluto may have overlooked advising his godson Tabaco to seek the support of the Church, but it was soon given him by the high dignitaries of the Church. Perhaps the king of the lower world had foreseen this.

Tobacco was not introduced into France from Spain. The unremitting wars between the two countries made trade difficult, and their contacts were not intimate. I have already mentioned that a friar who later renounced his vows, A. Thevet, proudly credited himself with having first brought tobacco to France from the lands of Brazil, years before the doctor Jean Nicot, who is generally held to have introduced tobacco at the court of Catherine de' Médicis, after having made its acquaintance in Portugal (José Rivero Muñiz: *"¿Quién fué el primero, Thevet o Nicot?" Revista Tabaco,* Havana, Year VIII, No. 81, p. 9).

In any case so high-ranking a member of the clergy as the Grand Prior, Cardinal François, Duke of Lorraine, to whom Nicot sent seeds and plants from Lisbon in 1560, played an important part in the introduction of tobacco into France (some say only of snuff); in his honor the American plant received the name of *herbe du Grand Prieur*. This dignitary, to use Brooks's phrase, was "the missionary of tobacco."

In Italy the use of tobacco must have begun slowly in Sicily, Naples, Milan, and other regions under Spain's domination at the time. Soldiers, sailors, and members of the Church must all have done their part in spreading it. But tobacco reached the Papal States thanks to two cardinals. Tobacco was taken to Rome before 1585 by Cardinal Prospero de Santa Croce, who had learned to use it in Portugal. For this reason tobacco was known there as *erba Santa Croce,* according to Castore Durante, who states this in his *Herbario Novo* (Venice, 1602). Another wearer of the purple, Cardinal Crescensio, is credited with having introduced Rome to smoking and having taught Pope Urban VII to take snuff on his return to Rome from England in 1590 (Antonio Nezi: *Vicende storiche di una pianta conquistatrice*). This may be the pontiff J. J. Ampére has in mind (*Promenade en Amérique,* Paris, 1855, Vol. I) when he relates that a pope offered his snuffbox to the head of some religious order, and the latter declined to take a pinch of snuff, saying: "Your Holiness, I do not have that vice," to which the Pontifex Maximus replied: "If it were a vice you would have it." The same anecdote is told in Cuba, where it is attributed to Archbishop Claret and a canon of the cathedral of Santiago, sometimes in connection with snuff and sometimes with chocolate. In Italy tobacco was also known as *erba Tornabuona,* because its cultivation had been started in the Papal States by Cardinal Niccolò Tornabuoni in 1574. As can be seen, tobacco was a cardinal's plant, which says much in its favor, for from times immemorial the term "fit for a cardinal" has been applied to the most exquisite and selected delicacies that would appeal to the sensual appetites.

Tobacco was bringing the peoples of all lands under its spell. As the Cuban poet Narciso de Foxá says:

> *Tobacco! Its delightful aroma*
> *Charms the wise man and makes the fool mad.*

Everybody gave himself over to this "dry drunkenness," as it was then termed. It was "the drunkenness of the temperate man" (*The Wandering Jew,* London, 1640). Kings, statesmen, theologians, and moralists thundered anathemas in vain and set their executioners to work on the adherents to that delightful temptation of the spirit that had come in from the Indies. Other objective and open minds, like those of doctors, naturalists, and seamen, or idealists like poets and dreamers, defended with their tales and poems this smoke that had come to them from the pagan New World to titivate the brain. It was another episode in the eternal conflict between error, which parapets itself behind theology and the power of the state, and truth, which seeks to find itself in experience and free thought.

The doctors were the stoutest defenders of tobacco. To all the therapeutic virtues attributed to tobacco by the healing magic of the American Indians, the physicians of Europe added new ones, some real, some imaginary, which they had derived from empirical applications and from literary channels. There were no limits to the lucubrations of those who regarded it as a panacea. It would take too long to mention the countless ailments for which tobacco was prescribed. There is no doubt that tobacco had medicinal applications that were known not only to the Indians but to the whites in America who learned their value, and the latter transmitted their knowledge and enthusiasm regarding its marvelous curative properties to their countrymen in the Peninsula. Even the clergy had no hesitation in making them known. Father Toribio de Benavente, known as Motolinía, in his *Historia de los Indios de Nueva España* (Barcelona, 1914, p. 78), written about the middle of the sixteenth century, alludes to the medicinal qualities of *picietl,* which was the name given tobacco by the Aztecs. There comes to mind the encomiastic chapter on the subject from the pen of Father Bernabé Cobo, of the Company of Jesus, in his *Historia Natural,* published in Seville toward the close of that century.

The laity of all Europe, and particularly the doctors, credited tobacco with marvelous cures. For example, during the great plague in England, in 1665, tobacco was believed to provide immunity against contagion, and this contributed greatly to its fame. Dr. Monardes was of the belief that certain ointments of tobacco placed on the navel and lower abdomen alleviated the discomforts of pregnancy and childbirth. Since in those days all medicine, both the professional and the home-remedy variety, set great store by miracle-working ointments, a tendency Cervantes satirized in Don Quixote's famous balsam, tobacco arrived just at the right moment to play its part in these magic salves. Those doctors convinced of tobacco's great value administered it in every kind of compound, poultice, infusion, powders, for fuming, for chewing, for inhaling.

In certain countries of Europe curious phenomena of transculturation sprang up through the channels of medicine. Certain doctors came to recommend the introduction of tobacco smoke into the body, not through the mouth, but from the opposite end. Enemas of smoke were known in Switzerland, Germany, and other countries of Europe (Brooks, op. cit., Vol. I, p. 55), probably as the result of a vague recollection of certain practices of the Indians. As late as 1844 in Scandinavia in certain illnesses the nose of the patient was stuffed full of tobacco (Brooks, I, 19, note).

Unquestionably tobacco was discovered by Europeans in an epoch that was propitious toward receiving it as a panacea. The superstitious beliefs in miracles and magic of the Middle Ages had not yet disappeared, and the experimental curiosity of the Renaissance was just coming into flower, although it had not yet expressed itself in scientific formulæ. And tobacco was at one and the same time a thing of wonder and a thing of science, a substance that attracted interest as much by reason of its exotic mystery and semifabulous origin as by the strangeness of its workings and the unexplored field of its possible applications, all of which offered countless opportunities for experimentation to curious physicians and for the snares of charlatans and quacks. Moreover, in connection with the most usual methods of taking tobacco—that is, in smoke or in powder—the medicine of Europe in those days was greatly given to the provoking of sneezing and nasal irrigations and fumigations "to clear the head"—*caput purgia,* it was called in Latin by the doctors. Powders to induce sneezing were very much in vogue, powdered pepper, myrrh, sneezewort, white hellebore, spurge, and so on, and these powders were introduced into the nose through the quill of a feather or in some similar way. This can be seen in the prescriptions of the celebrated Paré, sixteenth-century doctor (J. F. Malgaine: *Œuvres complètes d'Ambroise Paré,* Paris, 1841, Vol. III, p. 586). Snuff and cigars from the Indies chose the right moment to invade Europe.

It was a Spanish doctor, Francisco Hernández de Toledo, who recommended tobacco to King Philip II of Spain. According to certain authorities, he brought the seeds back with him from Mexico. It was another man of science, Dr. Nicolás Monardes, the physician of the Archbishop of Seville, who defended tobacco there in his book on medicine (*Segunda Parte del Libro de las cosas que se traen de nuestras Indias Occidentales que sirven al uso de la Medicina,* Seville, 1571). Perhaps Monardes was tobacco's greatest advocate. His opinions were translated and adopted by many physicians of other countries. Another doctor, a courtier and ambassador, Nicot, introduced it into France and gave it his name, from which the word *nicotine* has come. And it was an English doctor, Cheynell, who defended it against James I, not behind his back, but to his face during the visit of the monarch to the University of Oxford. To flatter the King the faculty of the university had declared itself against tobacco and had received with reverence the King's adverse and unqualified decision, when that bold doctor got to his feet and, with a pipe in his hands, extolled the virtues of the American plant. In Italy the doctor Castore Durante, praising the introduction of the tobacco plant by the Cardinal and Nuncio Prospero de Santa Croce, said that His Eminence would be rewarded for this act, just as his ancestors had been for having brought to Rome the wood of the cross on which Jesus Christ died.

The doctors were not impressed by the diatribes of kings and theologians branding tobacco as infernal and all its uses as nefarious. When Dr. William Barclay published his encomiastic work *Nepenthes, or the Virtues of Tobacco,* in Edinburgh in 1614, he dedicated it to "My Lord Bishop Murray," and he not only defended the aromatic plant, but referred to the land where it was grown as "the country God has honored and blessed with this happy and sacred plant."

Depending on the occasion, friends or enemies of tobacco predominated, but until now the thing that has triumphed over all is money, that spirit "of all the devils," which perturbs scientists and philosophers alike. Rightly enough Father Labat said:

". . . when it was a question of tobacco, the doctors did not forget to take advantage of the right they had acquired to pass judgment on all things. Even though they had never seen tobacco nor heard its name, not on this account did they refrain from discussing its nature, its properties, and its virtues. . . . To be sure [adds the ironical Dominican friar], reasoning as doctors do, without bases for their opinions, they almost never come to an understanding." (*Nouveau Voyage aux isles de l'Amérique,* Paris, 1713, Vol. VI, p. 276.)

But with equal justification the doctors could have retorted that the clergy is very given to preaching on the subject of things it has never seen, without excepting Father Labat himself, whose account contains crass errors. If it is not true that tobacco received its name because it first came from the island Tabaco or Tobago, as Dr. Monardes maintained, from whom Labat took the statement, neither was the plant or its name discovered in Tabasco, or in Yucatán about the year 1512, as the friar mistakenly contends—an error still repeated by certain dictionaries and historians who have followed a false trail.

Among the doctors it was customary to denounce the abuse of tobacco and to recommend that this *"sana sancta"* plant should not be employed without a doctor's prescription, to which the habitués of the weed answered that this was professional interest. From the middle of the seventeenth century there were satires written against the doctors who profited economically by the fad for tobacco.

Dogmatists and scientists bowed to the diabolical spirit of tobacco when, despite the martyrdom imposed upon its users, it managed to make its way among the highest and lowest classes, and became the source of lucrative taxes, tithes, revenues, and monopolies, not only for the functionaries of the crown but for those of the altar, too. And here also this was a result of the power of money, which the outspoken archpriest

Juan Ruiz had already observed in the court of king and pope
and commented sarcastically upon. When the royal councilors
realized how easy it was to levy taxes upon tobacco as an article
of pleasure, the persecutions came to an end, the moralists
ceased to inveigh, and consciences grew calm, while the devil-
spawned tobacco of the idol-worshipping natives of America
was allowed to contaminate the world in exchange for the payment of high taxes to their exalted governments. It was then
that the unspeakably cruel Sultan of Turkey, his eyes having
been opened to the economic advantages of tobacco, nullified
the *irade* ordering smokers impaled, and the fury of the
Ulemas died down. One Grand Mufti had ordered the persecutions in the name of God; later another Grand Mufti ordered a change of attitude, whether inspired by Allah or not it
is impossible to say. The same thing happened with coffee, first
condemned as being contrary to the laws of the Koran, and
then extolled as the "wine of Islam" as against the "wine of the
Christians." If earlier there was a legend to the effect that coffee was a drink brewed from the excrement of cloven-footed
devils, later there came a new legend, of Persian origin, piously
explaining that once when Mohammed was overcome with fatigue and sleep, God revived his prophet, sending him by the
Archangel Gabriel a drink heretofore unknown, as black as
the holy stone in the Kaaba at Mecca. Thus coffee had descended from heaven as a gift of Allah, and Turkey became an
outstanding country in the consumption of tobacco and coffee.

But it was not only in Constantinople and the Asiatic kingdoms that these changes occurred. Although King James I of
England had in 1603 published his violent *Counterblast,* or
Misocapnos, seu lusus reius de abusu tobacco, he ceased his attacks when tobacco began to bring him in revenue. And not
long afterwards, in 1628, another work was published in answer to the English King's furious pamphlet entitled *Anti-
Misocapnos,* on behalf of tobacco, its use and cultivation (R.
Brunet: *De la culture et fabrication du tabac,* Paris, 1903, p. 6),
under the auspices of the Polish Jesuits, or at least of certain
members of the Jesuit International living in Poland.

The Jesuits were now those who most favored the consumption of tobacco *ad majorem Dei gloriam,* with an eye to their
own plantations, particularly in America, which brought them

in great wealth through the exploitation of vast territories, with Negro slaves and Indian serfs to work them, and exempt from taxes, assessments, and levies of every sort. Ibáñez, an eyewitness, said of the Jesuit Kingdom of Paraguay that "from sunrise to sundown one hears them [the members of the Company] talk of nothing but plantations, ranches, herds, roundups, farm work, dealings in leather, yerba maté, tobacco, and cotton" (quotation from Father Miguel Mir: *Historia Interna Documentada de la Compañía de Jesús,* Madrid, 1913, Vol. II, p. 230). The German Jesuit poet Jacob Balde, published a satire against the abuse of tobacco on the part of doctors, and against the doctors believing in panaceas (*Satyra contra abusum Tabaci Medicina Gloria* (Munich, 1651), but not against the use of the plant itself, which he held in as high regard as did his Spanish fellow member Father Bernabé Cobo. It was these same Jesuits, those who had so upset orthodox theologians with their heretical "Chinese rites," who introduced into the Celestial Kingdom powdered tobacco and the fine art of taking snuff, under the Manchu dynasty, about 1715 (Laufer, op. cit., p. 32). The Chinese converts to Christianity were known among their countrymen, because of this vice, as "snufftakers." The marked leaning of the members of the Company of Loyola toward snuff, as well as the hatred they aroused because of their meddling in political affairs, was responsible for the general belief in "Jesuit snuff," by means of which they secretly poisoned their enemies, according to the legend that circulated against the Jesuits in Spain (B. E. Hill: *A Pinch of Snuff,* London, 1840, p. 39; Fairholt: *Tobacco, Its History and Associations,* London, 1876, p. 255).

Tobacco had become a powerful fiscal instrument, and a number of tobacco monopolies had been organized to assure the lean coffers of the state a handsome income. The Papal States made good use of the method. In Ferrara the pontifical authority, by order of Alexander VII, in 1657 set up a monopoly on tobacco *"in corda, in polvere,"* etc. This same Pope in 1660 ordered a similar monopoly in Bologna, and in 1665 it was established in Rome itself (Corti, op. cit.). In 1725 Pope Benedict XIII of his own accord abolished the dead-letter prohibition against snuff. His Holiness enjoyed the tickling and stimulation of the Indian weed, and the taking of snuff was permitted even at St. Peter's, not in the scandalous, unabashed manner that had prevailed earlier, but "with discretion" (Brooks, op. cit., Vol. I, p. 81). Snuff had become so much a part of the clergy's customs that they were unable to do without it even during the liturgical rites, and they even put their snuffboxes on the altar, according to Brunet and others (cited by Brooks, Vol. I, p. 157). In 1779 Rome opened a tobacco factory, and the Pope gave a certain German merchant the privilege of manufacturing cigars in the Spanish manner, known as *bastoni di tabacco,* in the States of St. Peter (Brooks, Vol. I, p. 167).

Father Labat had shrewdly observed of tobacco that "never was a thing so universally accepted, in spite of the contradictions, obstacles, and enemies that attempted to smother it in the cradle" (op. cit., p. 286). Never was greater propaganda carried on in the world than that in favor of tobacco except the apostolic *"propaganda fide"* of the Catholic religion. By the beginning of the seventeenth century an English poet (Joshua Sylvestre: *Tobacco Battered,* London, 1616, p. 17) said that "Don Tobacco" had more followers than Christ, and that smokers, or tobacconists, in their fanaticism were like the Jesuits, "real idolaters and enemies of the State."

Through the all-powerful intervention of money, peace was made with the devil everywhere. And kings and prelates smoked fine cigars without stigma or heresy, and surely without even sin if they were Havana cigars, always considered "the supreme good." Before the end of the seventeenth century "priests' cigars" were being manufactured in the Indies, and were the most expensive and desirable (A. O. Exquemeling: *Buccaneers of America,* London, 1684). The best Havana cigars were set aside for the King of Spain (Fairholt, op. cit., p. 217). From this came their name *"regalias"* which means "privileges of the king," and on the best brands of Cuban cigars such names as Regalia, Crowns, Scepters, Imperials, Queens, Princes, Royals, Kings, Queens, Isabelas abound, and other similar allusions to the royal prerogatives. But the clergy did not want to be left out, and there were special brands for them, too. "Havana cigars," said Fairholt in 1859 (ibid., p. 217), "vary in length and thickness; there is one kind that is particularly thick and of superior quality that is made for

LATE NINETEENTH-CENTURY CIGAR-BOX LABEL

REPRESENTING A TOBACCO WORKER, A PLANTATION,
AND BALED TOBACCO

TWENTIETH-CENTURY CIGAR-BOX LABEL
REPRESENTING VARIOUS MANUAL OPERATIONS
IN THE MAKING OF CIGARS

priests, out of selected leaf, for members of the clergy and manufactured by monks." It should be noted, however, that these
sacerdotal cigars had no special brands to indicate their purpose. There were regalias, but there were no pontificals; coro-
nas could be bought at the tobacconist's, but not mitres. We had
Queens and Princes, but no "cardinal's delight" nor "canon's
breva." This was probably due, as I have already indicated, to
the bar sinister of its original demoniacal origin, which tobacco
never wholly lost. For this reason, among the numberless commercial trade-names invented to distinguish brands, shapes,
manufacture, there have never been any Catholic denominations, as has been the case with so many other articles of trade.
The "Powders of St. Augustine" we knew in Havana were
tooth-powder, for the powders of tobacco never received saints'
appellations, in spite of their great use and enjoyment by
church dignitaries and devout Catholics.

From 1851 the papal court did away with the last trace of
suspicion against this plant which was called by some "devil's
plant" and by others "holy plant." Cardinal Antonelli, the com-
petent fiscal administrator of the Papal States, ordered earthly
imprisonment, not threats of other-world punishments, which
did not always carry too much weight, against anyone combat-
ing the use of tobacco by writing or rumors (Corti: *A History
of Smoking,* London, 1931). And so, in the end, kings, divines,
merchants, doctors, poets, and devils were all satisfied.

Perhaps certain traces of its religious origin still survive in the
use of tobacco. The symbolic offering of tobacco, employed
among the Indians before the discovery of America in sign of
peace and friendship, is still practiced among white, Christian
peoples, especially in America, like the exchange of incense and
kisses administered in Catholic churches with the pyx and the
ritual *pax vobis.* The dandies of the European courts of the
eighteenth century offered one another snuff with great polite-
ness. Even today it is a rite of cordial welcome to offer a cigarette or a cigar. And if it should be a Havana cigar, the gesture
acquires a certain symbolic distinction and solemnity, like
drinking a glass of champagne. In the manufacture of tobacco,
admixtures of other ingredients are employed, not precisely to
improve the product, but to differentiate it, to distinguish it
and give it mysterious and exciting properties, which brings to

mind those mixtures of powders, alkalies, and perfumes employed in the Indian rites. Centuries ago, when snuff was manufactured on a large scale, perfumes, spices, and essences were added to the ground Nicotiana; and when tobacco was manufactured in quantities for chewing, each maker had his own formula, like a magic secret, to give his plug tobacco a special taste and effect. The boy who smokes his first cigar experiences a certain solemn emotion, as though it were a vestige that had survived of the rites of initiation into manhood. Certainly in the nervous and psychic catharsis induced by tobacco there is that same ineffable element as of yore which is the essence of religion. But there are no longer anathemas against tobacco, nor is it linked to supernatural powers except in poetic and folkloric flights of fancy. There remain a few stubborn opponents among certain religious sects and doctors of alarmist tendencies. These are the same who formerly defended it to the death and who now, from time to time, with the same liberty exercise their right to warn us against its evils, as real or as imaginary as were its virtues. This is the privilege of science, always on the alert and always receptive to doubt.

But it is said that the devil has not died, that he lives and moves, and whatever he does is deviltry. Not long ago a cigarette was invented in Hungary that when lighted gives off smoke of different colors, so ladies can match it to their dresses. Thus Madame will have to select not only hat, slippers, purse, and gloves to match her gown, but cigarettes as well. This is something new the devil has thought up to tempt mortals to sin, but he knows what he is doing and keeps his eye on his work. It would seem that he had tempted a contributor to a certain English Catholic journal (D. W., "Talking at Random," copied from the *Tablet* by the American religious journal *Orate Fratres,* October 1939) who suggests that this idea of cigarettes burning with polychrome smoke might be utilized to beautify the ceremonies of the Church, having the incense change color, like the chasubles, according to the service: red on martyrs' days, white for virgins, purple during Lent, and even green on occasion. Who can say where the transculturation of tobacco may yet end? With a little more imagination and candor it might be possible to use Havana tobacco in rites demanding the use of incense. Perhaps the day will come when

in the diocese of the Amazon or the Orinoco leaves of tobacco
will be burned on the coals of the newly lit fire for the resurrection ceremonies of Holy Saturday instead of pellets of incense. And if it is inspired by faith, there will be no sacrilege
in this, only the syncretic adaptation of an aromatic token of
reverence that was formerly consecrated to their gods by the
infidels. This has been happening for a long time in certain religious missions of Africa where the drums of the sacred *bembé*
are used to accompany hymns at the altar of the cross, and the
disguised devils dance their mysteries to give greater solemnity
to the sacramental ceremonies of baptism and confirmation,
which are the magic initiations of the Christian neophytes.

The devil has conquered. Tobacco holds sway over all mankind. "This Indo-Antillean word has made its way into all the
languages of the world and is understood everywhere" (Laufer,
op. cit., p. 65). Today tobacco everywhere is a luxury that is a
prime necessity. What a devilish paradox! There are lands that
do not eat bread and where wheat is unknown, but none that
are unacquainted with tobacco or do not know how to smoke.
And famished people have been seen begging for tobacco, often
more eagerly than for food. This was observed during the last
fratricidal war in Spain, where it was diagnosed as a "tobacco
psychosis." Christians devoutly ask divine providence for their
"daily bread," but there are other human beings, and even they
themselves, who, as an added blessing, ask more fervently for
their daily tobacco, as though reciting in their subconscious a
blasphemous Lord's Prayer to Our Lord Almighty Devil.

Brooks quite rightly is of the opinion that among mankind
"tobacco is the most social of its appetites." And the most egali-
tarian. During the days of snuff-taking, these doggerel verses
were popular in England:

> *What introduces Whig or Tory*
> *And reconciles them in their story*
> *When each is boasting in his glory?*
> *A pinch of snuff.*

In the service of tobacco strangers are brought together and
help one another out, like Masonic brethren of smoke. And the
attacks against the democracy of tobacco are resented more
deeply than others, to the point of making history. The anger

aroused among the masses of the people against the tobacco monopolies, which were generally in the hands of rich aristocrats or Jews, and against the abuses of their despotic holders was such that it provoked riots in various European cities, and is credited with having been responsible for the first uprisings against the aristocracy in the eighteenth century (Brooks, op. cit., 146, 158). Steinmetz (*The Smoker's Guide,* 1878, p. 13) says that Jean Bart, the French naval hero, "by smoking in the presence of Louis XIV carried out an act of such audacity and implication of equality that it may be regarded as the real beginning of the French Revolution." From this point of view the riots of the tobacco-growers and the Cuban middle class against the tobacco-monopoly holders might be considered the beginnings of national consciousness, which paved the way for the people of Cuba in their rebellion against other trade, political, ecclesiastical, and social monopolies.

Today tobacco belongs to the people; its use is an inalienable right of the individual. Tobacco now is so widespread and such a necessity among all nations, races, sexes, religions, and classes that Laufer, who has contributed so much to the historical study of this plant, has arrived at the following political conclusion:

"Of all Nature's gifts, tobacco has been the most powerful social factor, the most efficient pacifier, the great benefactor of humanity. Tobacco has made all the world of the same lineage, and has united it in a single bond. Of all luxuries it is the most democratic, and the most universal; it has played a great part in making the world democratic."

The very evolution of the different types of smoking would seem to be due in part to environment and not entirely to economic factors and religious beliefs, as though the rhythm of social life had had an influence on the customs of smokers. The pipe is more frequent in cold countries and indoors for traditional ceremonies of peace and religion. The cigar is rather the companion of walkers in warm lands and for active magic operations, relaxation, or celebrations. The cigarette, being of paper, shortlived and light, is the offspring of a miscegenation of culture, a transcultural product engendered in times and habits of tension and haste. "The automobile is the enemy of smoking," says Jose Aixalá, and rightly (*Diario de la Marina,* Havana, December 9, 1939); but tobacco, which interferes with the tense moments of activity of the present social rhythm, fills all its pauses. In modern existence, with its breakneck pace and machine rhythm, tobacco would have been frightened away if it had not been upheld by the cigarette, oiling its pistons and valves and renewing its energies. The pipe has been called by an Englishman "sedentary"; the cigar was dubbed by another author "ambulatory." We now read that the cigarette is called "impatient." These adjectives are very expressive. The pipe tends to repose and evocation of the past; the cigar is more a pathway and the enjoyment of the present; the cigarette, haste, nervous release, and hope for the future. These times we are living in, with so much restlessness, haste, worry, and aspiration, are the age of the cigarette, which everyone seeks as a source of relaxation and stimulation at one and the same time.

During these past convulsive years tobacco has been regarded as a "weapon of war," like the oil and grease that power the machinery of combat. Tobacco, it has been said, is a great aid to the morale of the fighting man, and no army goes into action without it. But no, tobacco continues to be an instrument of peace, indispensable to keep up in each soldier submitted to the unprecedented and terrible nervous tension of war that minimum of independence, personality, comfort, dreams, and hope without which society would disintegrate and the human being go mad. Tobacco, which is a social rite of peace and friendship, is the soldier's most faithful friend and in war is always and at every moment "his peace." It is the temporary redoubt where his individuality can defend and comfort itself when, as a prisoner of Mars, he draws a free breath as he smokes and creates in himself the illusion of having recovered for a moment the enjoyment of his personal sovereignty.

All this bears witness to the constant and intimate power of tobacco, its modern social functions, its triumph, its universal transculturation.

8

On the Beginnings of the Sugar-Producing Industry in America

In order to clear up certain specific points that have been in dispute concerning the beginning of the sugar mills in the West Indies and particularly in Cuba, I shall quote from a number of important works bearing on the problem.

The first text is from Oviedo, and the unmistakably capitalistic character of the sugar industry is manifest to the reader. It also contains extremely interesting information regarding sugar's beginnings in Hispaniola. It is Chapter viii of the fourth book of the first part of the work by Captain Don Gonzalo Fernández de Oviedo y Valdés: *Historia General y Natural de las Indias, Islas y Tierra-Firme del Mar Océano* (*General and Natural History of the Indies, Islands and Mainland of the Ocean-Sea*), and it was composed in the year 1546, the date which appears at the end of the chapter in question:

"Which Deals with the Mills and Sugar Plantations on This Island of Hispaniola, Whom They Belong to and How This Rich Industry Developed in These Regions, and First on This Island

"Now sugar is one of the richest crops to be found in any province or kingdom of the world, and on this island there is so much and it is so good although it was so recently introduced and has been followed for such a short time; and even though the fertility of the land and the abundant supply of water and the great forests that provide wood for the great and steady fires that must be kept up are all so suitable for such crops, all the more credit is due the person who first undertook it and showed how the work should be done.

"Everyone was blind until Bachelor Gonzalo de Velosa, at his own cost and investing everything he had, and with great personal effort, brought workmen expert in sugar to this island, and built a horse-powered mill and first made sugar in this island; he alone deserves thanks as the principal inventor of this rich industry. Not because he was the first to plant sugar cane in the Indies, because some time before his coming many had planted it, and had made syrup from it; but, as I have said, he was the first to make sugar on this island, and following his example, afterwards there were many others who did the same. As he had a large plantation of sugar cane, he built a horse-powered mill on the banks of the Nigua River, and brought out workmen from the Canary Islands, and ground the cane and made sugar before anyone else.

"But investigating the matter further, I have learned that certain reputable old men, who at present live in this city, say differently, and they maintain that the person who first planted sugar cane on this island was one Pedro de Atienza of the city of Concepción de la Vega, and that the Mayor of Vega, Miguel Ballester, a native of Catalonia, was the first to make sugar. And they say that he did this more than two years before Bachelor Vellosa; but at the same time they say that the Mayor made very little, and that the source of both the one and the other was the cane of Pedro de Atienza. So, whichever version one wishes to accept, it would seem that the original basis or origin of sugar in this island of the Indies was the cane of Pedro de Atienza, and from these beginnings it grew and multiplied until it became the industry it is today, and each day it increases and grows, for although in the last fifteen years certain *ingenios* have failed or deteriorated for reasons I shall set forth in their proper place, others have been perfected. Let us return to Bachelor Velosa and his mill.

"As he came to understand the business better, the Comptroller, Christobal de Tapia, and his brother, the Warden of this fortress, Francisco de Tapia, went into partnership with him, and the three of them together established an *ingenio* in Yaguate, a league and a half from the banks of the Nizao River, and some time ago they had a disagreement and the Bachelor sold out his part to the Tapias. Later the Comptroller sold his to Johan de Villoria, who afterwards sold it to the Mayor, Francisco de Tapia, who thus became the sole owner of the first *ingenio* that existed on this island. As in those early days people did not understand as well as they should have the need of great areas of land and accessible water and wood and other supplementary things for such an industry (of which there was not so much as was necessary there), the Mayor, Francisco de Tapia, abandoned that *ingenio* and moved the copper or boilers and equipment and everything he could to another, better location right on the banks of the Nigua, five leagues from the city, where, until he died, the Mayor had a very good *ingenio,* one of the most important in this island.

"In order not to repeat over and over what I shall say now, the reader should bear in mind that what is said of this *ingenio* applies

to all the others of the same sort, that each of the important and
well-equipped *ingenios,* in addition to the great expense and value
of the building or factory in which the sugar is made, and another
large building in which it is refined and stored, often requires an investment of ten or twelve thousand gold ducats before it is complete and ready for operation. And if I should say fifteen thousand
ducats, I should not be exaggerating, for they require at least eighty
or one hundred Negroes working all the time, and even one hundred and twenty or more to be well supplied; and close by a good
herd or two of a thousand, or two or three thousand head of cattle
to feed the workers; aside from the expense of trained workers and
foremen for making the sugar, and carts to haul the cane to the mill
and to bring in wood, and people to make bread and cultivate and
irrigate the canefields, and other things that must be done and continual expenditure of money. But it is a fact that the owner of an un-
encumbered and well-equipped *ingenio* has a fine and rich property,
and one that brings in a great profit and return to its owner.

"So this was the first *ingenio* established in this island, and it
should be observed that until sugar was produced on it the ships
sailed back to Spain empty, and now they return loaded with sugar,
carrying a bigger cargo than they brought out, and a more profitable
one. And as this plantation was started on the banks of the Nigua, I
wish to mention the others that are located along the same river.

"Another important *in'genio* on the same bank of the Nigua River
is that of the treasurer, Esteban de Passamonte, and his heirs, which
is one of the best and most important on the island, both in build-
ings and in other equipment, and it has a fine supply of water and
woods and slaves and all that is required; it is seven leagues from
this city.

"On the same bank of the Nigua, farther down, is another very
good *ingenio* built by Francisco Tostado, six leagues from this city,
which was left to his heirs, and is a very fine property and has everything it needs.

"On this same bank of the Nigua there is another *ingenio,* one of
the best and most important of this island, which is close to the out-
let to the sea, four and a half leagues from this city of Santo Domingo, which belongs to the secretary, Diego Caballero de la Rosa,
alderman of this city; it is a beautiful and valuable property, both
because of its location and for the other features it possesses.

"Above the Nigua, on a river called Yamán, eight leagues from
this city, there is another fine *ingenio,* established by the late Johan
de Ampies who was commissioner of Their Majesties and alderman
of this city; it now belongs to Doña Florencia de Avila, and the heirs
of the commissioner.

"Another central, one of the finest of the island, belongs to the Duke Admiral, Don Luis Columbus. Because this industry of sugar and the mills manufacturing it began on the banks of the Nigua River, in order to mention all those located on it, which are the five foregoing, that of the Admiral was not put first, as is fitting that in everything pertaining to the Indies his name should precede all the rest, for all who live off these properties and everything they have earned from them owe him first consideration, for it was through his grandfather that these lands became known, for he discovered them for all who are now enjoying them. But as I have said, to keep things in their proper order, it was necessary to speak first of the *ingenio* of the Mayor, Francisco de Tapia, and after him to list the others as I have done, because when the Admiral's *ingenio* was established, there were already others on the island. This was built and founded by the second Admiral, Don Diego Columbus, four leagues from the city, at a place called New Isabela; and afterwards his wife, the Vicereine, Doña Maria de Toledo, moved it to where it is now, which is a better location and closer to this city, and in three or four hours they bring the sugar down-stream in boats and load it on the ships; it is finer and better than any other *ingenio* here.

"Another *ingenio* was started by Licentiates Antonio Serrano, formerly alderman of this city, and Francisco de Prado, which later became the property of the paymaster Diego Caballero, one of the aldermen of this city, who now, by the grace of his Cæsarean Majesty, is marshal of this island. As he decided to go to Spain, he neglected this *ingenio* and lost it; because it was founded by lawyers, and nothing had been said in their law texts about such matters, they did not know what they were doing, for they did not understand the requisites of such an undertaking, nor were their purses long enough to keep up and outfit the *ingenio*. Moreover, because of the unhandy location of the property, their expenses were greater than their profits, and as the second owner of the property understood things better, he got rid of it after he had made what he could out of it, selling off the Negroes and cattle, as well as part of the equipment, prudently preferring to lose a part rather than the whole.

"Another *ingenio* was founded three leagues from this city, and for a time it was thought to be very good, because it seemed so and ground a quantity of cane; but it, too, was founded by lawyers, near the bank of the Hayna, by Pedro Vázquez de Mella and Esteban Justinian, a Genoese; and after their death it went to their heirs and it was lost because of the lack of a dam, and in attempting to bring in the water from the Hayna River they wasted a great deal of time and money. And so the heirs decided to divide up the lands and the slaves and the cattle and all the implements they could make use of, and gave up the making of sugar in order not to lose everything they had in the enterprise. But later on, Juan Bautista Justinian fixed it up and kept the house and built a horse-powered mill, and now he is grinding sugar and it will improve every day and be a rich property if he can get a sufficient supply of horses.

"Another *ingenio* was established by Christobal de Tapia, the inspector of the gold-smelting of this island and alderman of the city, since deceased. It was left to his son Francisco de Tapia, and is located four leagues from the city, at a place called Itabo, which is a brook. After the time of Christobal de Tapia his son Francisco de Tapia could not keep it up, and he left it, because the expense was greater than the profits, and this *ingenio* was lost, like others I have mentioned.

"The heirs of the treasurer, Miguel de Passamonte, have a very fine *ingenio* on the bank of the Nizao River, eight leagues from this city of Santo Domingo. It is one of the best on the island and one that has lasted; it could be included among the eight most important.

"Alonso de Avila, who was Their Majesties' paymaster on this island, and alderman of this city, established another very good *ingenio,* eight leagues distant on the banks of the Nizao, which he left to his son and heir, Esteban Dávila, and his sister, and it is a very fine property.

"Another very good *ingenio* was established by and belongs to Lope de Bardecia, a resident of this city. It is on the banks of the Nizao, nine leagues distant from the city of Santo Domingo, and is one of the very good properties of this sort.

"Another *ingenio,* and one of the best and most important on the whole island, was established by the Licentiate Zuazo, who was one of judges of the Royal Tribunal established in this city by Their Majesties. It is along the banks of the River Oca, sixteen leagues from this city of Santo Domingo, and is one of the fine properties of these parts, and it was left after the days of the Licentiate to his wife, Doña Phelipa, and his two daughters, called Doña Leonor and Doña Emerenciana Zuazo, along with much other wealth and property. And it is the opinion of some who are versed in this industry that this *ingenio* alone, with its Negroes and cattle and equipment and land and appurtenances, is now worth over fifty thousand gold ducats, because it is very well fitted up. And I heard Licentiate Zuazo say that every year he had an income of six thousand gold ducats from this *ingenio,* or more, and that he thought it would bring in much more in the future.

"The secretary, Diego Caballero de la Rosa, in addition to the *ingenio* mentioned before on the banks of the Nigua, has another very good one twenty leagues from this city near the town of Azua. This *ingenio* is on the banks of the river called Capizepi, and is a very fine property and very productive. Jácomo Castellón set up another very good *ingenio* near the town of Azua, on the banks of the river called Bia, twenty-three leagues from this city of Santo Domingo; and after his death this *ingenio* and all his other possessions went to his wife, Doña Francisca de Isasaga, and his children; and it is a very fine property and profitable, in spite of the fact that this *ingenio* has not been looked after as it should be since the death of Jácomo de Castellón.

"Fernando Gorjón, resident of the town of Azua, has another *ingenio* in the town itself, twenty-three or twenty-four leagues from this city of Santo Domingo. This property is very useful and profitable to its owner, and greatly admired.

"The precentor, Don Alonso de Peralta, of this holy church of Santo Domingo, set up a horse-powered mill in the town of Azua itself, and after his days it was left to his heirs. These mills are not so strong as those powered by water, and are more expensive because instead of the water turning the wheels to grind the sugar, it must be done with the power of the many horses required for this work; and this estate went to the heirs of the precentor and to Pedro de Heredia, who is now the Governor of the province of Cartagena on the mainland.

"There is another horse-powered mill in the same town of Azua which belongs to an honorable man who lives there, by name Martín García.

"In the town of San Juan de la Maguana, forty leagues from this city of Santo Domingo, there is another important *ingenio* which belongs to the heirs of a resident of the town, whose name was Johan de León, and the German company of the Welsers, which bought half of this *ingenio*.

"In the same town of Sanct Johan de la Maguana there is another very good and powerful *ingenio* which was established by Pedro de Vadillo, and the secretary Pedro de Ledesma and the Bachelor Moreno, since deceased. It was left to their heirs and is a very fine and rich property.

"Eleven leagues from this city, along the banks of the river called Cazuy, the late Johan de Villoria, the elder, and his brother-in-law, Hierónimo de Agüero, set up a very good *ingenio*. This property was left to the heirs of the two, and also to the heirs of Agostín de Binaldo, the Genoese, who had a share in this *ingenio*.

"This same Johan de Villoria set up another *ingenio,* which is one of the best on the island, on the banks of the river called the Sanate, twenty-four leagues from this city of Santo Domingo, near the town

of Higuey, which after his days went to his heirs and to Doña Al-
donza de Acebedo, his wife, and is a fine property.

"The Licentiate Lucas Vázques de Ayllón, who was one of the
judges of the Royal Tribunal of Santo Domingo, and Francisco de
Ceballos, both since deceased, established a very good and important
ingenio in the town of Puerto de Plata, which is forty-five leagues
from this city on the northern coast. This property now belongs to
their heirs.

"Two hidalgos from the city of Soria, by name Pedro de Barrionuevo and Diego de Morales, residing in the town of Puerto de
Plata, established another very good *ingenio* in that town, and it is
a very fine property.

"In this same town of Puerto de Plata there is a good horse-
powered mill built by Francisco de Barrionuevo, former Governor
of Castilla del Oro, and Fernando de Illiescas, who live in that town,
and it is a very good property.

"In this same town of Puerto de Plata Sancho de Monasterio, of
Burgos, and Johan de Aguillar have another horse-powered mill,
and it is a fine property.

"In the villa of Bonao, nineteen leagues from this city of Santo
Domingo, there is another good sugar *ingenio* which belongs to the
sons of Miguel Jover, the Catalonian, and Sebastian de Fonte, and
the heirs of Hernando de Carrión, and it is a good property.

"The Licentiate Christobal Lebrón, who was judge of this Royal
Tribunal, established another *ingenio* in a very well-located and
fruitful spot, ten leagues from this city of Santo Domingo, known
as Arbol Gordo. It is a very good property, and went to his heirs.

"Hernando de Carbajal and Melchior de Castro started another
good *ingenio* on the banks of the Quiabón River; twenty-four leagues
from this city of Santo Domingo, on a very fine spot; but it was
never finished, because they went out of partnership, and because it
was far away, and because it seemed to them that it would take too
much capital to get it going; in a word, it did not last.

"Now, summing up the account of these mills and rich sugar
plantations, there are on this island twenty important *ingenios* completely installed and four horse-powered mills. And there is the opportunity to set up many others on this island, and there is no island
or kingdom among Christians or pagans where there is anything
like this industry of sugar. The ships that come out from Spain return loaded with sugar of fine quality, and the skimmings and syrup
that are wasted on this island or given away would make another
great province rich. And the most amazing thing about these great
plantations is that during the time that many of us have lived out
here, and of those that have spent over thirty-eight years here, all

these have been built up in a short time by our hands and our work, for there was not a one to be found when we came out here. And this is sufficient for sugar and the mills producing it, and it is worth bearing in mind in connection with the comparison I made before between the island of Hispaniola and its fertility and those of Sicily and England.

"There are other *ingenios,* though not many, on the islands of Sanct Johan and Jamaica, and in New Spain, which I shall mention in their proper place. Each arroba of twenty-five pounds, each pound weighing sixteen ounces, brings a price of one peso in this city of Santo Domingo, and at times over a peso and a ,half of gold, or a little less. In other parts of the island it is worth less because of other expenses and cartage to bring it to port, in this year of 1547 since the Nativity of Christ, Our Redeemer. And this brings to an end Book IV, because the account of other matters will be continued in this *Natural e General Historia de Indias."*

The second text I have drawn upon for the origins of the sugar industry in Hispaniola is from the pen of the famous Father Bartolomé de las Casas in his work *Historia de las Indias.* It reads as follows:

"The inhabitants of this island entered upon another industry, which was to seek a way of making sugar, in view of the great abundance of sugar cane in this land. And it is said that a resident of Vega, by the name of Aguilón, was the first to make sugar on this island, and even in all the Indies, using certain wooden instruments with which he squeezed the juice out of the canes, and although it was not very well made because he did not have the equipment, it was, nevertheless, real and almost good sugar. This must have been about the year 1505 or 1506. Afterwards a resident of the city of Santo Domingo, the Bachelor Vellosa, attempted to make it, and he was a surgeon, and a native of the town of Berlanga. This was around the year 1516, and he was the first in that city to make sugar. He had better apparatus, and so the sugar was whiter and better than that first made in Vega, and sugar candy was made from it and I saw it. Vellosa devoted himself to this industry and finally set up a *trapiche,* which is a mill moved by horses, in which the canes are crushed or pressed and the sweet syrup from which sugar is made is extracted from them. . . .

"When the Hieronymite friars who were there saw the success with which the Bachelor was carrying on this industry, and that it would be very profitable, in order to encourage others to undertake it, they arranged with the judges of the Tribunal and the officers of the crown to lend 500 gold pesos from the funds of the royal treasury to anyone who should set up a mill, large or small, to make sugar, and afterwards, I believe, they made them other loans, in view of the fact that such mills were expensive. . . .

"In this way and on this basis a number of the settlers agreed to set up mills to grind the cane with horse-power, and others, who had more funds, began to build powerful water-run mills, which grind more cane and extract more sugar than three horse-powered mills, and so every day more were built, and today there are over thirty-four *ingenios* on this island alone, and some on Sant Juan, and in other parts of these Indies, and, notwithstanding, the price of sugar does not go down. And it should be pointed out that in olden times there was sugar only in Valencia, and then later in the Canary Islands, where there may have been seven or eight mills, and I doubt that that many, and withal the price of an arroba of sugar was not over a ducat, or a little more, and now with all the mills that have been built in these Indies, the arroba is worth two ducats, and it is going up all the time." (*Historia de las Indias,* Madrid edition, Chapter cxxix, Vol. III, p. 249.)

The texts of the two chroniclers, as well as various others that could be cited, are somewhat confused, and contradict each other in places. A serious critical interpretation is needed, in addition to certain corrections, on which I am already working, but which I am reserving for an extensive work now in preparation, *Introduction to the Economic-Social History of America.*

I should merely like to observe here that the early chronology of sugar in America is probably as follows:

1493 (December): Introduction and first planting of sugarcane roots in Hispaniola by Christopher Columbus.

1501 (approximately): Planting of first canefield by Pedro de Atienza.

1506 (or the previous year): First sugar produced by Miguel Ballester or by Aguilón or Aguiló.

1515 (or before): The first grinding by the first mill, by Gonzalo de Velosa.

1516: The establishment of the first *ingenio,* by Gonzalo de Velosa and the brothers Francisco and Cristobal Tapia.

This brief chronology requires some explanation. The chronicler Francisco López de Gómara, who wrote as follows in

1552, seems to have hit upon the truth, or to have come close to it:

"What has multiplied greatly is sugar, and there are close to thirty prosperous *ingenios* and mills. The first to plant sugar cane, before any other Spaniard, was Pedro de Atienza. The first to extract sugar was Miguel Ballestero, a Catalonian, and the first horse-powered mill was set up by the Bachelor Gonzalo de Velosa." (*Hispania Victrix, Historia General de las Indias*. Biblioteca de Autores Españoles, Vol. XXII, p. 177.)

There is often considerable confusion among the historians of these Antilles who have not occupied themselves with examining all aspects of the sugar production of those times; some of them adduce information dealing with the early and successful production of sugar, while others deny the existence of *ingenios* at that time, or else cite the accounts of the chroniclers on the operation of different mills and yet fail to recognize the real facts of the sugar trade.

In the Spanish Antilles the production of sugar existed before there were *ingenios*. The term *ingenio* was used to mean "industry, skill, craft," aside from the procedures that were considered natural or invariable, to secure some new mechanical effect, such as a mill to crush cane, to bring water from a river, or smelt metals. And before such mechanical devices existed here, syrup was extracted from cane by the use of the simple Indian *cunyaya,* or a lever press. It was in this way that Miguel Ballester and Aguiló, too, probably extracted the syrup in 1505 or 1506. That is to say, they produced sugar before mills existed. Bartolomé de las Casas bears this out very clearly when he says that Aguiló made sugar "with certain wooden instruments with which he squeezed the juice from the canes." Gonzalo de Velosa or Vellosa also made sugar this way, but later on he used "other more suitable instruments" until he himself, persisting in his determination to manufacture sugar, finally set up a machine, about the year 1515, with which the first sugar-grinding in America took place.

But there is still a basic point to be cleared up. Was this really the first *ingenio* that was established, this one in 1515 by Bachelor Vellosa? It must be borne in mind that the first mills were driven by living power, that of slaves, horses, or oxen who turned a main wheel by moving steadily around the machine, as in the old chain pumps the Arabs brought into Spain, or the set of wheels of such mills was moved by hydraulic power, like the famous "artifice or device of Juanelo," famous in those days, which raised the water of the Tagus River in Toledo. Oviedo alludes to the fact that Velosa's mill was "horse-powered." And Las Casas is equally precise, saying that Velosa "finally set up a *trapiche,* which is a mill moved by horses." As is evident from this paragraph of Las Casas, the word *ingenio* had a generic meaning and included the simple mills known as *trapiches,* "which grind the cane by horse-power," and the "powerful mills," which were water-powered.

But before long the word *trapiche* came to be generally applied to the mills moved by animal power, and the word *ingenio* came to signify those employing hydraulic power. In this paragraph of Las Casas's the distinction is already apparent; but in Oviedo's long account, which I have reproduced, this division between *trapiches* and *ingenios* is very plain, according to the differing structure, motor power, and grinding capacity of each. Oviedo says that in Hispaniola there were "twenty *ingenios* and four horse-powered *trapiches.*" López de Gómara, who was never in Hispaniola and who took his information from Oviedo and other sources, made proper use of this terminology when he said that it was Velosa "who had a horse-powered *trapiche.*"

López de Gómara could have added that it was this same Gonzalo de Velosa who likewise had the first *ingenio* of America, the first of the "powerful water-run mills." Oviedo himself speaks of it in detail in the chapter quoted. According to Oviedo, Velosa, "as he came to understand the business better," the business of making sugar, and with "the Comptroller, Christobal de Tapia, and his brother, the Mayor of this stronghold, Francisco de Tapia, went into partnership with him, and the three of them together established an *ingenio* in Yaguate, a league and a half from the banks of the Nizao River." Oviedo continues his account of this partnership up to the point where Francisco de Tapia had become the sole owner of the "first *ingenio* there was on this island." It is clear that *ingenio* meant "a powerful mill," because of its hydraulic power, which was far greater than that of man or beast; one of those "powerful water-run *ingenios* grind more cane and produce more sugar than three *trapiches*," according to Bartolomé de las Casas. So it was Gonzalo de Velosa in company with the Tapia brothers who set up the first "powerful *ingenio*." Juan de Castellanos in his *Elegías* confirms this fact, alluding to the "great industry brought about by setting up powerful *ingenios* to grind sugar," and says:

> *The first inventor of this device*
> *Who first brought it to perfection*
> *Was, they say, Gonzalo de Velosa,*
> *A man held in esteem for his learning.*

The distinction between *ingenio* and *trapiche* is made even clearer in a report presented to the King in 1561 (*Cartas de Indias,* Arch. Hist. Nac., Madrid, Box 2, Number 12, quotation from Silvio Zabala: *Revista de Historia de América,* Mexico, December 1938, p. 214). In it the Licentiate Echagoian says to the King that in Hispaniola there were more than thirty sugar mills, "some of which were *trapiches* whose wheel was moved not by water but by horses."

So, in conclusion, it appears clear that if a Catalonian, Miguel Ballester, was the first to make sugar in the West Indies, an Extremaduran or Castillian, "the wise Velosa," as Castellanos calls him, set up the first *trapiche* and later the first *ingenio*— that is to say, the first sugar mills of America. I might remark in passing that Ballester, Aguiló, and Velosa were probably of Jewish origin, as was, in the opinion of many, Christopher Columbus.

As for the difference in industrial capacity between a *trapiche* and an *ingenio,* aside from Las Casas's calculation that an *ingenio* did the work of three *trapiches,* we have another more exact calculation, taken from the history of sugar in Brazil, where the same mechanical differences existed. They calculated there that in twenty-four hours a *trapiche* or horse-powered mill in the sixteenth century could grind 25 to 35 cartloads of cane and extract 840 pounds of sugar, while a hydraulic mill could grind from 40 to 50 cartloads and extract 1,120 to 1,960 pounds (Hermann Watgen, *O Dominio Colonial Holandés no Brasil,* Rio de Janeiro, 1930, p. 427). Putting this into modern terms, the *trapiche* produced less than three sacks of sugar (holding 325 pounds each) and "the powerful *ingenio*" could secure at most double that amount per day, or only six sacks of sugar. It will be seen from this information how limited the production of sugar in Hispaniola was during the sixteenth century, in spite of its many and early mills, and particularly in Cuba where because of lack of water the modest *trapiches* predominated until the nineteenth century, when the steam engine was introduced.

In legal documents this distinction between *ingenios* and *trapiches* was preserved. All the Royal Letters Patent granting special privileges to foster the sugar industry say without exception that the benefits of royal favor are for those who set up or possess *ingenios,* and never once make mention of the *trapiches.* (See the Royal Letter Patent quoted in Part II, Chapter x of this study.)

At the end of the sixteenth century, in the Royal Letter Patent of December 23, 1595 (*Recopilación de Indias,* Law VIII, Section XIII, Book VI) ordering that Negroes, not Indians, were to be employed in certain tasks and in the sugar industry, the text reads "sugar *ingenios* and *trapiches.*" And this distinction continued until the nineteenth century, as can be seen in a Letter Patent of 1804, and a Royal Ordinance of July 17, 1848.

With these antecedents it should be possible to bring order into the confusion so often encountered in the history of Cuba when one learns that at an early period sugar was already being manufactured and "was on the increase," and at the same time one learns that mills did not yet exist and subsidies were requested of the King to establish them. The contradiction is only seeming, not actual, for sugar was made with *trapiches,* and at the same time requests were being made for land, money, slaves, and special privileges in order to set up *ingenios* to make more and better sugar. In Cuba the water-powered *ingenios* were of necessity few, compared with those in Hispaniola, and they always were because of the relative scarcity of hydraulic power. For this reason when the sugar production of Cuba increased, nearly all the plants manufacturing sugar were *trapiches,* and these in turn received the generic name of *ingenios.* It was not until the introduction of steam and the great mechanical development that followed that the time-hallowed *trapiches* came to be contemptuously regarded, and were given the name of *cachimbos,* while the steam-powered sugar factories, incomparably more powerful, kept the name of *ingenios;* until finally another name for the ultrapowerful *ingenio,* "the even more powerful," came into being, which is that of *central.*

Ingenio, therefore, has always been the accepted generic denomination, which comprehended, in different epochs, the specific meanings already indicated, and other supplementary and varying, such as *trapiche, cachimbo,* and *central.* The understanding of this makes it easier to grasp the history of the Cuban sugar industry.

9

"Cachimbos" and "Cachimbas"

I have used the term *cachimbo* in these pages. The word *cachimbo* is met with in Cuba and in other regions of Latin America and is used with different meanings in the two principal fields of our agrarian economy, tobacco and sugar.

In tobacco terminology the word *cachimba* means pipe; in that of sugar, with a masculine declension, *cachimbo* means a little sugar-grinding mill. It is used contemptuously, comparing the humble plant with its prominent puffing smoke-stack to a pipe or *cachimba.*

10

On How the Sugar Ingenio has Always Been the Favored Child of Capitalism

The sugar-cane roots brought out to the Indies by Christopher Columbus on his second voyage flourished from the moment they were put into the ground, to the delight of the Admiral, who shrewdly discerned the brilliant future in store for sugar cane and the sugar industry in these lands. "They will not be overshadowed by those of Andalusia," Columbus wrote the sovereigns, with prophetic foresight. When on January 30, 1494 Columbus drew up in Hispaniola a memorandum for the monarchs to be delivered to them by Antonio de Torres, he already spoke in glowing terms of the economic future awaiting "the sugar cane, to judge by the way the few that were planted are growing," comparing it with that of Andalusia and Sicily.

But it is one thing to have cane and another to produce sugar on a commercial scale. Between the raising of the cane, which experience had shown to be merely a question of man power, and the commercial production of sugar, which in Europe had a steady and growing market, stood the problem of industrial production, which demanded machinery and technicians that did not exist here, and of necessity had to be imported from Europe. In a word, capital was needed to buy slaves, to bring in experts and skilled workers and all the machinery for milling, boiling, evaporating, and refining. Even aside from the land required, the production of sugar was perforce a capitalist enterprise.

Columbus and the rulers of Spain took the production of sugar in the newly discovered islands very seriously, but some time went by before the sugar industry became organized. Hispaniola was the starting-point.

Sugar became an industry in Hispaniola, and the same thing took place in the other countries when the chief obstacle, which was the lack of capital, was overcome. "Masters in sugar," or technicians, as we would say in modern times, were also needed. History has preserved the names of some of these masters, for their calling was an important one. But the experts to *manificar* the sugar could be supplied if there were money enough to induce them to leave the Canary Islands or the Portuguese islands and offer them a tempting salary. For this reason the sugar industry must have been the first of those undertaken by Europeans in America that demanded a joint, unequivocal association of labor and money—that is, the social and economic organization of a company. With the exception of the different conquests, most of which were undertaken under carefully worked-out agreements by real companies, either of the type known today as co-operatives or stockholders, besides special types organized to meet the complex requirements of those extraordinary undertakings involving discovery, conquest, settlement, and development. For example, there is a case on record showing that before 1515 there existed in America, on the mainland, in the settlement of Darien (Castilla del Oro), a "company of sugars," constituted by three partners, Francisco de Arcos, Luis Fernández, and Pedro Hortís, as evidenced by a notary's document drawn up in Seville on the 19th of April of that year, in the office of Mateo de la Cuadra (Card 155 of the *Catálogo de los Fondos Americanos del Archivo de Protocolos de Sevilla,* Vol. III, p. 44). We have already seen that the first *ingenio* set up in America was founded in Santo Domingo, shortly after 1515, by a society consisting of three partners, Gonzalo de Velosa, who was the experienced expert and founder of the first *trapiche* in Hispaniola, and the brothers Tapia, one the Comptroller and the other the Warden of the fortress, which gave them the backing of money and of authority.

Nevertheless, the enterprise was too big for individual initiative, even for "companies of sugars." The experts were very few, and the available capital very limited. Moreover, the temptations the Indies offered to bold promoters in that early period of the sixteenth century were almost irresistible, when new explorations, conquests, and business ventures were being undertaken on all sides. One can cite as an example the economic preparations of Francisco Fernández de Córdoba (1517), Juan de Grijalva (1518), and, especially, Hernán Cortés (1519) in Cuba for their expeditions to Yucatán and Mexico. Capital was scarce and was not loaned out at interest, for this was in those days a sin condemned by the Church. And men engaging in these ventures were not satisfied with wages or salaries, but wanted a share in the riches that might accrue, for, as López de Gómara says, "in the Indies everyone wants to ascend in rank or win great wealth." Under such circumstances it was very hard to find capital for long-term investment in real estate or agro-industrial enterprises, such as the sugar *ingenios* represented. And, unable to secure sufficient capital, either by loan or by stockholding companies, the royal exchequer became the only source.

I shall not review here the different phases and vicissitudes through which the establishment of the sugar industry passed in America, as a result of its basically capitalistic nature. In Part II, Chapter viii I have quoted passages from Oviedo and Las Casas dealing with this point, which are sufficient.

It is not true, as Herrera claimed and others have repeated after him, that it was the Hieronymite friars who ruled the Indies through the Regent, Cardinal Cisneros, nor that they established the first sugar *ingenios* in America. But in this epoch those wishing to set up *ingenios* enjoyed special favor in the matter of money loans. And the magnates of Hispaniola took advantage of this situation, and with the land grants they received, the forced labor of the Indians, and the King's money they soon became sugar barons, real feudal lords, for the *ingenios* were true fiefs, with serfs attached to the soil.

Letters patent issued by the crown in Castile in the year 1517 and the following year granted fiscal favors to plantation-owners and those who established *ingenios*—that is, "powerful *ingenios*" run by water.

In 1517 a moratorium was granted to the conquerors and settlers of Cuba, and the Hieronymite friars were authorized to act in the matter of debts owed by the inhabitants of that island (published in the *Colección de Documentos de América,* Second Series, *Documentos de Cuba,* Vol. I). The text reads as follows:

"The King and Queen.—Reverend Father and devoted friars, etc. —Pánfilo de Narvaez, in the name of the island Fernandina, formerly known as Cuba, has informed us that the island has been recently settled, and that those who have conquered it have got into debt buying certain supplies from our stores and from other persons, and that as they have discovered little gold they are in need and in difficult circumstances; in view of which he has asked us to order that these debts owing us be collected with certain moderation from the persons who owe them, and that they be allowed to pay what they owe other people little by little, or that We take such steps in the matter as may seem to us advisable; and after consulting with our advisers it was agreed that We should issue this decree to you upon this matter, and We found this good; therefore We order and charge you to look after the above matter, and according to the instructions you have received from us to take care of it and remedy it as seems best for our service and for the good and welfare of the island in question and the dwellers therein that they may not be too heavily burdened. Dated, in Madrid, on the . . . day of the month of . . . . . of the year MDXVII.—F., *Cardinalis.*"

The following year, 1518, another royal decree was issued, containing instructions to the Licentiate Rodrigo de Figueroa, with new instructions about loans and moratoriums. It reads as follows:

"Zaragoza, December 9, 1518.—The King.—What you, Licentiate Rodrigo de Figueroa, who are going out as our resident judge to the island Spañola, should know:

"In addition to what has been previously said, know that the Catholic Queen, my lady, and I are very desirous that the aforesaid island Española be settled and ennobled with every kind of plant and other commodities, as are these our kingdoms, which, according to the information we have received concerning the fertility and abundance of that island, would be the case if the governors and authorities it has had, and the inhabitants and dwellers there, had taken thought of it and had wished to remain and live there, especially as regards the sugar which has been made and is made there. I order you to use all diligence to see that the residents of the aforesaid island set up *ingenios* of sugar, and that you assist and favor in every way you can all those who wish to do so, both lending them funds from our treasury to help them establish these *ingenios,* as well as giving them privilege and use of the lands, and that you do the same with all other residents and settlers who are willing to work and wish to remain there and build, settle, and plant and do such other things as are required for the good and ennoblement and settlement of these lands.

"Likewise, because it is my will that in all that can conveniently be done they be relieved and unmolested, and being informed that many of them owe debts to one another, and if these were demanded and collected from them much harm would result and the loss of their property, therefore I charge you that if the persons who owe debts on the aforesaid island cannot conveniently pay them, you arrange with their creditors that, being assured that they will pay them within a certain length of time, they wait for them, and in this you will do me a service.

"This and whatever else you should see that may redound to our service I order and charge you to do with the fidelity, care, and diligence that I expect of you. Dated in Zaragoza, IX of December of the year 1518. I the King. Countersigned by Covos. Witnessed by

the Chancellor and the Bishop of Burgos or of Badajoz and of
Zapata."

These orders can be interpreted as antedating the installation
of the *ingenios* of sugar—that is, the hydraulic power mills.
At the beginning of 1518, almost a year before the decrees
transcribed above were issued to Licentiate Figueroa, an eyewitness had sent the Emperor Charles V a report from Hispaniola on the fine cane plantations and the *ingenios* that were
being set up. A part of the enthusiastic letter of Licentiate
Alonso de Zuazo to the Emperor referring to the island of Hispaniola runs as follows:

"Herds of cows numbering thirty or forty and branded wandered
off, and in three or four years they were found in the hills to the
number of three hundred or four hundred. The same thing happens
with swine and sheep and mares and other livestock. . . . The hill-
sides are covered with cotton, and now they are setting up gins to
clean it. . . . There are also fields of sugar cane that are wonderful
to see, the cane as thick as a man's wrist, and as tall as the height of
two men of medium stature. And mills are also being set up to make
sugar, which will be a source of very great wealth. . . . Santo Domingo, January 22, 1518." (*Colección de Documentos de América,*
Vol. I, p. 292.)

A petition to the King from Hispaniola dated the same year,
1518 (September 14), contains the request that the tithes to be
paid the Church by the planters of cane and the manufacturers
of sugar should be reduced to the thirtieth part; this request
was refused by the government (*Indice Gen. de los Reg. del
Consejo de Indias, de 1509 a 1608;* cited from J. A. Saco). This
petition was very important, for the tithes were very burdensome. As Mariano Torrente pointed out (*Bosquejo económico
político de la Isla de Cuba,* Vol. II, Havana, 1853, p. 351): "every plantation obligated to pay the tithe lost the amount represented by the tithe in question." The tithe was a tax paid to the
Church to use for its own purposes. It represented the tenth
part of all the fruits or vegetable products of the land, as well
as of all fowl and quadrupeds that were raised, of the milk,
butter, cheese, honey, and wax.
The ecclesiastical tax of the tithe and first fruits was set up

by the Catholic kings in a royal order of October 5, 1501, which shortly afterwards became Law II of Section XVI of Book I of the Laws of the Indies. In this levy sugar was not expressly mentioned, according to several of the published editions of the *Recopilación*. It is true that in those texts a tithe of the "green barley that may be sold" is included, and that in the copy of this tariff law included in the *Cedulario Cubano* of José María Chacón y Calvo (p. 35) to the word "green barley" there is added "(*sic*) sugar." But this note is a mistake, for the word in question, *"alcacer"* does not mean "sugar," but "green barley," according to the *Dictionary* of the Spanish Academy. Nevertheless, it would seem that in the original levy of 1501 the tithe on sugar was clearly included, as appears from the copy of the same included in the above-mentioned *Cedulario Cubano,* in which the paragraph reads as follows (this text is taken from an unpublished volume of the *Colección de Documentos de la Real Academia de la Historia,* Vol. V, p. 23, according to Chacón y Calvo):

"A tithe is to be paid on sugar cane, one cane out of each ten. If those to whom the tithe is due should ask those in charge of collecting the levy to grind the cane they had received in tithes, they are to do so at once. And if any difference should arise between those in charge of the tax-gathering and those receiving the tithe, the former alleging that there was more cane for grinding than that received in tithes, they are to grind the cane without charging anything for it."

These church taxes of tithes and first fruits had to be paid in kind, not in money. And with regard to the tithe of sugar cane, it is evident from the text quoted that not only did the Church receive ten per cent of the sugar cane, but it was to be ground free of charge. The tithe on sugar cane really amounted to a tax equivalent to the payment in sugar of ten per cent of all the cane of each harvest, without being able to discount anything for the work of grinding and the manufacture of the sugar. As a result, the ecclesiastical tithe weighed very oppressively on the new sugar-raisers, who had to pay other taxes as well, and for this reason they complained to the King. If the planters of today, even the most devout Catholics, had an ecclesiastical tax of ten per cent of their gross production levied against them, in addition to the taxes imposed by the state, they, too, would "cry to high Heaven." Heaven was a little deaf to the cries of the sugar-growers four centuries back. The Church was very powerful in Spain at that time and could hardly be expected to give up without a murmur this rich contribution from the Indies, and the King, who was entitled to the tithes of the Indies "by reason of apostolic concessions from the Pope" (Law I, Section XVI, Book I of *Leyes de Indias*) continued to demand the payment of "all tithes due" in order to provide the churches with clergy, ornaments, and other things necessary to the cult.

But the sugar-planters kept on complaining and did not pay the tithes, or cheated the Church and the King in their payments, until finally the good sense of the exchequer prevailed and brought about a compromise. The Church had to yield to the planters, who were the magnates of the colonies, and certain reductions were finally granted them. The Emperor Charles V on February 8, 1539 issued the following order:

"We hereby order and command that in order to avoid defraudation of the Church, before any division such as is customarily made between the planters and cultivators of sugar and the owners of mills making white sugar, refined sugar, skimmings, second skimmings, loaf sugar, muscovado, brown sugar, clarified sugar, and syrups, tithes be paid in our Indies and adjoining islands in the following form: on the first white, solidified, purified sugar a tithe of five per cent is to be paid; and on the refined, skimmings, loaf sugar, muscovado, brown sugar, clarified sugar, and syrups at the rate of four per cent, and the same on all other products, every year, and all those having sugar mills are obliged to pay this tithe, unless a different custom prevails somewhere."

This legal arrangement later became Law III of Section XVI of Book I of the *Leyes de Indias,* and it bears the following epigraph: "The Emperor Don Charles on February 8, 1539. And in Madrid, on September 19 of the same year. The Emperor and the Cardinal Governor, there, on July 15, 1540. And in Talavera, on April 11, 1541. And the Prince Governor, in Madrid, on May 31, 1552." Which shows how many times this privilege of the sugar-planters was ratified by the kings.

These measures were favors granted by the government of Castile to its subjects in the Indies who had embarked on industrial ventures. The royal concessions followed one after another. A royal order of 1519 sending sugar experts to Hispaniola reads as follows:

"Lope de Sosa, our general envoy and Governor of Castilla de Oro: know by these presents that our representatives on the island of Hispaniola have written me that on that island there is a great need of masters and workmen to set up sugar mills and to manufacture sugar because every day many *ingenios* are being established there, and the place is very well fitted and suitable for this, and I have been informed that in the Canary Islands there are many masters and workmen who would go out to the island if it were not that certain people put obstacles in their way, and because you, while you are preparing to undertake your trip to serve us in that capacity, can do much to attract the aforesaid masters and workmen, and on the way, since you will be stopping at the aforesaid island, you can take them with you there, I am sending you with this the enclosed letters so the governors of those islands may not put any obstacle in the way of their departure, but rather help and favor them in it, and for this reason I order you to present these letters to them, and do everything you can to induce as many masters and workmen to go out to the island as is possible, etc."

"In Barcelona, 16 of August of 1519. I, the King. Countersigned by the secretary Cobos, and witnessed by the abovementioned."

(*Archivo de la Casa de Contratación, Archivo de Indias,* 139–1–6.)

Another royal order was issued in Valladolid by the Prince in the name of his mother, Doña Juana, and reads as follows:

"To our officials residing in Hispaniola (the branch there of the Casa de Contratación of Seville) and our customs officers and taxcollectors of the aforesaid island: You know the desire my lady, the Catholic Queen, and I have held in regard to the welfare, settlement, and growth of the aforesaid island, and the steps we have taken to secure these ends. I have been informed that one of the principal activities that have been begun and are being carried on there is the manufacture of sugar, and, God be praised, there are sugar mills in abundance, and Licentiate Antonio Soriano in the name of this island has reported to me that because the construction of such mills is very expensive, and the materials and machinery required for them brought out from this Kingdom or realms near the aforesaid island cost too much for people to be able to buy them, and this will prevent this activity from developing further, he implores us to order that the machinery, materials, and other things taken out from this Kingdom for the construction and work of these mills shall not pay tax or other duties, or whatever my pleasure should be. And I, for the reasons set forth, found this good. Therefore, I order you, etc."

And the order follows to the effect that the materials in question be allowed to enter duty-free.

A letter from Licentiate Figueroa, the royal judge in Hispaniola, to the Emperor, dated July 6, 1520, reads as follows:

". . . plans are under way to build forty *ingenios* and more; and most of these by obligation, for some have received allotments of Indians, and others loans of money from Your Majesty for two years. Advise Passamonte to be openhanded with these loans, for this is what will revive the island, and this island is the support of everything else in these parts. . . . Negroes are very much in demand; none have come here for nearly a year." (*Colección de Muñoz,* Vol. LXXVI; quoted from Saco.)

In this résumé of the privileges and protection dispensed to sugar in the Antilles mention must also be made of another great economic and social boon conferred upon the sugar-manufacturers by the sovereigns, which was allowing and helping them to bring in Negro slaves for three hundred years. Cogent data has been adduced to show the strong link that existed between the slave trade and the development of the sugar plantations.

In those days the settler who wanted to raise sugar could ask the King for whatever he needed, aside from the audacity and enterprise supplied by himself; he could ask for land grants for plantations, money for mills, subsidies for bringing in expert workmen, tax exemptions, and, above all, the privilege of importing Negro slaves. Nearly always the planter got what he asked for, and when in spite of all this aid he found himself in financial difficulties, he was helped out with moratoriums and cancellation of debts, and even with *a priori* immunity from any action against him for debts.

By March 1522, 2,000 arrobas of sugar were being shipped from Hispaniola to Seville, and its production increased until it had become, to quote Father Bernabé Cobo "the principal industry of these islands [Santo Domingo and Puerto Rico] because people have developed such a taste for sweets" (op. cit., Vol. I, Chapter xxxii).

The friar and royal judge of Hispaniola, Licentiate Luis de Figueroa, in 1523

". . . asked that at the expense of the Royal Exchequer some sugar mills be built in that island and in Cuba, Jamaica, and San Juan de Puerto Rico where those settlers who are unable to set up mills of their own could take their cane to be ground, paying a fair price for the grinding. The government agreed to this request, ordering a mill to be built in each of the four islands at the expense of the Royal Exchequer." (José Antonio Saco: *Historia de la Esclavitud de la Raza Africana en el Nuevo Mundo y en especial en los países Américo-Hispanos,* Havana, 1938, Vol. I, p. 216.)

Father Luis de Figueroa also requested that the inhabitants of these islands be allowed to sell their sugar and other products freely, wherever they wished, even outside the realms of Charles V, but this petition was not granted. It ran counter to the monopoly of the Spanish state and the vested interests that profited by it, and it was not until three centuries later that they agreed to forgo their privileges, in the face of graver threats.

Charles V conceded one of the most important privileges enjoyed by the owners of sugar mills in the Indies: namely, that such mills could not be attached, seized, or sold by law for debt. The royal decree bears the date of January 15, 1529, in Toledo, and became Law IV of Section XIV of Book V of the *Leyes de Indias,* and runs as follows:

"There can be no attachment of sugar mills.—We order that there can be no attachment of sugar mills in any part of the Indies, or of slaves or other things necessary for their sustenance and milling, unless it be for sums due to Us, and We allow payment to be made in sugar and products of the mills, and this privilege cannot be renounced by the owners, nor is the renunciation valid if they make it. And it is also our will that the notaries drawing up contracts and deeds insert no renunciatory clause, under penalty of being suspended from office, and that the officers of the law shall not carry it out."

This royal decree exempts the mills and all their equipment from attachment except for debts owing the King—that is to say, for loans from the exchequer and taxes, allowing only the seizure and public sale of their industrial products, such as sugar and syrups. This privilege was so binding that it could not be renounced even by those benefiting from it, and it was repeatedly ratified in subsequent royal decrees: by Charles himself in Palencia on September 20, 1534, by the Empress in Valladolid on May 4, 1537, by Philip II and the Princess Regent on March 30, 1557 in Valladolid, and later in Madrid on August 3, 1570, and in San Lorenzo on September 28, 1588. And by Philip III in Olmedo as late as October 2, 1605.

With this privilege the sugar-millers were converted into lords exempt from all legal action against the source of their wealth, which was the industrial totality of the sugar-producing machinery: land, machines, slaves, livestock, utensils. To be sure, there were the industrial products—sugars, syrups, molasses—but these were of relatively slight value and not always sufficient to cover the expenses of upkeep and running of the mill, and of the lavish life of the planter. Be this as it may, the mill of the planter could not be attached, which is another way of saying that he could not be deprived of his privileged economic position, of his "fine property."

It soon became clear that this privilege was excessive and, above all, that it prevented the establishment of new mills, if these were to be built with borrowed capital. Nobody, aside from the Royal Exchequer, was willing to lend his money to build a mill if there was no way of holding a lien upon it to guarantee payment and it could not even be attached for debt. For this reason the above-mentioned royal decree of 1529 was modified by Charles V himself in another royal decree of November 8, 1538, which became Law V of Section XIV of Book V of the *Recopilación de Indias,* which reads as follows:

"There may be attachment of any mill smelting metals or manufacturing sugar if the debt amounts to the full value of the mill.— Our purpose in ordering that there may be no attachment against mills smelting metals and manufacturing sugar, with their slaves, instruments, and machinery, was that they might not cease to prosper for the common good of these kingdoms and of the Indies, because if this were done, the results would be very injurious, and neither the plantiff nor the defendant would benefit thereby. And because it is necessary to take the claims of creditors into account, we declare and order that if the debt is so large that it amounts to the full value of the mill, with slaves, equipment, and all its fittings, and the debtor has not other property to satisfy the creditor's claims,

the whole mill, with slaves and equipment, may be attached in payment of the debt, and the person attaching the property must give
guarantees that he will maintain it in working order, as the debtor
had it."

This royal decree was reissued by Philip II at the Pardo, on
March 13, 1572, and by its terms, when the planter owed as
much as the value of his mill or more, and had no other property, he could be deprived of it.

The settlers of Cuba also received royal concessions in their
sugar-manufacturing enterprises. The first dates from the year
1523.

About this time and at the request of Diego Velázquez, the
first Governor of Cuba, the chronicler I have so often mentioned, Gonzalo Fernández de Oviedo, had himself carried to
the King several samples of the sugar cane raised in Cuba with
great success. According to Herrera, even before this date Cuban sugar was on the increase.

On February 13, 1523 a royal decree of importance for Cuba
was issued ordering that all those who wished to establish
sugar mills on that island be helped in their undertaking with
loans. The royal decree reads as follows:

"The King: our officials on the island of Fernandina (Cuba), because they have seen from their own experience how, after Our Lord
was pleased to have the industry of sugar begin, the aforesaid island
has prospered and flourished, from which it is hoped that great utility, improvement, and perpetuity will redound to the inhabitants of
that island, and Juan Mosquera has informed me, in the name of
the others, that many inhabitants and residents of the island wish to
build mills and follow that industry, and that because the building
of the mills in question is very costly, as is their operation, the industry cannot be undertaken nor kept up if We do not order the loan to
certain persons of a sum of money . . . and he has asked me to order this done, if this is my pleasure. And because I have a great desire to favor the inhabitants and settlers of this island (Cuba) in
every way possible, and that they be favored and helped in this,
which is a very necessary thing, I am pleased to do so, and to this
end I order you to inform me as to what people there are in this island who have a way or desire to set up mills of sugar, and who
are not able to do so with their own resources, and who are trust-
worthy people, in your opinion, and to those having these qualifica-

tions, etc." (*Archivo de la Casa de Contratación,* Archivo de Indias, 139–1–6.)

The special dispositions follow, in keeping with the foregoing exposition.

Special favor was also shown the Cuban sugar industry in the matter of acquiring slaves. A royal decree of April 4, 1531, issued from Ocaña, authorizes the loan of money to the settlers of Cuba for a period of two years from the revenues due the crown in Cuba that year so they may purchase slaves to help develop the sugar mills. In the year 1532, when there were already almost five hundred Negroes in Cuba, according to a letter of Licentiate Vadillo, the Council of Havana requested His Majesty to ship them more slaves. At the end of 1534 the agent Fernando de Castro requested permission of the King to bring in fifty Negroes in order to set up a sugar mill. In 1535 the council reported that there were some thousand Negroes on the island, and Gonzalo de Guzmán asked His Majesty to exempt him from the payment of taxes on the fifty slaves he had been granted to set up a sugar mill. There was no end to the requests: governors, bishops, monks, municipalities, landowners, and merchants asked for slaves and more slaves. And this went on for over three centuries.

To give an account of the successive royal concessions to the sugar-planters would be to travel the main highway of Cuba's history during its life as a colony. I shall cite just one, one of the most striking, which it is not altogether inopportune to recall at the present moment we are traversing.

Although almost all the royal decrees referred to were, in general, issued for the Indies, especially those referring to the exemption from attachment and seizure from 1529 to 1538, toward the end of the sixteenth century when plans were under way for setting up mills in the vicinity of Havana, the same privilege was requested. The city council of Havana and the Governor in 1595 requested His Majesty, in order to foster the establishment of mills, that they be decreed exempt from attachment for debt, their lands as well as their equipment, their slaves, and their livestock. And His Majesty the King replied to this request with a royal decree that was published in Cuba on October 23, 1598, which had been issued by the King on

December 30, 1595. Moreover, during those years the King had ordered loans made from the exchequer to the amount of forty thousand dollars to builders of mills, with which the industry grew. This is recorded in the proceedings of the Council of Havana of February 3, 1601, according to the historian Urrutia, (Vol. II, p. 96). These concessions were really efficacious, for by 1610 there were records to show that sugar was being exported from Havana. The historical importance of this special privilege has been very great for the Cuban people.

These royal decrees of exemption were the Magna Charta of the plantation-owners having sugar mills in Cuba; they remained in force for centuries, and upon them was built the whole juridical and social structure of the sugar capitalism of the island. It is not until the second half of the nineteenth century, when the growth of economic liberalism made radical changes necessary, that this privilege of the plantation-owners, which had been the basis and the support of the sugar aristocracy of Cuba, disappeared. As late as the year 1833 a royal decree was issued expressly stating that those royal decrees of the sixteenth century, which had lasted for three hundred years, continued in force unchanged. On July 17, 1848, the Council of Authorities of the island of Cuba accorded that the historic privilege of exemption from attachment should be renounceable; but it continued in effect until 1865, for by a royal decree of April 2, 1852, issued by Isabel II, it was declared to be in force still, even though its reform had been begun. The new mills established were to be subject to the "common law," but the three-century-old grants of privilege were still to obtain for the mills existing prior to the year 1865.

A few years earlier, in 1858, as though to define the social-economic problem that then existed in Cuba, which, in his opinion, made a national revolution predictable and explained the threat of socialism, a Cuban statesman as sensible and conservative as the Count of Pozos Dulces, referring to the sugar-planter, said: "He has importance and wealth in a land where no one else has it," and he even added: "he has the law on his side, and, unfortunately, the power as well" (letter from Paris, June 18, 1858, to the Havana newspaper *Correo de la Tarde,* in his *Colección de escritos,* Vol. I, Paris, 1860, p. 372).

By this time even conservative opinion was condemning

these exceptional prerogatives and calling them immoral, and blaming them for the dissipation, the frauds, the usury, and the "gross profiteering of certain capitalists" (M. Torrente, op. cit., Vol. II, p. 355). The land-holding and industrial privileges of the planter declined then, slavery and slave-dealing declined; liberalism came into power, stimulated from abroad by the industrial revolution and the capitalism of finance. Spain could not understand this and she set herself mulishly against economic progress. She refused to accept reforms, and in 1868 the revolution broke out which, at times favored, at times hampered by international contingencies, culminated in 1902 in our independence, and the Republic of Cuba entered upon another era of its economic, political, and social history, which, according to the most cautious judgments, will soon be reaching its end.

II

The First Transatlantic Shipments of Sugar

It is on record that on the 29th of June 1517 there arrived in the port of Seville in the cargo carried by the ships of Juan Ginovés and Jerónimo Rodríguez a "little box" of the first sugar made in these Antilles, in Hispaniola, which was sent by the Hieronymite friars who at that time constituted the theocracy, or, to be more exact, the hierocracy, that ruled the Indies by order of Cardinal Francisco Ximénez de Cisneros. In the name of Joanna the Mad he was absolute ruler of Castile, backed by the power of his political, ecclesiastic, and feudal pre-eminence, his economic position, and his stubborn and rigidly authoritarian character.

It has been assumed that prior to this date there were no shipments of sugar, but there may have been, even though, rather than to Castile, they were sent to the rest of the Indies discovered up to then, which were provisioned from Hispaniola and Cuba. We have only to recall what has already been said about the different methods of producing sugar, and how there may have been and surely was sugar in the Antilles before there were ingenios there.

This can be inferred from the royal decree of February 13, 1523, dealing with the island of Fernandina or Cuba (see text on page 279 of this book). The King stated clearly that "they have seen from their own experience how, after Our Lord was pleased to have the industry of sugar begin, the aforesaid island has prospered and flourished, from which it is hoped that great utility, improvement, and perpetuity will redound to the inhabitants of that island. . . ." The date of this royal decree is of the beginning of the year 1523, and it is evident from the text that before this year Cuba had "prospered and flourished" as a result of the inception of the sugar industry, and there was every reason to augur great profits and increasing wealth for the settlers of that island. And all this before the existence of *ingenios,* with only horse-powered grinding mills. This is strengthened by Oviedo's and Herrera's observations on the sugar of Cuba. From all this it is evident that the industry of sugar had been started in Cuba before 1523, and this meant that sugar was being produced not only for domestic consumption, but for export; otherwise this new source of economic prosperity could not have developed so rapidly, this "flourishing and abundance" to which the King alludes. The conquest and organization of Yucatán and Mexico were then at their peak, and Cuba was the principal base of supplies for these enterprises. This island provided cassava bread, smoked meat, and these were undoubtedly accompanied by sugar and syrups.

12

How Havana Tobacco Embarked upon Its Conquest of the World

Tobacco was a little slower than sugar in becoming an article of export from America. But as far as Cuba is concerned, tobacco and sugar were almost contemporary; one might even say that tobacco was the more important of the two. When sugar was still of slight importance among Cuban exports, tobacco had already won a place and a name for itself in Europe.

Who first exported Cuban tobacco to Europe? It must have been Christopher Columbus and his companions on their return from their first voyage, to judge by the impression made on them by the "highly esteemed plant" of the Indians and the lighted "rolls" they discovered in use on the islands of the archipelago. But Columbus, in all probability, did not foresee the commercial possibilities of tobacco as he did those of sugar on his second voyage.

The belief that Father Ramón Pané sent tobacco seeds to Charles V has no basis in fact. It is probable that the friars sent out years before by Cisneros to acquaint themselves with the peoples and products of the Indies, their possibility of development, and the rule and behavior of the Admiral, Don Christopher Columbus, informed the mighty Castilian Cardinal regarding the *cohobas, tabacos,* and other typical customs of the Indian inhabitants. Fray Francisco Ruiz brought Cisneros many different Indian products, such as cassava bread, hammocks, and "a case or two of idols made of different woods," all of which the Regent placed in the University of Alcalá he had founded, thus setting up in Europe the first museum of American ethnography. Some time later Alejandro Geraldini, Bishop of Hispaniola, who died there in 1524, sent Pope Leo X many gifts of Indian objects, among them several of the idols used by the *behiques* in their rites. It seems likely that along with the images, masks, and belts Cisneros and Leo X received rolls of tobacco, the forked tubes, and the polished dishes used in the cohoba rites in which the discoverers and chroniclers were so interested, but, of course, there is no proof of this. And still less can it be said that from that moment the use and cultivation of tobacco began in Europe.

Passing from Indian to Negro to white man, the transculturation of tobacco in these Indian lands was fairly rapid. The proof of the rapid spread of tobacco among the white conquerors is the fact that by 1535 they had learned to distinguish between the tobacco of the Antilles and another species found on the American continent. In 1535 the first botanical selection of tobacco began in the Antilles; the first selection in its history, when the species *Nicotiana tabacum* was brought in from Yucatán (Brooks, loc. cit., p. 19), that with the little red flower having the gamopetalus corolla and five points like a star, which was preferred to that which grew in Cuba, the *Nicotiana rustica,* smaller in size, stronger and more bitter, and having a flower with ruffled edge and of a greenish-yellow color. This proves beyond a doubt how quickly the Europeans in the Indies became aware of the charms of tobacco and realized that selection and elimination were very important factors in its cultivation.

But although tobacco was making its way quickly in the American colonies, going from Indian to Negro and then to white man—that is to say, working its way up—the realization of its possibilities as an article of commerce of economic value for which there was a transatlantic market did not come so rapidly. As a matter of fact, it was not until the second decade of the sixteenth century that the economic colonization of America began. Until that time the settlers had hardly done more than establish themselves in Hispaniola, get a certain amount of gold, manage to find food, and make voyages of exploration to Cuba and the other islands of the Antillean archipelago and to a few regions of the mainland. After this period a permanent economic basis began developing for transatlantic trade, the conquest was spreading to the islands and coasts of the continent, and the sailors on their home voyages and the Spaniards returning to their native land introduced the exotic novelties of America to the port cities of the Old World.

It was in the second half of the sixteenth century that tobacco acquired an economic standing in world commerce and in the agriculture of certain countries, both American and European. The tobacco of Cuba was raised for the market and was much sought after. In the ports of the islands, which had natural defenses against the tempests of the sea, but which lay practically open and defenseless against man's aggression, the prized native tobacco could be obtained by freebooters and smugglers in connivance with the settlers, all of whom resented the monopolies established by the Spanish government. This clandestine trade was large. The port of Havana became the center for the distribution of tobacco because of the circumstance that it was at its roadstead that the Spanish ships assembled with their crews of rowdy sailors and their rich passengers for the return trip to the Guadalquivir. For centuries Havana was of great importance as the center for the Spanish fleets from Cartagena, Nombre de Dios, Portobello, Vera Cruz, Campeche, and Santo Domingo, all of which gathered in its spacious bay, protected from hurricanes and pirates and situated at the mouth of the Bahama Channel, which was the return route they were obliged to take because of currents and winds. From there these fleets set out in convoy under the protection of the armadas on their way to Seville. So large was this floating population gathered in Havana for months that its entertainments and diversions had to be organized for those thousands of people, some of whom had made fortunes in the Indies, while others had become degraded by their sordid existence as galley slaves or by the quarrelsome, dissipated life of the soldier. As a consequence of this wealth and extravagance that came in with the fleets, in June 1581 one of the governors of Havana wrote:

". . . this place [Havana] is the most expensive in all the Indies; this is because of the great number of ships that pass through here, and the people traveling on them who cannot refrain from spending even if they wanted to" (letter from Gabriel de Luján to His Majesty the King, *Papeles del Archivo General de Indias,* Academia de la Historia de Cuba, Vol. II).

It is on record that at this time the public sale of tobacco was a business in Havana. The first legal measures taken by the Spaniards against the plant date from this time. Cuban tobacco appears in our legislative history because of the restrictions against it and not, as in the case of sugar, because of the favors lavished upon it. The first restriction is that of the year 1557 against its sale in Havana; the second is of 1606, against its planting in Cuba and part of Hispanic America; the third is of 1614, against international trade in it. Under penalty of flogging, confiscation, and death.

By an ordinance of the city council of Havana, dated May 14, 1557, Negress slaves were forbidden to have taverns or inns, and specifically to sell wine or tobacco, under penalty of fifty lashes. This was a measure of marked racial discrimination on the part of the government. The Negresses carried on a thriving business in Havana during the months the fleets plying in and out of the Indies were in port. These "tripe-selling Negresses," as they were called, slaves or manumitted, entertained the seafarers in their establishments, where they ate, drank, gambled, danced, sang, played the guitar, and amused themselves in every imaginable way. In these centers of dissipation tobacco was one more pleasure. Without doubt these Havana taverns were schools for smokers. And the Negresses must have sold a great deal of tobacco for immediate consumption, for the voyage, and for future sale in Seville to make the long-headed aldermen, pricked on by envy and greed, abolish their lucrative business. Since then the business of tavern-keeper has been the prerogative of the white man from the Peninsula. And trade in tobacco with returning Spaniards, troops, and crews of the fleets and armadas of His Catholic Majesty and even with the other Atlantic ports of the Spanish-American empire was one of the characteristics of this port.

By that time regular commercial relations had been established between the different Spanish colonies in the Indies, and there was an inter-American trade in tobacco which the interests of the metropolis were already hedging around with restrictions and exactions. A royal decree issued by Philip II at the Escorial on September 16, 1586 included among other commercial restrictions on distilled liquors the following on tobacco:

". . . no store or tavern-keeper or other person of whatever state or condition may sell, give, or take to the aforesaid city [Panamá] or any part of its environs or territory, publicly or secretly, any tobacco, much or little, or plant it or have it, even though he says that he intends to take it somewhere else, under penalty of a fine of fifty gold pesos the first time, and the confiscation of the tobacco, which shall be burned in public as a plant harmful and forbidden in the aforesaid city and its territory, and the second time the fine doubled and exile for life from the Kingdom; and if the person should be a Negro or Negress, free or slave, any of these penalties shall be doubled, and in addition they shall receive two hundred lashes in the public streets. And we authorize each apothecary to have two pounds and no more in his shop, with the authorization of the courts, the city council and administration, making declaration thereof to them."

Commercial activities were springing up outside Havana, between the other Cuban ports and the traders of heretic nations hostile to Spain, even though, because of their clandestine nature, they left few documentary traces on the Spanish side. The cultivation of tobacco must have been on a considerable scale by this time in Cuba, which received frequent visits from freebooters and filibusters flouting Spain's monopoly on trade. As early as 1504 there were French corsairs along the coast of Brazil and perhaps in the Caribbean. In 1523 one of them seized off the Azores the treasures of Montezuma which Hernán Cortés was sending to Charles V. In 1527 the first English ship to come to America appeared off Santo Domingo, probably captained by John Rut, and the crew came ashore and seized the provisions they needed. And shortly afterwards the attacks and raids on Cuban ports began. In 1531 a fleet of thirteen ships and three thousand men was fitted out in Dieppe to seize Havana, but the expedition changed its plan and proceeded against the Portuguese instead. In 1536 Havana was sacked by a Frenchman; the next year, Santiago de Cuba. In 1554 the French attacked the latter city once more, and secured a booty of eighty thousand dollars. In 1555 another Frenchman, the Lutheran Jacques de Sores, seized, pillaged, and burned Havana. In 1562 the French Huguenots under Gaspar de Coligny established themselves in Carolina and in 1564 in Florida. This was the heyday of the great French, English, and Dutch corsairs in the waters of the Caribbean. In 1586 Drake raided the city of Santo Domingo, and shortly afterwards sacked the rich city of Cartagena de Indias. This bold sea-dog even raided the city of Cádiz in Spain itself, as a prelude to the victory over King Philip's "Invincible Armada," which took place the following year. In all these contacts of the foreign aggressors, a part must have been played by tobacco leaves, with which the "Lutheran" invaders had undoubtedly made acquaintance in the "Papist" ships they had seized and the cities they had occupied. But in the exportation of Cuban tobacco smuggling played a more important role than foreign piracy. Smuggling in Cuba was not a chance or sporadic affair, but a normal and highly profitable business. As Pezuela points out:

"Smuggling, which was almost continuous from the end of the sixteenth century and throughout the seventeenth, laid the basis for even richer fortunes than the establishment of Havana as a center for the fleets; and these fortunes, assured of reproduction and increase by the same methods, were influential in gradually extending the cultivation of tobacco" (Pezuela: Introduction to his *Diccionario,* p. 31).

In such manner, licit and illicit, peaceful and violent, Cuban tobacco made its way to Europe. Probably not a single ship left Cuba from the time of its discovery that did not carry some tobacco in its cargo, in the sailors' trading stores or duffel. Trade in tobacco was interoceanic: from the West Indies it was taken to Europe across the Atlantic and to Asia across the Pacific. And the tobacco of the New World had not only become a desirable article of trade; it had transmigrated to the Old World and was being sown and cultivated wherever there was a climate that even remotely resembled its tropical birthplace. The universal diffusion of tobacco, so amazing because of its rapidity and its defiance of the severest official restrictions, was possible from the material point of view because of the nature of the seeds of this plant. I have already mentioned how numerous they are; a single plant can produce over a million seeds, says W. G. Freeman. They are so tiny and numerous that when the Cuban *guajiro* sows them, he broadcasts them without opening his hand. The seeds are so small that they go through the cracks of his fingers into the air to drop into the furrows. Tobacco is sown with a clenched fist, like the symbolic Communist gesture. It is evident from this how easy it was to transport these seeds, even with the greatest clandestinity. One handful was enough to plant a whole plot, and in a couple of years to cover a whole region with tobacco.

This tobacco brought in from America was being planted all over the world. But the demand for Cuban tobacco persisted, despite the relentless persecution of which it was the object, and even after its cultivation had grown greatly in those same far-off lands that had formerly proscribed it. The reason was that, owing to special conditions, Cuban tobacco was and continues to be insuperable for the knowing smoker's taste. It should be pointed out in this connection that by the end of the sixteenth century Cuban tobacco was held in high esteem in England, even before that country had colonies in America. There can be no doubt that the tobacco of the greatest island of America was surreptitiously brought into the greatest island of Europe in smugglers' ships.

Cuban tobacco probably reached Europe in the forms in which it was used here by the Indians and later by the Spanish settlers—that is to say, as snuff, plug, and, above all, for smoking, as it was discovered in 1493 in that typical twisted form which Father Las Casas compared to *"mosquetes."* This inference is inescapable because of the fact that the Indians of the West Indies did not smoke pipes, which the Spaniards came to know some time later on the mainland, especially in Brazil and North America. In Spain cigars were smoked as in the Antilles. When the Spaniards referred to them they called them by the generic name of *tabacos,* or *tubanos,* or, more popularly, cigars, and the cigarettes they used, wrapping the shredded tobacco in paper, after the usage of Mexico, were known as *papeletes* or some similar name. All these terms clearly refer to tobacco in rolled or wrapped form for smoking. There were no pipes for tobacco in Spain until they were introduced by the Portuguese.

Proof that in the rest of Europe, and particularly in the Latin countries, tobacco was first smoked in the form of a cigar is the fact that in the first descriptions of this exotic American pleasure of smoking reference is made to the rolled-up leaves. We see this in Thevet, in Benzoni, in Lobel, and in De Lery.

The work of André Thevet (*Singularités de la France Antarctique,* Paris, 1558) and its anonymous English translation (*The New Found Worlde, or Antartike,* of 1568) are considered the first published work containing a graphic description of the smoking of a cigar. Thevet's observations dealt with Brazil. His book contains an engraving by the artist Jean Cousin which is an attempt to interpret the text of Thevet. It is clear that this draftsman, who was very accurate in matters of human anatomy and in details of landscape, was less so in regard to certain exotic objects that he did not know by sight, like the cigar smoked by one of the Indians in his drawing, but he was attempting to represent graphically the rolled-leaf cigar. In the work already mentioned by this same author, Thevet (*La Cosmographie universelle,* 1575), there is a reproduction of another drawing of Indians smoking, revealing more assurance of design, but an even greater disproportion in the dimensions, which I alluded to earlier. In any case, the twisting of the leaves is even more marked here.

As for the work of Benzoni, who lived for almost a year in Havana, in 1554, the drawings of the cigar (see Figure H, p. 159) are not in any way influenced by Thevet's work. The proportions are more exact, and the rolling of the leaves is clearly apparent. The detail the artist has omitted is the smoke. This work is of the year 1572 and was published in Venice.

The author Matthias Lobel, de l'Obel, or Lobelius has inserted in the work he wrote with Pierre Pena entitled *Nova Stirpium Adversaria* (Antwerp, 1576) an engraving (see Figure K) showing a head smoking a big cigar rolled in a tube, with this inscription beside it: *"Nicotiana inserta in fundibulo ex quo hauriunt fumum Indi et naucleri."* This was, said Lobel, "the same as Columbus had seen in the mouth of the natives of San Salvador," and, he goes on to say, what the sailors on the ships plying to the West Indies had later made great use of. "You will observe," says Lobel, "how the sailors as well as the others who returned from there (that is to say, America) use little tubes made of leaves or palm fiber, the end of which they fill with the crumbled dried leaves of this plant we are talking about (that is to say, *petun*). They then light it and, opening their mouths as much as they can, draw in the smoke with their breath." Fairholt, who reproduces this curious drawing (*Tobacco: Its History and Associations, including an account of the plant and its manufacture, with its modes of use in all ages and countries,* London, 1859, p. 16), observes in a note that "the reader must bear in mind that the original artist paid no attention to the size of the tube in proportion to the head of the smoker, the first being excessively long."

Therefore one has to query the authenticity of this type of cigar, which Lobel must have drawn from imagination, without ever having seen it, on the strength of what he had read and the equally farfetched drawings of Benzoni and Thevet, especially the latter, which served him as model. And one must also deny the assertion that this was the way in which the Indians of Guanahani smoked tobacco, for Lobel had no basis for this statement, even though he could have inferred it with reasonable assurance if he knew how the Indians of the Antilles smoked tobacco, in the form of rolls of twisted leaves.

The Frenchman J. de Léry (*Histoire d'un voyage fait à la terre du Brésil, autrement dit Amérique,* Rochelle, 1578), speaking of the Brazilian Indians, relates that they used the Nicotiana plant or *petun* in this fashion: "After gathering little bunches of leaves and drying them in their houses, they take four or five of them and roll them in another large leaf of a tree, like a cornucopia of spices. They light this cone at one end and inhale its smoke." Speaking of certain ceremonies among the Caribs of

FIGURE K. Act of smoking a cigar, according to Lobel: *Nova Stirpium Adversaria,* Antwerp, 1576

Brazil, this same de Léry tells that he saw them "take a cane about four or five feet long, at the end of which was some of the *petun* plant, dried and burning. Dancing and blowing the smoke from it in every direction over the other savages, they said to them: 'May you all receive the spirit of strength to enable you to overcome your enemies.' " This latter description by the French author is not very clear, for it might be interpreted as being of a pipe rather than a cigar of the kind mentioned by Thevet and Lobel; but the first description coincides with that of these other authors, even as to the enormous size of the cigar, which might well be the exaggeration of this exotic object discovered among the savages of America, with that lack of exactitude of which travelers and chroniclers of those times were often guilty, as we have earlier observed, or the description of a cigar of special size used in ceremonial or tribal rites, as Brooks supposes (op. cit., Vol. I, p. 283).

Pipe smoking became known among Europeans later than the cigar, and was introduced by the Portuguese, who discovered it in Brazil, and by the English, who made its acquaintance in Florida and Virginia. Perhaps the first published drawing of an Indian smoking a pipe is that contained in the work of de Bry: *Historia Brasiliana* (1590), even though earlier descriptions of the use of the pipe among the Indians of America already existed, as, for example, in the history of Hawkins's second voyage, which took place in 1565, and refers to Florida (*Hakluyt's Voyages,* Vol. X, p. 57), though this account was not published until years after the afore-mentioned work by Thevet.

In the sixteenth century in England two ways of smoking tobacco were known, the cigar and the pipe, and the former must have been the earlier, taken from the Antilles. In the later years of that century the custom of smoking had become so widespread in England that a German traveler wrote: "Wherever they happen to be, the English are continually smoking the Nicotiana plant, which in America is known as *Tobaca*" (P. Hentzner: *Itinerarium or Journey to England,* 1598). By the beginning of the seventeenth century the chemists' shops in England could not take care of the trade, and special shops had to be opened. In these the plant and the accessories for smoking it were sold, and there the users assembled to chat while indulging in their vice without hindrance, just as opium-smokers gather furtively today. Tobacco was a positive rage in England at that time. The word *tobaccanalia* was invented. The use of tobacco was a mark of the highest social distinction. The habitual use of tobacco lent a man rank and prestige. Even today these psychosocial effects can still be observed among budding smokers.

"The beardless adolescent endures stoically the disagreeable initiation into the habit of smoking. He candidly believes that tobacco, by some magic power, will confer upon him the attributes of manhood. The same phenomenon can often be observed among weakly men of small stature, who for this reason are victims of a marked inferiority complex. These little men often smoke big, strong cigars, for to them, as to the adolescent, tobacco is a symbol of virility. This may perhaps explain the liking of certain women for tobacco." (Miguel A. Manzano: *"Fumar,"* Revista de Agricultura, Industria y Comercio, San Juan de Puerto Rico, 1940, p. 237.)

In all these examples the conspicuous act of smoking is an attempt at social elevation; the smoker wants to show that "he is a man" or that she has progressed beyond the social taboos limiting the feminine personality. Something of the same sort took place in Europe when this startling, exotic, expensive, and new vogue of smoking was introduced and became known. Smoking then was "the smart thing," what was done by the best people. The dandies of London went to the theater so people could see them smoking, and they became expert in achieving amazing effects with the smoke. And as often happens in these periods of transcultural transition, tobacco was eagerly adopted not only by people of rank, but by those who without having social rank, or being able to pay for such luxuries, were novelty-seekers and vain, and wanted to seem more than they were. Tobacco was fashionable, but the excesses in which people indulged made it ridiculous and in poor taste. In the London of those days there were smoking masters, just as there were dancing masters, and masters of any other "liberal art." In Germany, too, there were similar displays of this showy faddishness among smokers, and the university students spent part of their time learning the most elegant manners in smoking (Fairnolt, op. cit., p. 54). The rage for this new American pleasure acquired such proportions that English satirical literature seized upon the theme of tobacco and the follies of the smoking fops and the "professors of whiffing."

It is to these English satirists that we owe the first mention of the smoking of Cuban tobacco. Its reputation was so great that the term *Cuban* was applied to a certain fashionable manner of smoking. In the year 1599 a play of Ben Jonson's, *Every Man out of His Humor* (London, 1600), was put on in London. Several of the characters, among them some who bring to mind the popular figures of the Italian *comedia dell' arte,* such as Puntaruolo and Carlos Buffone, hold forth upon the fashionable habits in vogue at the moment, and particularly upon the subject of tobacco. Referring to the "mysteries" or most exquisite and refined forms of the art of smoking, as it was taught by the "professors," one of Jonson's characters mentions three of these "mysteries" or latest refinements, which bear the technical names of "Cuban ebolition, Euripus, and Whiffe."

There is no mystery today connected with the last term. *Whiffe*, today written *whiff*, means a puff or a mouthful of smoke. The word perhaps meant at that time emitting the smoke all at one time, opening the mouth wide. The smoke was drawn in and then emitted as something no longer of any use, as one spits out a gargle after it has served its purpose.

As Brooks says, the first two of these "mysteries" taught by the professors of the art of whiffing "have baffled nearly all the critics of English literature who have tried to explain this passage." Euripus, says Brooks, was "an old channel, famous for its strong currents and the violence of its tides"; therefore the application of this word to the art of smoking would seem to suggest "a rapid exhalation of smoke." This emission of the smoke inhaled by the smoker must have been through the nasal passages, as through two springs or fountains flowing intermittently, two spigots opened in a barrel of smoke. This manner of smoking was a kind of drainage system. It was the most usual method of smoking. Laufer (op. cit., p. 34) states that in England at this time "the usual thing was to exhale the tobacco smoke through the nose."

But what did "Cuban ebolition" mean? The meaning of Cuban is clear; but what did the other signify? *Ebolition* is an archaic form of *ebullition*. Brooks suggests that it might imply a "rapid, uneven emission of the smoke," and adds that one dictionary interprets Jonson's expression in that sense (*A New English Dictionary*, ed. by Sir James A. H. Murray et al., Oxford, 1888–1928).

The expression "Cuban ebolition" was not the whimsical or passing metaphor of a poet, for it persisted in the jargon of English smokers at least as long as those fads and exotic fashions in smoking endured. In a play by an anonymous author (perhaps John Day) acted by the students of Cambridge in 1602 (*The Return from Parnassus, or the Scourge of Simony*, London, 1606) the phrase appears again, though in a somewhat garbled form. One actor speaking of another named Pródigo says that he is an "excellent" person because "he takes the Gulan ebullitio in an excellent manner." And the expression is to be found once more in a satirical work of the year 1617 by Richard Brathwaite: *A Solemne Ioviall Disputation,* etc. (London, 1617). In this work the typical manners of smoking are mentioned, the whiffe, the gulp, the retention, and the Cuban ebolition. The first, the whiffe, has been explained; the gulp implies swallowing the smoke, or smoking "in swallows"; and the retention probably means the smoker's method of holding in the smoke.

There were still other fashionable manners of smoking, such as the "Receit Reciprocall," which in Brooks's opinion (op. cit., Vol. I., p. 376) suggests "the blowing out of the smoke in rings, making each of these larger or smaller." This was one of the manners in vogue in Germany at the time. This whimsical fashion consisted in emitting the smoke in the shape of a ring, and before it had time to dissolve blowing another similar ring, but at right angles through the first. This is the description given in Dekker's *The Gull's Horn-book* (1602). Perhaps the "Receit Reciprocall" may have been a method employed by pairs of smokers, in which one emitted the smoke following the fashion set by the other.

There is no clear explanation of these capricious manners of smoking, especially the "Cuban ebolition." By elimination, it can be affirmed that ebolition did not resemble the other fashions of smoking. I do not think it is too farfetched to suggest that the ebullition or boiling of the smoke must have referred to a way of exhaling the smoke from the mouth in a column, with a certain pressure and bubbling, in the same way steam comes out of a closed kettle of water through a spout or when the lid is raised a little. It would seem that this manner of emitting the smoke must have been the last word in elegance and distinction, to judge from the fact that on the title page of certain books published in England in defense of tobacco there are figures of smokers with their heads turned skyward, releasing from their mouths a kind of plume of smoke—in ebolition. (See illustration opposite.) This example is typical and was published on the occasion of a satire against Sir John Suckling and his followers (*The Sucklington Faction; or, Suckling's Roaring Boyes,* 1641). Smokers today are still aware of these and other ways of emitting tobacco smoke in their hours of deep thought, when

THE SVCKLINGTON FACTION:

OR

(SVCKLINGS) Roaring Boyes.

Much meate doth gluttony produce. — Hee needes no napkin for his hands
And makes a man a swine — His fingers for to wipe
But neuer a temperate man indeed. — Hee hath his kitchin in a box
That with a leafe can dine — His Roast meate in a pype

...ere sits the prodigall Children, the younger brethren *Luk. 15. 12.* acting 2 parts Cavaliers and disguised ...s of the world, as one that is into a farre Countrey. *Nisi placuerit Deo, non erit eo*. Because his father humors ...ith the Idolatrous Ceremonies an errand Peripatetick ...follow Popish Innovations in a dudgeon and discontent ...Gods houshold, and consequently the Almighties direction. Not having God ...le, hee hath the Devill to ...ter, walking now not only ...sts of the flesh, and of his ...illing the desires of both; ...e Prince of the ayre, the ...now worketh in the children of disobedience. With the desperate attendants of these lascivious living times, he drawes his ...y through his throat, being the creatures to consumption ...a sumation of his intemperancie, delicate luxury, and ...odigality, spending all either in his belly or his backe, sole ...pe out, aporth, and feske, and ...ashions of the times, to ...selfe a painted Puppet on ...t vanity.

...a, *Venus, tribus his suum ...situs egeunt.*

...h wine and women, horses, ...whores, dancing, dicing, ...rinking, may the prodigall ...I am brought unto a mor ...d, yea unto the very huskes Pride of spirit makes him ...n Alehouse, and therefore ...er eagernesse bee daily ...avernes : where some ...s by his liquor, and bloud of the Vine, and the spirits of the ...hausting, and infusing them unto mad ebriety : thus drinking ...ne mensura, whole ones, by measure without measure, like ...unt through the juice of Mulberries, he is enraged unto bloud, ...amnable resolutions and designes, terminated in the death and ...n of the next man he meetes, that never did, neither thought ...e. Or having a noyse of renegado Fidlers, Musicke-abusers, ...him, and he with them, sings and danceth, danceth ...ike a Nightingale *, or Canarie bird. He is pro ...vish.

...Donec deceptus & exspes,
...quicquam fundo suspiret nummus in imo,
...paring till all be spent, dancing, and drinking away both wit

...and wealth. Now he acts his ryots, anon his revels, and forthwith ser ...ries to a Play-house, or Bawdy-house, where the woman with the attire of an harlot kissing him, al ...lures this simple sot, voyd of understanding, to solace himselfe, (ver. 18.) *and take his fill of love untill the morning*. Lust leades him to dalliance, till a dart (*Ver. 23.*) strike thorow his liver, untill bee be cast downe and wounded, yea and slaine by her.

This notorious good-fellow (corruptly so called) being a confederate of the Greekes, *Titere tu*, or joviall roaring Boyes, is of the Poets mind, when he said ;

Fæcundi calices quem non fecere disertum? Whom hath not wine made witty? He drinkes that he may bee eloquent and facete, after his cup of *nimis*, he harps on *Barnabies* Hymne, or *Bacchus* his inebriating Catch, bousing verily, and chanting on this wise merrily :

*Æsculapi tandem sapi,
quid medelas blateras?
Mithridatum est potatum
inter vini pateras.
Ad liquores & humores
tædium crescunt salices
Si quis ægrotet, mox epotet
decem vini calices.
Quid problema, aut poema
vult acute texere,
Ordiatur, vino satur,
& uvarum nectare.
Nil acute, nil argute,
dictum sine delio;
Audivi sales, nunquam tales
ait in œnopolio.
Quorsum plura, hæc figura
satis rem nobilitat:
Vas rotundum totum mundum
plenè consignificat.*

These are children of spirituall fornication, such as goe a Who ...ring from God after the idols of their owne braines : *Hos.* 1. 2. such are superstitious Romanists, ...tutoured by their Ghostly Fathers, to beleeve In grosse as the Church beleeveth, which (as *Luther* saith) is grosse Divinity. These fall not onely from piety to impuritie, but also from Christian verities, to Antichristian vanities, fopperies, and trumperies.

FINIS.

Printed in the Yeare, MDC. XLI.

THE ELEGANT MANNER OF SMOKING
AND BLOWING OUT SMOKE CALLED "CUBAN EBOLLITION"
(*from an English broadside of* 1641)

LONDON TOBACCO SHOP,
SHOWING THE CHARACTERISTIC SIGN OF A
NEGRO SMOKING A LARGE PIPE
(from Richard Brathwait's *The Smoaking Age*, 1617)

tobacco is their companion and traces in the air the cryptic signs
of their speculations.

With reference to the Cuban ebolition, not so many years ago
a historian of tobacco asked, without being able to find the answer: "Why Cuban?" (G. L. Apperson: *The Social History of
Smoking,* New York, 1916, p. 49). Why was this fashionable
manner of smoking called Cuban? From the antecedents alluded to, it is possible to infer the reason. This typical style of
smoking had been learned from the Cubans by English smokers. Sailors and freebooters probably picked it up from the
traders or inhabitants of Cuba with whom they had dealings,
just as the early Spaniards had learned it from the Indians. This
style of smoking must have been one of the supreme rites of
the *behiques,* perhaps in honor of the god Huracán, the deity
of the terrible storms, the thunder, the winds, the rains, and in
general of agricultural fertility; and the rite that produced gusts
and "ebullition" of clouds of tobacco smoke must have been
connected with him, just as among the Aztecs, according to
Father Gerónimo de Mendieta, tobacco was the goddess spouse
of Tlaloc, the god who smoked and made clouds, the god of
rain and of the agro-sexual rites.

Blowing the smoke out of the mouth as in a boiling column
must have been the *behiques'* way of directing the magic smoke
of the tobacco toward their idols, their sick patients, the heavens,
and such other objects as they wished their magic to affect. This
simple or combined method of exhalation in smoking is the
only way of sending the smoke in a certain direction, as though
to carry out some transcendent plan. In this way the smoke
emitted from the body after the tobacco has been consumed is
not the mere residue of something that is no longer of use, but
continues to be a substance that has not yet lost its supernatural
efficacy, and if it first communicated the sanctifying grace of the
deities to the believer, it now returns to these powers with a
mystical message, or is transmitted to surrounding objects to
bestow upon them this mysterious sacred power. Thus the ritual
efficacy of smoking is greatly increased: first through the inhalation of the smoke the divine essence penetrates to the remote meanders of the brain and the other internal organs of the
body where the personality has its seat, and then this super-

natural stimulus is deliberately given forth so that it may exercise its influence upon other surrounding beings. In other simpler methods of smoking, the individual plays a more passive role; he alone receives the mystery of the gods, and the effects come to an end in him. Through the "Cuban ebolition" the smoker becomes the owner of the sacripotent smoke and "socializes" it, emitting it toward the gods, his fellow men, or objects with the purpose of establishing a social relationship. In other methods of smoking, those which are merely inhalatory, the smoker is the only one who absorbs into himself the smoke or powder of tobacco and then gets rid of them when their usefulness to him is over, with which the rite comes to an end and the participant awaits such subjective effects as may ensue. By the other method the rite does not conclude with the smoker; on the contrary, it would seem that the inhalation of the smoke is only the first step, and the rite reaches its fulfillment with the exhalation of the sacred essence, which will accomplish its effects outside the smoker. Smoking here is no longer merely the inhalation of a magic substance by the smoker to achieve a vomitory catharsis and spiritual cleansing of himself, but the preparation and control of the supernatural essence by an individual in order to make ritual use of it afterwards and set it to work upon others. In this way the smoker becomes an active officiant in the purification rite. Cuban ebolition must have originally been a highly typical liturgy of the Cuban Indians, and particularly of their priests or *behiques*. And for this reason this ritual, whose traditions and meaning were functionally of a priestly character, aside from its higher æsthetic form and more complicated cathartic process, was naturally received by European smokers as the most refined and aristocratic manner of smoking. This attitude of the smoker, launching his message upwards, is the most hieratic and the most religious. It is the rite of the priests of Tlaloc sending their god the magic cloud of tobacco silently to implore the blessing of rain from this deity. It is like sending a breath of the soul to the skies as is done in the *ecobio* of the Ñañigo Negroes of Cuba.

It is the manner of smoking of the thinker. As Eça de Queiroz, the great Portuguese novelist who disliked Cuba, where he was consul, and Havana so heartily—in his opinion it was nothing but "a tobacco warehouse"—said, "thinking and smoking are two identical operations; both consist in launching little clouds into the air" (*Epistolario de Fadrique Mendes,* Chapter viii). If tobacco, in the judgment of Victor Hugo, was "the plant that turns thoughts into dreams," this Cuban manner of smoking was that of the thinker, delighting himself with a melody written in smoke spirals that express thoughts as ineffable as dreams.

Moreover, the Cuban ebolition demanded greater liberty and elegance of manners on the part of the smoker; it allowed for greater expression of the personality, as in the manipulation of a fan or a lace handkerchief or the courtier's handling of an exquisite snuffbox. And with regard to the intimate psychological effect of smoking, these "directed puffs" opened an unlimited horizon to the imagination where the spirit could take refuge from the tensions that beset it.

At any rate, this terminology in the language of the English smokers indicates that by the end of the sixteenth century the Cuban manner of smoking tobacco was sufficiently well known there, and so distinctive and highly esteemed as to be the symbol of aristocratic elegance. And if the Cuban manner of smoking was considered supreme, it was only logical that the tobacco of Cuba should be equally esteemed. We have only to recall what took place with regard to the development of a taste for tobacco, particularly the Cuban, and its rapid diffusion, to understand how this local mannerism of the smokers of Cuba reached England, along with its tobacco, to become there the most elegant manner of smoking. This was due to the fact that it was the most exotic, the most traditionally aristocratic, the most communicative and social, and, above all, that which brought to mind the presence of the most highly esteemed, the most expensive, and the most exquisite tobacco in the world, which was the Cuban.

Among those dandies, those exhibitionistic smokers in the Cuban ebolition manner, the pipe predominated; but they also took snuff and smoked cigars (Brooks, Vol. I, p. 52). The Cuban ebolition did not demand the cigar or Havana tobacco; it worked equally well with the Nicotiana plant in a pipe. And many illustrations of the books on tobacco published in England at this time show smokers with a pipe in the hand. But if this special manner of smoking called, and rightly so, Cuban, was famous there, is it not logical to assume that the tobacco of the island of Cuba was not only known but esteemed in all its merit? And granted all this, is it not likely that what the English have always called Havana cigars were known there, too?

It was not the Havana cigar that the English were in the habit of smoking in those days, but the pipe. It has been held that the English adopted the use of the pipe because they learned it from the Indians of Virginia (Ralph Linton: *Use of Tobacco among North American Indians,* Chicago, 1924, p. 8). However, the English were smoking long before that colony was settled. The Portuguese carried pipes to Europe from Brazil before the English came to know them from the North American Indians. They were the ones who taught the African Negroes of Guinea, Congo, Angola, and Mozambique to smoke this way in their dealings with them as slave-traders, and the Asiatic peoples in their explorations and trading in the East Indies. It was likewise the Portuguese who spread pipes and tobacco through Persia, Arabia, and Turkey. On the other hand, the sailors, who were the ones who introduced smoking, preferred to chew tobacco, for which they required no light. Their pleasure could be satisfied in this simpler manner without the dangers incurred by the use of fire on those long sailing voyages. But when he came ashore, and even on board, despite the risks, the sailor smoked a pipe, which was more practical than a cigar.

It is not to be wondered at that the cigar fell into disuse in Europe and was generally replaced by the pipe. In order to smoke the typical cigars of the Antilles one either had to have an abundant supply of them, which, for economic reasons, was difficult unless one were rich, or one had to have the dry, whole leaves, properly cured, to roll and wrap them every time one wanted to smoke, which, in addition to being expensive and requiring considerable effort, was complicated and not at all easy unless one had acquired the skill. With the pipe this and other difficulties were obviated. Nevertheless, sailors were smoking cigars in Europe before 1565 (Brooks, Vol. I, p. 44). The almost continuous wars between England and Spain from 1588 to 1604 did not greatly hamper the trade in Spanish-American tobacco with London, for it was carried in French and Flemish boats (Brooks, Vol. I, p. 87), and widespread smuggling flourished, even though there was an increase in prices. The possibility should not be excluded, therefore, that England made the acquaintance of the cigar during that period. It was probably brought in from the American lands where it was produced, though in such small quantities as only to whet the appetite.

Other information, taken from these same works of English literature, bear out the presence of the Cuban type of cigar in England at this time. Ben Jonson speaks of "making of the patoun." *Patoun* or *petun* was the name given tobacco by the Portuguese, and "making a petun" was the same as "making a cigar," the expression used later. To take tobacco in snuff, or for chewing or for smoking in a pipe it was not necessary to "make it." Making could only be applied to the cigar, which had to be twisted and rolled. This is how Brooks interprets it (op. cit., p. 376), saying that the expression "making of the patoun" used by Jonson meant "making and smoking a cigar, which at that time was beyond doubt a mystery for the gentlemen of the Elizabethan court, accustomed to the pipe."

Another author already quoted, Richard Brathwaite, is even clearer on this point. His work *The Smoking Age* (1617) has a frontispiece with an engraving (opp. p. 302) of a little Negro smoking a big cigar and carrying another under his arm. This was the sign of the tobacconists. This is proof that the most typical, the prototypical form of Cuban smoking—that is, the cigar of the Indians—was already known in London to the point where it was used as the unmistakable trade-mark of the best tobacco to attract smokers. There can be no doubt that the Cuban cigar at that time must have been rare in London and for this reason was more eagerly sought after by the fashionable dandies and voluptuaries.

This does not mean that the cigar reached England only from Cuba; it may have been taken from other parts of the Indies, especially Trinidad and other of the Caribbean islands where the sailors of Great Britain and Holland carried on their smuggling trade. The Jesuit Bernabé Cobo in his work *Historia del Nuevo Mundo* says:

". . . the tobacco plant is very well known not only in the Indies, but in Europe as well, where it has been taken from these lands, and is highly esteemed because of its many and excellent qualities. . . . There is no place in all the Indies today where there are not many people who take the smoke of tobacco. . . . So great is the amount of tobacco that is used in the Indies and shipped to Spain that there are whole provinces where the sole occupation and activity of their inhabitants is the cultivation and care of this plant, and that of certain regions brings a higher price than that of others." (Op. cit., p. 147).

We have the word of the King of Spain himself for it that the tobacco of Cuba and other countries of America was eagerly sought by the European trade. A royal decree issued on August 26, 1606, at San Lorenzo del Escorial, alludes to the abundance of tobacco cultivated in the Indies "because it is the principal crop the natives possess," and to the number of foreign ships calling at their ports to buy it because this tobacco "was highly esteemed and sought after in the aforesaid countries," as well as mentioning the fact that it was the royal wish that the inhabitants should cultivate other crops and work the mines. In view of all these circumstances, the cultivation of tobacco was forbidden for ten years in Santo Domingo, Cuba, Margarita, Venezuela, Puerto Rico, Cumaná, and Nueva Andalucía.

This edict, which was aimed principally at the Dutch, must not have been very effective, or, to be more exact, it was found to be extremely harmful, and King Philip III rescinded it, permitting the cultivation of tobacco, but repeating that its sale to foreigners was illicit and would be punished by death.

This royal decree of Philip III was, according to Bennett, in answer to the measures taken by James I of England, who ordered a tax on foreign tobacco in order to favor that of the colonies, and the Spanish King hoped that by placing severe restrictions on smuggling between England and his American possessions he would force the English to go to Seville and pay a fancy price for this most highly esteemed tobacco of the Indies.

About this period, says Pezuela:

". . . from 1610 to 1620 the cultivation of this valuable plant was begun in the bottom lands along the Almendares River near Havana, and the Arimao, not far from Trinidad. This productive branch of agriculture had developed in a few years to the point where its smuggled exportation to foreign countries was several times undertaken, and it was sold in Havana to the passengers on the fleets returning to Spain." (Pezuela: *Diccionario de la Isla de Cuba,* 1863, Vol. I, p. 32.)

Tobacco from the Indies brought "its weight in silver" in London at this time. Alfred Crowquill said it was worth "its weight in gold" (quoted by Steinmetz and Brooks). The amount of tobacco entering England was very considerable. From the island of Trinidad (which the European reader should not confuse with the region of Trinidad in the island of Cuba) and the basin of the Orinoco alone tobacco to the value of 200,000 pounds sterling was imported into England in the year 1615, and its trade "was almost completely in the hands of Spaniards." By 1620 the demand for Spanish tobacco was growing so in England, according to Edward Bennett, that as a result new colonial establishments were springing up in the Hispanic Indies. The dealers made jokes about it, saying that they paid the English for their merchandise "with smoke." Nor must the contraband trade in tobacco be overlooked, either that carried on in the Indies or that in the Peninsula itself, which was not at all negligible. Even the friars were accused of being smugglers, and their convents were subject to search (Brooks, Vol. I, p. 144).

By the second decade of the seventeenth century the tobacco of Cuba and other Hispanic countries had to compete with that of Virginia and Bermuda, as well as of the other Antilles and the mainland and even certain countries of Europe. Brathwaite in the second part of his work, published in 1617 under the title of *The Smoking Age,* refers to the tobacco of Bermuda, Trinidad, and Varinas. Bermuda, Trinidad, Caracas, Varinum were terms employed by those same smokers of London who went into ecstasies over the "Cuban ebolition." The use in England of the tobacco of Barinas or Varinas, a remote inland region of Venezuela, goes to show how frequently the monopoly the Spanish laws tried to set up was flouted. In connection with this smuggling, it must be pointed out that some Havana tobacco must have reached England as Bermuda tobacco. Bermuda, as everyone knows, is a little island not far from Havana at the end of the channel where the European-bound fleets changed their course. This island was famous among navigators as a very dangerous spot, so much so that Shakspere refers to it in this sense in *The Tempest;* this was due not only to its sudden squalls and its reefs, but because it was one of the favorite stations of the English corsairs, from which they emerged to fall upon

the Spanish galleons. Bermuda, which by this time was occupied
by the English, and the Bahamas with their barren keys were
lairs of smugglers, pirates, and wreckers who preyed upon the
Spanish merchant ships engaged in trade with Cuba and those
that crossed the two Bahama channels. To a certain extent,
though to a lesser degree, these islands were to Cuba what
Tortuga was to neighboring Hispaniola, the hiding-place of the
enemy waiting to pounce. In this period trade between the heretical English and Cuba was strictly forbidden, but it was carried on quite regularly and more or less openly, and the origin
of the products of the Antilles was concealed, especially those
from Cuba and the city of Havana, where a brisk smuggling
trade was carried on in the open bay of Matanzas and from the
near-by islands. Bermuda did not begin to export tobacco until
1614, and its output was always small (Brooks, Vol. I, p. 87),
but its clandestine dealings with Cuba were more than occasional. By 1616 Virginia tobacco was being consumed in England. This was the American colony on which the English
based their tobacco production, and this single-crop system became so dangerous that it was known as "the colony founded
on smoke" (W. Bullock: *Virginia Impartially Examined,* London, 1649).

It is well to recall this trade in tobacco with the coastal regions of South America here, for it had great historic consequences; because of it three European nations, England, France,
and Holland, embarked upon the colonization of the Antilles.
These tropical colonies of America were the offspring of tobacco. About the middle of the first decade of the seventeenth
century groups of adventurers from each of these nations made
their appearance at the same time along the coast of the Guianas, with a view to establishing tobacco plantations to be
worked by white laborers.

Tobacco was the chief incentive in these first colonizations
attempted in America during the early years of the seventeenth
century by nations hostile to Spain who paid no attention to the
solemn papal bulls issued by Alexander VI giving the Catholic
kings and their successors the dominion of the New World. In
1606, with the founding of Jamestown, the real colonization of
Virginia by England began. During this period the English,
French, and Dutch had staked out their claims in Surinam,

Cayenne, and Essequebo, respectively. In 1623 England seized San Cristobal, an island of the Antilles, to which they gave the name of St. Kitts; in 1624, Barbados, and so on. In 1632 Holland occupied the little islands of St. Eustatius and Tobago, and in 1634 the more important one of Curaçao. France laid hold of Guadeloupe and Martinique in 1635. In all these colonies the principal economic basis of their settlement and wealth was tobacco. All these new tobacco lands supplied the distant European mother countries with tobacco. Nevertheless, the demand for Cuban tobacco continued. Among knowing smokers "Cuban ebolition" never lost its supreme charm.

The first notice of great exportations of tobacco from Havana appears in 1626 in secret charges brought against Governor Cabrera, who was accused of having sent to the Canary Islands, on his own responsibility and without permission from the clearing house of Seville, a ship carrying a cargo of tobacco worth 200,000 pesos (J. de la Pezuela: *Historia de la Isla de Cuba,* Madrid, 1868, Vol. II, p. 51). Even though the ports of the Canary Islands were not free, as they later became, they were very well suited for smuggling on a large scale, by transshipment, and for this reason the cargo sent out by Cabrera amounted to illicit dealings in tobacco with Lutheran foreigners.

By this time the King of Spain had discovered that the tobacco of Cuba could supply goodly sums to the royal coffers, and the exchequer began to wrap its tentacles about the tobacco industry. In either 1632 or 1636 the levying of duty on tobacco entering Spain was no longer sufficient, and its cultivation was declared a "regalia of the crown"; in 1634 a monopoly on it was established in Spain; in 1665 there were tobacco companies acting as lessees; in 1670 the first snuff factory was set up in Seville, with raw material coming from the Indies. The pressure of the exchequer was increasing. The monopoly was reaffirmed in 1719 and 1726, and infraction of it was punishable by death.

In the seventeenth century the fame of Havana tobacco grew greatly after the end of the bloody Thirty Years' War, in 1648. Havana had become a great producing and shipping center for tobacco, not only to Spain, but to all Spain's American empire, Mexico, Costa Rica, and the countries of the Pacific (A. O.

Exquemeling, *Buccaneers of America,* London, 1684). According to documents of the period, tobacco was said to be "the most important fruit" of Havana in the seventeenth century (J. Le Riverand: *Los molinos de tabaco hasta 1720,* Habana, 1940).

Taking these antecedents together with those previously outlined dealing with the transculturation of tobacco, one has a schematic outline of the struggle in which the tobaccos of America had to engage to conquer the world. At the head of them marched that invincible leader of all, Havana tobacco.

During the seventeenth and eighteenth centuries the cigar was almost wholly restricted to the Antilles, the other Spanish colonies, the Peninsula, and certain regions of Asia and the Philippines where it had been taken by the Spaniards. As England's colonies in North America grew and could supply its smokers with tobacco from Virginia and later from the Carolinas and Maryland, the taxes on Havana tobacco grew heavier and heavier until its use became restricted to those wealthy enough to pay for such a costly luxury. Most Europeans used tobacco only in plug, for chewing, in twist or cut for pipe smoking, or as snuff. In the eighteenth century when English writers used the word *cigar,* they felt obliged to explain its meaning to their readers. J. Cockburn, alluding in 1735 to three friars of Nicaragua, speaks of the "seegars" they were smoking and says: "These gentlemen gave us some cigars to smoke. These are tobacco leaves rolled together in such a way that they can be used in a pipe or smoked by themselves; it is the only fashion they know here, for in all New Spain there is not a pipe to be found."

With the conquest of Havana in 1762 by the English, the Havana cigar in turn conquered the ships' crews and the British and North American regiments that occupied the city for a year, and was brought back to the ports of England and its colonies in America on the thousands of ships that called at Havana during that year of free trade with cargoes of merchandise from Europe and America and slaves from Africa. But the English gave up Havana, and a few years later a new war broke out that gave the United States its independence, and in which the Spaniards and Creoles had a share. Havana fitted out several expeditions to help the rebels, not to mention that of the French under Rochambeau, which played such an important part in the final defeat of the British at Yorktown. The war came to an end, but new conflicts arose, and other peace treaties, which disturbed the commercial relations between Cuba and the Anglo-Saxon peoples (Herminio Portell Vilá: *Historia de Cuba en sus relaciones con los Estados Unidos y España,* Vol. I, Havana, 1938, Chapters ii and iii).

During this long period Havana cigars were becoming known abroad, but the trade in them continued illicit. It is strange to observe that in the commercial documents of Cuba, tobacco is hardly mentioned as an article of export except to Spain and its viceroyalties of America. This was because at the time trade in tobacco was doubly contraband; it was forbidden to be taken out of Cuba by Spain, and its importation into other countries was forbidden either because of their own tobacco production or because of a state monopoly.

Before the end of the eighteenth century the Havana type of cigar was coming to be known in England and the rest of the Continent, apart from Spain, where it had always been used. By this time the habit of smoking had waned considerably in England. Dr. Samuel Johnson stated in 1773 that smoking had disappeared, but this was an exaggeration, even though snuff had come to predominate over the pipe. In 1779 the papal government gave Peter Wendler, a German, permission to manufacture *"bastoni di tabacco"* in Rome. Cigar factories were also opened in France, modeled after that of Seville. And with the spread of the habit of smoking cigars manufactured in Europe, the taste for real Havana tobacco grew.

In Germany the cigar of the Havana type soon was in demand. In 1788 one Schlottmann set up a factory in Hamburg to manufacture them of leaf imported from Cuba. And in time Hamburg became a great center of the tobacco trade; buyers were sent out who settled in Havana, opening warehouses, cigar factories, and even banking houses to finance their operations. But throughout Europe Havana tobacco met great obstacles at that time, for just as the cigar was beginning to spread over that continent the era of the great revolutions, starting in 1789, and wars which shook that part of the world for a quarter of a century began.

It is said that it was in the year 1801 that the manufacture of Havana-type cigars was begun in the United States, in Connecticut, and that in 1810 genuine Havana cigars were first imported (Carl Werner: *A Textbook on Tobacco,* New York, 1914, p. 22). It seems more likely, however, that these events took place in inverse order: first the importation of Havana cigars, followed by their local manufacture. Probably this date refers only to their official importation, for long before this Cuban cigars must have been smuggled into Anglo-America.

In spite of the wars of Europe, which hampered economic prosperity and normal trade relations for years, they nevertheless in a measure stimulated acquaintanceship with the Havana type of cigar because of the role played by the Spaniards and Portuguese in certain episodes of these conflicts. In Austria the cigar was introduced about 1805, thanks to a Spanish diplomat who presented some to Marshal Windisch-Graetz. The gift must have been of Havana tobacco, in view of the rank of the two men. In England and in general in the rest of Europe it was not until the Napoleonic Wars and the uprising of the Spanish people that the Havana cigar came into its own. In those war years Spain, whose monarchists and reactionaries had again and again asked for the assistance of foreign troops, was crossed and recrossed by English and French armies, those of Wellington, of Bonaparte, of Angoulême, and the returning soldiers carried back with them, along with their laurels, their wounds, and their campaign memories, the Spanish custom of smoking cigars and cigarettes manufactured in that country and in its colonies. As a result of the Peninsular and American wars in England, from 1814 on, the custom of cigar-smoking had a vigorous resurgence. It was an expensive luxury, which was an added reason for its revival. The moralists once more attacked the "Spanish vice"; prim Queen Victoria, who came to the throne in 1837, hated tobacco as much as James I did, and forbade its use at court. At her insistence, Wellington in 1845 had to order his officers to refrain from using it, but tobacco triumphed against this royal hostility. In 1823 England imported only 26 pounds of cigars; a year later the amount had increased to 15,000, and every year Havana cigars were winning more addicts in Great Britain. Among the countries of the Continent whose troops had been in Spain the use of the Havana type of cigar was spreading. Napoleon did not smoke, but he took a great deal of snuff, and his officers learned to smoke cigars *"a la española,"* and the fashion spread through the rest of Europe.

But there was still a serious obstacle in the way of victory for Cuban tobacco or Havana-type cigars.

With the spread of the use of the cigar came the custom of smoking in the street. Tobacco became ambulatory and was regarded by the authorities as a sign of revolutionary tendencies and liberalism as compared with the conservative pipe, which preferred limited confines, quiet and sedentary habits (Corti, op. cit., p. 246). For this reason in the absolute monarchies of Europe smoking in the street was forbidden until the Revolution of '48 won this new "personal liberty" for the people. First in the Anglo-Saxon countries and later in the constitutional monarchies of Europe the cigar came to predominate, almost completely replacing the pipe and snuff-using, which were relegated to certain circles. As civil liberties triumphed and political constitutions were guaranteed, the cigar came into the ascendancy once more, coinciding with the advent of economic liberalism in Cuba, which threw the port of Havana open to all nations. And in this atmosphere of free industrial and commercial enterprise Havana tobacco, by the unanimous plebiscite of the world, was awarded the imperial scepter of the tobacco world. Havana tobacco from then on became the symbol of the triumphant capitalistic bourgeosie. The nineteenth century was the era of the cigar. The ground is now being cut from under its feet by the democracy of the cigarette. But cigars and cigarettes are now being made by machines just as economy, politics, government, and ideas are being revised by machines. It may be that many peoples and nations now dominated by the owners of machines can find in tobacco their only temporary refuge for their oppressed personalities.

Appendix

PRAYER OF THE RIGHTEOUS JUDGE[1]

"Righteous Judge, King of Kings and Lord of Lords, who reigns always with the Father, the Son, and the Holy Ghost, help me, protect me, and deliver me, by sea or by land, from all who seek to harm me, as Thou delivered the Apostle Saint Peter and the Holy Prophet Jonah from the belly of the whale; O Almighty Lord, help me, for I am Thy slave, in everything I undertake and in every kind of game such as cock-fights and card-playing, through the grace of the Holy Righteous Judge, whence comes the Holy Trinity. Let these great powers, these great relics, and this holy prayer help me to defend myself against all evil, so that I may find buried treasure, however difficult it may be, without being disturbed by spirits and ghosts, and so that when I am attacked or on the battlefield neither bullets nor knives may prevail against me. Let the arms of mine enemies shiver in their hands, and their guns be powerless, and let mine be successful and never overcome; let mine enemies fall before my feet like the Jews before Jesus Christ; let prison doors, handcuffs, knives, chains, keys, locks, bars break apart. And Thou, Righteous Judge, who wast born in Jerusalem and sacrificed between two Jews, if mine enemies should pursue me, let not their eyes see me, nor their lips speak to me, nor their hands lay hold upon me, nor their feet overtake me. I shall be girded about with the arms of Saint George, I shall be locked in the cave of the Lion with the keys of Saint Peter, in Noah's Ark shall I be sheltered; with the milk of the Virgin Mary shall I be sprinkled, with Thy precious blood shall I be baptized, and in the name of the Lord's Prayer Thou hast repeated, and the three holy wafers Thou hast consecrated, I implore Thee, O Lord, to walk with me and to enter and dwell in my house with joy and comfort. May the Righteous Judge protect me, the Blessed Virgin cover me with her mantle, and the Blessed Trinity be my shield upon my right side. Amen."

(From *Broad and Alien is the World,* by Ciro Algería, translated from the Spanish by Harriet de Onís, copyright, 1941, by Rinehart & Company, Inc., and reprinted with their permission.)

[1] See page 204.

Glossary

abridora—stripper of tobacco leaves
andullo—plug tobacco
apartadora—selector of tobacco leaves
behique—Indian medicine-man
bohío—rustic hut of poles thatched with palm leaves
bonchero—unskilled cigar-maker
breva—type of cigar
buhuitihú—Indian medicine-man
cajetilla—pack of cigarettes
cancháchara—drink of sugar and water
capa—wrapper
capadura—second growth of tobacco
cartabón—mold for making cigars
cemí—Indian idol
cepo—mold for rolling cigars
cibucán—press for draining liquid from grated yucca for making cassave bread
cigarros—cigarettes
corona corona—type of cigar
cuaba—torch of resin wood used by Indians in certain religious rites
cuje—pole for drying tobacco leaves
cunyaya—pump-handle device for squeezing juice from sugar cane
curros—free Negroes who came to Cuba from Spain. They considered themselves an aristocracy among Negroes.
enmatulado—bale of tobacco
enterciadores—tobacco-packers
escogedor—tobacco-sorter
gavilla—"hand" of tobacco
guajiro—Cuban countryman, small farmer
hacendado—sugar-planter
mabinga—poor grade of tobacco
mambises—Cuban rebels against Spanish government

mamones—suckers
mancuerdas—pairs of tobacco leaves
manojo—four gavillas
matul—bale of tobacco
mazo—bale of tobacco
picadura—fine-cut tobacco
pilón—sweating process
regalía—type of cigar
repasadora—tobacco-selector
rezagador—tobacco-selector
rueda—bunch of tobacco leaves
sitiero—small farmer
tabacal—tobacco patch
tabaquito—small cigar
tarea—hank of tobacco leaves
tripa—filler
vega—bottom land where tobacco is grown
veguerio—group of vegas

INDEX

[i

Fernando Ortiz (1881–1969) was one of Cuba's most influential public intellectuals. His vast work on Cuban culture includes the five-volume work *Los instrumentos de la música afrocubana*. Fernando Coronil is Assistant Professor of Anthropology and History at the University of Michigan.

Library of Congress Cataloging-in-Publication Data
Ortiz, Fernando, 1881–1969.
[Contrapunteo cubano del tabaco y el azúcar. English]
Cuban counterpoint, tobacco and sugar / by Fernando Ortiz ; translated from the Spanish by Harriet de Onís ; introduction by Bronislaw Malinowski ; prologue by Herminio Portell Vilá.
Includes index.
ISBN 0–8223–1616–1 (paper)
1. Tobacco industry—Cuba—History. 2. Sugar growing—Cuba—History. 3. Cuba—Civilization. I. Title.
HD9144.C90713 1995
338.1'7371'097291—dc20   94–38200 CIP

Please remember that this is a library book,
and that it belongs only temporarily to each
person who uses it. Be considerate. Do
not write in this, or any, library book.